Geoscience Canada, Reprint Series 1

Facies Models

Reprinted with revisions from a
series of papers in
Geoscience Canada, 1976-1979,
published by the Geological Association of Canada.

Edited by

Roger G. Walker
Department of Geology
McMaster University
Hamilton, Ontario L8S 4M1, Canada

May, 1979

Additional copies may be obtained by
writing to:
Geological Association of Canada
Publications
Business and Economic Service Ltd.
111 Peter Street, Suite 509
Toronto, Ontario M5V 2H1

ISBN 0-919216-15-3

Geological Association of Canada
Department of Earth Sciences
University of Waterloo
Waterloo, Ontario N2L 3G1, Canada

Printed by Ainsworth Press Limited,
Kitchener, Ontario

Contents

Preface

The series of articles on Facies Models in Geoscience Canada was initiated by editor Gerard V. Middleton, partly in response to a suggestion from Erich Dimroth that "brief, well-written, concise articles, outlining the techniques of rock interpretation, and the concepts and criteria for regional evaluation would be an enormous help to the general purpose field geologist". Each author was asked to prepare a review that covered the basic ideas, without unnecessary jargon, in terms that the student or "general purpose field geologist" could understand. Most of the illustrations in the articles have been prepared with this objective in mind. The articles have attracted sufficient attention that the Geological Association of Canada has decided to reprint them here.

Many of the older articles have been revised for this volume, and some new illustrations and references added. In addition, three contributions appear here without prior publication in Geoscience Canada, in order to increase the scope of this volume. It is not intended to be a complete survey of all facies, and some facies (such as iron formations and volcaniclastics) have been included here because of their importance in a Canadian context. We hope, however, that we have achieved an international coverage of ideas and examples.

I have re-arranged and re-numbered the articles to give a more logical sequence, from alluvial fans to deep basins, and from carbonates, through evaporites into iron formations. The volume is completed with the volcaniclastic and seismic-stratigraphic contributions.

At an early stage of planning, I asked Noel James to help select and advise on the carbonate and evaporite contributions. Maureen Czerneda, as Managing Editor for Geoscience Canada, has been a great help in editing and communicating our wishes to the printer. I thank the Geological Association of Canada, and Glen Caldwell as chairman of the publications committee, for their help and support. The sketches that head each article are the work of Peter Russell, and the cover design is by Peter Russell and Dave Bartholomew. I particularly thank many authors and journals who have kindly allowed us to use their illustrations. These are acknowledged as "from, or after, A. N. Author, 1984" in the captions. Finally, I thank the authors for meeting my deadlines—it made my job so much easier.

Roger G. Walker
McMaster University
May 1979

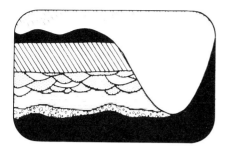

Facies and Facies Models 1. General Introduction

Roger G. Walker
Department of Geology
McMaster University
Hamilton, Ontario L8S 4M1

Introduction
In this general introduction, considerably expanded from its first appearance in Geoscience Canada, I will comment on three ideas – facies, facies sequences and facies models. The intent is to simplify and de-mystify, and to return a useful concept from the metaphysicist to the geologist.

Facies
Ever since the term *facies* was introduced by Gressly in 1830, there has been discussion as to the meaning and use of the term (see Middleton, 1978). In particular, arguments have focussed on: 1) whether the term implies a "set of characteristics" as opposed to the rock body itself; 2) whether the term should refer only to "areally restricted parts of a designated stratigraphic unit" (Moore, 1949) or also to stratigraphically unconfined rock bodies (as originally used by Gressly and other European workers), and 3) whether the term should be purely descriptive (e.g., "sandstones of Facies A") or also interpretive (e.g., "fluvial facies"). Succinct discussions of these problems have been given by Middleton (1978) and Reading (1978) – I will use the term in a concrete sense rather than abstractly implying only a set of characteristics, and will use it in a stratigraphically unconfined way. Middleton (1978) has noted that:

"the more common [modern] usage is exemplified by De Raaf *et al.* (1965) who subdivided a group of three

formations into a cyclical repetition of a number of facies distinguished by "lithological, structural and organic aspects detectable in the field". The facies may be given informal designations ("Facies A", etc.) or brief descriptive designations (e.g., "laminated siltstone facies") and it is understood that they are units that will ultimately be given an environmental interpretation: but the facies definition is itself quite objective and based on the total field aspect of the rocks themselves . . ,The key to the interpretation of facies is to combine observations made on their spatial relations and internal characteristics (lithology and sedimentary structures) with comparative information from other well-studied stratigraphic units, and particularly from studies of modern sedimentary environments."

Defining Facies
Breaking a rock body down into constituent facies is in part a classification procedure. The degree of subdivision is first and foremost governed by the *objective of the study*. For example, if the objective is the description and interpretation of a particular stratigraphic unit, a fairly broad facies subdivision may suffice. Alternatively, if the objective is the refinement of an existing facies model, or setting up an entirely new model, then facies subdivision in the field will almost certainly be more detailed.

The *scale of subdivision* is dependent not only upon one's objectives, but on the time available, and the abundance of physical and biological structures in the rocks. A thick sequence of massive mudstones will be difficult to subdivide into facies, but a similar thickness of interbedded sandstones and shales (with abundant and varied examples of ripples, cross bedding and trace fossils) might be subdivisible into a large number of distinct facies. As a general rule, I would advocate erring on the side of oversubdividing in the field – facies can always be recombined in the laboratory, but a crude field subdivision cannot be refined in the lab.

Subdivision of a body of rock into facies ideally should not be attempted until one is thoroughly familiar with the rock body. Only then will it be apparent how much variability there is, and how many different facies must be defined to describe the unit. In the field, most facies studies have relied on distinctive combi-

nations of sedimentary and organic structures (e.g., De Raaf *et al.*, 1965; Williams and Rust, 1969; Cant and Walker, 1976). Statistical methods can also be used to define facies, especially where there is considerable agreement among workers as to the important quantifiable, descriptive parameters. In carbonate rocks, percentages of different organic constituents, and percentages of micrite and/or sparry calcite have been used as input to cluster and factor analyses, with the resulting groupings of samples (in Q mode) being interpreted as facies (Imbrie and Purdy, 1962; Klovan, 1964; Harbaugh and Demirmen, 1964; see also Chapter 7 of the book by Harbaugh and Merriam (1968) on Computer Applications in Stratigraphic Analysis – Classification Systems). Unfortunately, statistical methods are unsuited to clastic rocks, where most of the important information (sedimentary and biological structures) cannot readily be quantified. Readers unfamiliar with the process of subdividing rock bodies into facies should consult the papers listed in the annotated bibliography, to see how the general principles briefly discussed here can be applied in practise.

Facies Sequence
It was pointed out by Middleton (1978) that "it is understood that (facies) will ultimately be given an environmental interpretation". However, many, if not most facies defined in the field have ambiguous interpretations – a cross-bedded sandstone facies, for example, could be formed in a meandering or braided river, a tidal channel, an offshore area dominated by alongshore currents, or on an open shelf dominated by tidal currents. The key to interpretation is to analyze all of the facies communally, in context. The sequence in which they occur thus contributes as much information as the facies themselves.

The relationship between depositional environments in space, and the resulting stratigraphic sequences developed through time as a result of transgressions and regressions, was first emphasized by Johannes Walther, in his Law of the Correlation of Facies (Walther, 1894, p. 979 – see Middleton, 1973). Walther stated that "it is a basic statement of far-reaching significance that only those facies and facies-areas can be superimposed primarily which can be observed beside each other at the present time". Careful application of the law therefore suggests that in a vertical sequence, a *gradational* transition from one facies to another implies that the two facies represent environments that once were adjacent laterally. The dangers of applying the Law in a gross way to stratigraphic sequences with cyclic repetitions of facies have been emphasized by Middleton (1973, p. 983). The importance of clearly defining *gradational* facies boundaries in vertical section, as opposed to sharp or erosive boundaries, has been emphasized by De Raaf *et al.* (1965) and Reading (1978, p. 5). If boundaries are sharp or erosional, there is no way of knowing whether two vertically adjacent facies represent environments that once were laterally adjacent. Indeed, sharp breaks between facies, especially if marked by thin bioturbated horizons implying non-deposition, may signify fundamental changes in depositional environment, and the beginnings of new cycles of sedimentation (see De Raaf *et al.,* 1965; and Walker and Harms, 1971, for examples of sharp facies relationships accompanied by bioturbation).

The first formal documentation of relationships between facies was published by De Raaf *et al.* (1965, Fig.22) in a diagram resembling a spider's web. An example of what is now referred to as a "facies relationship diagram" (F.R.D.) is given in Figure 1. In the diagram, the number of transitions from one facies to another, observed in the field, is shown by numbered arrows. Note that gradational relationships can easily be distinguished from sharp or erosional contacts.

In the 15 years or so since the first F.R.D., I believe that one method for handling the data has emerged as significantly better than others. It is based upon a method suggested by Selley (1970).

First, an F.R.D. is constructed, and the number of transitions from one facies to another are tabulated (Fig. 2A) and converted to probabilities (Fig. 2B). I will use the data of Cant and Walker (1976) from the Devonian Battery Point Sandstone of Quebec to illustrate the method. In Figure 2A (for example) Facies F is followed once by Facies G, and twice by SS, hence the observed transition probabilities p_{ij} are 0.333 and 0.667 respectively in Figure 2B. Each row in Figure 2B must total 1.0.

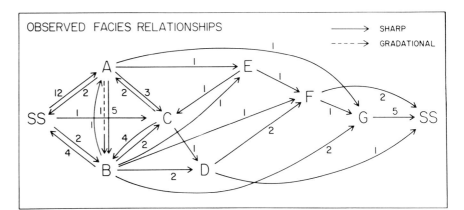

Figure 1

Example of a Facies Relationship Diagram, showing the observed number of sharp and gradational transitions between facies. Data are from Cant and Walker (1976), and refer to facies in the Devonian Battery Point Sandstone of Quebec. SS - scoured surface; A - *poorly defined trough cross bedding; B - well defined trough cross bedding; C - large planar tabular cross bedding; D - small scale planar tabular cross bedding; E - isolated scours; F - trough cross laminated fine sandstones and shales; G - low angle stratification.*

Figure 2A. Observed number of transitions between facies.

	SS	A	B	C	D	E	F	G
SS		12	2	1				
A	2		6	3		1		1
B	4	1		2	2	1	1	2
C		2	4		1			
D	1						2	
E				1			1	
F	2							1
G	5							

Figure 2B. Observed transition probabilities.

	SS	A	B	C	D	E	F	G
SS		.800	.133	.067				
A	.154		.462	.231		.077		.077
B	.308	.077		.154	.154	.077	.077	.154
C		.286	.571		.143			
D	.333						.667	
E				.500			.500	
F	.667							.333
G	1.000							

Figure 2C. Transition probabilities for random sequence.

	SS	A	B	C	D	E	F	G
SS		0.320	0.245	0.151	0.075	0.038	0.075	0.094
A	0.280		0.260	0.160	0.080	0.040	0.080	0.100
B	0.259	0.315		0.148	0.074	0.037	0.074	0.093
C	0.237	0.288	0.220		0.068	0.034	0.068	0.085
D	0.222	0.270	0.206	0.127		0.032	0.063	0.079
E	0.215	0.262	0.200	0.123	0.062		0.062	0.077
F	0.222	0.270	0.206	0.127	0.063	0.032		0.079
G	0.226	0.274	0.210	0.129	0.065	0.032	0.065	

Figure 2D. Observed minus random transition probabilities.

	SS	A	B	C	D	E	F	G
SS		+0.480	-0.112	-0.084	-0.075	-0.038	-0.075	-0.094
A	-0.126		+0.202	+0.071	-0.080	+0.037	-0.080	-0.023
B	+0.049	-0.238		+0.006	+0.080	+0.040	+0.003	+0.061
C	-0.237	-0.002	+0.351		+0.075	-0.034	-0.068	-0.085
D	+0.111	-0.270	-0.206	-0.127		-0.032	+0.604	-0.079
E	-0.215	-0.262	-0.200	+0.377	-0.062		+0.438	-0.077
F	+0.445	-0.270	-0.206	-0.127	-0.063	-0.032		+0.254
G	+0.774	-0.274	-0.210	-0.129	-0.063	-0.032	-0.065	

Figure 2

Tabulation of facies relationships. See text for details, and see caption of Figure 1 for explanation of letters.

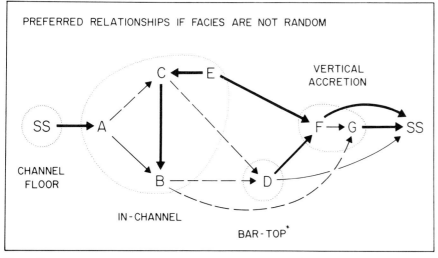

PREFERRED RELATIONSHIPS IF FACIES ARE NOT RANDOM

VERTICAL ACCRETION

CHANNEL FLOOR

IN-CHANNEL

BAR-TOP*

Figure 3

Simplification of Figure 1, based on transitions that occur more commonly than random. Heavy lines indicate much more commonly than random, and dotted lines indicate a little

more common than random. Light solid lines intermediate between heavy and dotted. Study Figure 2D to see how this diagram is derived. Letters explained in caption of Figure 1.

Second, another transition probability matrix is constructed on the assumption that all facies transitions are random. The probabilities depend only upon the absolute abundance of the various facies, and are given by:

$$r_{ij} = \frac{n_j}{N - n_i}$$

where r_{ij} is the random probability of transition from facies i to facies j, n_i and n_j are the number of occurrences of facies i and j respectively, and N is the total number of occurrences of all facies. This equation applies in all situations, regardless of whether the facies sequence is continuous or contains faults and covered intervals. The equations published by Miall (1973, p. 351) *only* apply to continous sequences of N occurrences of facies with N-1 facies transitions. Miall's equations, for example, do *not* apply to the original data of Cant and Walker (1976, p. 112), where there were 67 occurrences of the various facies, but only 60 facies transitions due to the presence of six covered intervals. The random transition probabilities are shown in Figure 2C.

Third, a difference matrix is calculated, showing the observed minus random probabilities.

$$p_{ij} - r_{ij}$$

The result is shown in Figure 2D, with a possible range of numbers from +1.0 to −1.0.

Fourth, we enter into the stage of understanding what the difference matrix tells us. It is noted that some values are high-positive (transitions much more common than if facies were random) and some are high-negative (transitions much *less* common than random). I suggest that a new F.R.D. be gradually constructed, putting on paper first only those transitions with "high" positive values (heavy arrows on Figure 3). What is included in the "high" category will depend on the range of numbers in the difference matrix - some trial and error is a good thing because it forces one to examine the data carefully, and to think what each more-common-than-random transition might mean geologically. It is also important to reexamine the original "spider web" (Fig. 1) when constructing the new F.R.D., because the absolute numbers of transitions can be shown on the new F.R.D. if desired. More importantly, the new arrows can be drawn to indicate dominantly sharp or gradational contacts. This overcomes most of the main objections that H.G. Reading expressed with respect to Selley's technique (see discussions of Selley, 1970, p. 575-581).

As the difference matrix is examined in detail, lighter arrows can be drawn on the F.R.D. (Fig. 3) to indicate transitions only a little more common than random (random being those transitions with values close to zero). The result of this method of data analysis should be a

better understanding of the structure of the raw data, and finally, an F.R.D. that can be used as a basis for overall interpretation (Fig. 3).

It is now important to distinguish between a single facies sequence, and repeated sequences (or cycles). The F.R.D. in Figure 3, along with basic facies interpretations, establishes the probable fluvial origin of the Battery Point Sandstone. The scoured surface SS can then be interpreted as the fundamental boundary between cycles, and hence individual cycles can be defined on the original complete stratigraphic section (Cant and Walker, 1976, Fig. 2). Using the F.R.D. as an idealization of all of the cycles combined (a local *model,* see below), each individual cycle can be compared with the model to identify points in common and points of difference. The reader may do this with the Battery Point cycles as an exercise (and see Cant, 1978).

Facies Models

The construction and use of facies models is one of the most active areas in the general field of stratigraphy. This emphasis is not new: many of the ideas were embodied in Dunbar and Rodgers' Principles of Stratigraphy in 1957, and were based upon studies dating back to Gressly and Walther in the 19th Century (Middleton, 1973). However, the importance of facies models at the present time is due to an increasing need for the models, and a rapidly increasing data base on which the models are formulated.

In this volume, facies models are expressed in several different ways - as idealized sequences of facies, as block diagrams, and as graphs and equations. The term model here has a generality that goes beyond a single study of one formation. The final F.R.D. for the Battery Point Formation (Fig. 3) is only a *local summary,* not a model for fluvial deposits. But when the Battery Point F.R.D. is compared and contrasted with F.R.D.'s from other ancient braided river deposits, and then data from modern braided rivers is incorporated (e.g. Cant, 1978), the points in common between all of these studies begin to assume a generality that can be termed a *model.*

A facies model could thus be defined as a general summary of a specific sedimentary environment, written in terms that make the summary useable in

4

at least four different ways. The basis of the summary consists of many studies in both ancient rocks and recent sediments; the rapidly increasing data base is due at least partly to the large number of recent sediment studies in the last 15 years. The increased need for the models is due to the increasing amount of prediction that geologists are making from a limited local data base. This prediction may concern subsurface sandstone geometry in hydrocarbon reservoirs, the association of mineral deposits with specific sedimentary environments (for example, uraniferous conglomerates), or the movement of modern sand bars in shallow water (Bay of Fundy, tidal power). In all cases, a limited amount of local information *plus* the guidance of a well-understood facies model results in potentially important predictions about that local environment.

Sedimentary Environments

There is now some agreement among sedimentologists as to how to subdivide up the depositional environments of the world into commonly recurring types. At a recent count (1972), there were 18 major environments, 40 sub-types, 14 sub-sub-types, and 20 sub-sub-sub-types. I deliberately do *not* cite this reference! Nevertheless, there is some agreement on a very basic subdivision based upon morphology, physical and chemical processes, and biological processes. The geologist involved with ancient environments would add the criteria of stratigraphic record and diagenesis to the above list. A typical set of environments that most sedimentologists would not object to is shown in Table 1. Aeolian and glacially-influenced environments might also be added to the list, and herein lies the beginning of confusion - some environments are being defined geomorphologically (e.g., alluvial fans) and others by

Table I *Major environments of deposition of clastic rocks*

Terrestrial:	Alluvial fans
	Rivers and their floodplains
Marginal-Marine:	Deltas
	Alongshore sand bodies
	(beaches, cheniers, barriers)
Marine:	Shelf
	Submarine fans - Turbidites -
	Abyssal plains

process (e.g., aeolian). Aeolian sediments can exist on their own (in many deserts) or can be blown into alluvial fan and fluvial environments, yet still be identifiable as windblown.

The point to make here is that our aim as geologists is not only to recognize environments, but to understand the range of processes that can operate within them. We must also be sure of why we want to identify environments in the first place. Is it to provide a name showing that we have thought about the origin of the unit we have mapped ("the Ordovician Cloridorme Formation consists of deep water turbidites"), or is it to provide a framework for further thought? It is the latter - the framework for further thought - that in my mind separates the art of recognizing environments from the art of FACIES ANALYSIS and FACIES MODELLING. The meaning and implication of these terms will become apparent.

Facies Models—Constuction and Use

The principles, methods and motives of facies analysis are shown in Figure 4, specifically for turbidites. This is to help link the general comments on facies models in this article with specific comments on the turbidite model discussed later. In Figure 4, the local examples 1 through 6 are turbidite examples; howere, I emphasize that the ideas embodied in Figure 4 can and should apply to facies models for all other environments. If enough examples of modern turbidites can be studied in cores, and if enough ancient turbidite formations are studied in the field, we may be able to make some general statements about turbidites, rather than statements about only one particular

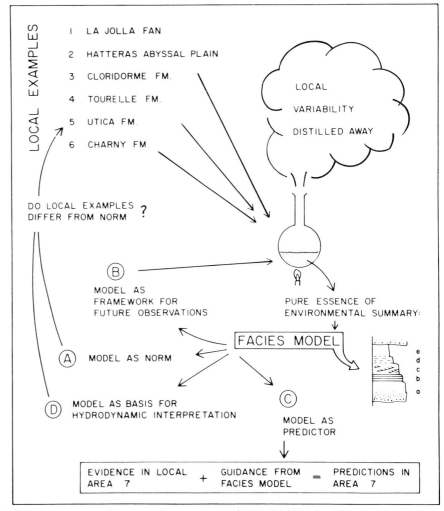

Figure 4
Distillation of a general facies model from various local examples, and its use as a norm, framework for observations, predictor, and basis for hydrodynamic interpretation. See text for details.

example. The process of extracting the general information is shown diagrammatically in Figure 4, where numbers 1 and 2 represent recent sediment studies (cores from, say, La Jolla fan and Hatteras abyssal plain) and numbers 3 through 6 represent studies of ancient turbidites (for example, the Cloridorme and Tourelle Formations of Gaspé, the Utica Formation at Montmorency Falls, and the Charny Formation around Quebec City). The entire wealth of information on modern and ancient turbidites can then be distilled, boiling away the local details, but distilling and concentrating the important features that they have in common into a general summary of turbidites. If we distill enough individual turbidites, we can end up with a perfect "essence of turbidite" – now called the Bouma model. But what is the essence of any local example and what is its "noise"? Which aspects do we dismiss and which do we extract and consider important? Answering these questions involves experience, judgement, knowledge and argument among sedimentologists, and the answers also involve the ultimate purpose of the environmental synthesis and summary. We will not consider the process of distilling the information here, but will consider each environment at its present level of understanding – emphasizing its beauty but pointing out its warts.

I pointed out earlier that the difference between the summary of an environment and a facies model perhaps depends mainly on the use to which the summary is put. As well as being a summary, a FACIES MODEL must fulfill four other important functions:
1) it must act as a *norm,* for purposes of comparison
2) it must act as a *framework* and *guide* for future observations
3) it must act as a *predictor* in new geological situations
4) it must act as a *basis for hydrodynamic interpretation* for the environment or system that it represents.
Figure 4 has been constructed to illustrate these various functions. Using the example of the turbidite model, the numbers 1 through 6 indicate various local studies of modern and ancient turbidites. There is a constant feedback between examples – in this way the sedimentologist exercises his judgement in defining the features in common and identifying "local irregularities". This

is the "distillation" process that allows the environmental summary (that will act as a facies model) to be set up.

Having constructed the facies model, it must act first as a norm (Fig. 4, A) with which individual examples can be compared. Without a norm, we are unable to say whether example 5 of Figure 4 contains any unusual features. In this example, Utica Formation turbidites at Montmorency Falls are very thin, silty, and many beds do not begin with division A of the Bouma model (Fig. 4); they begin with division B or C. Because of the existence of the norm (Bouma model), we can ask questions about example 5 that we could not otherwise have asked, and whole new avenues of productive thought can be opened up this way. Thus there is a constant feedback between a model and its individual examples – the more examples and the more distillation, the better the norm will be, and the more we must be forced into explaining local variations.

The second function of the facies model is to set up a framework for future observations (Fig. 4, B). Inasmuch as the model summarizes the important information, geologists know that similar information must be sought in new situations. In our example, this would include the individual characteristics of the five Bouma divisions. Although the framework ensures that this information is recorded wherever possible, it can also act to blind the unwary, who might ignore some evidence because it is not clearly spelled out by the model. This leads to imprecise interpretations, and would cause a freeze on any further improvement of the facies model – hence the feedback arrow (Fig. 4, B) implying that all future observations must in turn be distilled to better define the general model.

The third function of the model is as a predictor in new geological situations (Fig. 4, C). In example 7 (for example Archean rocks in the Manitou Lake area, N.W. Ontario) (Fig. 4, C) let us imagine that we have just enough evidence to suggest a turbidite interpretation. Because we have the turbidite model and (in an ideal world) understand its operation, we can take the combination of the model and the limited data from area 7 to make further predictions about area 7. This is obviously a vitally

important aspect of facies modelling, and good surface or subsurface prediction from limited data can save unnecessary exploration guesswork and potentially vast sums of money.

The fourth major function of a facies model is to act as an integrated basis for hydrodynamic interpretations (Fig. 4, D). Again, it is imortant to eliminate the local "noise" before looking for a general hydrodynamic interpretation, and again, there can be a feedback between the hydrodynamic norm and local examples (Fig. 4, D). This is indicated by the feedback arrow to example 5 (Fig. 4), implying the question "does the interpretation of example 5 differ from the idealized hydrodynamic interpretation?" If there *is* a difference (and there is), we can again ask questions that could not be asked if we had not used the facies model to formulate a general interpretation. This usage of the facies model is demonstrated particularly well by the turbidite model, discussed in a following article.

The turbidite / submarine fan model has been discussed because it is reasonably well understood, and because it demonstrates particularly well the four functions of a model illustrated by Figure 4. Some of the other models discussed in this volume are less well understood – because the environmental summary is weaker, so the functioning of the model is weaker. I emphasize that the construction and functioning of facies models is essentially similar for all environments, and that the turbidite example was discussed above to make the general statements about facies models a little more specific.

Basic Sources of Information
The following bibliography is not intended to be complete, and my annotations apply *only* to the aspects of the books or papers relevant to environmental summary and facies modelling.

Books Containing General Environmental Syntheses
Selley, R.C., 1970, Ancient sedimentary environments: Ithaca, N.Y., Cornell University Press, 237 p.

Selley introduces the volume as "not a work for the specialist sedimentologist, but an introductory survey for readers with a basic knowledge of geology". The book achieves this end very well – it summarizes, it leans on classical exam-

6

ples, and it very briefly indicates the economic implications (oil, gas, minerals) of some of the environments. This volume is a good place to start.

Spearing, D.R., 1974, Summary sheets of sedimentary deposits: Geol. Soc. Am., Map and Charts Mc-8.

A series of 7 large sheets with many line drawings, minimal text, and useful references on selected sandstone depositional environments. This is a quick way to get a feeling for alluvial fan, alluvial valley, aeolian, regressive shoreline, barrier island, tidal, and turbidity current environments, and the series also provides a very good entry to the recent literature.

Blatt, H., G.V. Middleton and R.C. Murray, 1972, Origin of sedimentary rocks: Englewood Cliffs, N.J., Prentice Hall, 634 p.

Chapter six, on facies models, is only 29 pages long, but summarizes concisely the general principles of facies and facies analysis, and briefly reviews alluvial fans, alluvial plains, deltas, barriers, offshore shoals and turbidites – deep basin environments.

Allen, J.R.L., 1970, Physical processes of sedimentation: New York, American Elsevier, 248 p.

Although slanted toward physical processes, the book contains chapters on winds and their deposits, river flow and alluvium, shallow marine deposits, turbidity currents and turbidites, and glaciers and glacial deposits. Each chapter begins with a discussion of processes, but ends with useful generalized descriptions of the environments. This volume would not be the place to start reading, but would be good followup material for readers wanting a better understanding of physical processes operating in various environments.

Pettijohn, F.J., P.E. Potter and R. Siever, 1972, Sand and Sandstone: New York, Springer-Verlag, 618 p.

Chapter 11 (p. 439-543) is a review of sand bodies and environment written at a fuller and more technical level than Selley, Spearing, or Blatt, Middleton and Murray. It considers Alluvial, Deltaic Estuarine, Tidal Flat, Beach and Barrier, Marine Shelf, Turbidite and Aeolian environments, with separate remarks on sand body prediction. Useful follow-up

reading after Blatt, Middleton and Murray (1972), Spearing (1974) and Selley (1970) in that order.

Rigby, J.K., and W.K. Hamblin, eds., 1972, Recognition of ancient sedimentary environments: Soc. Econ. Paleont. Min., Spec. Pub. 16, 340 p.

Contains separate papers on many important environments written at a technical level. Many of the papers are disappointing as reviews but there are excellent contributions on Alluvial Fans, Fluvial Paleochannels, Barrier Coastlines and Shorelines. Most of the authors present their environmental summaries but do not attempt to use them as models.

Reading, H.G., 1978, Sedimentary environments and facies: Blackwell, Oxford, 557 p.

An excellent compilation of data on depositional environments and facies models. An indispensable reference, and the best available summary of major depositional environments.

Reineck, H.E., and I.B. Singh, 1973, Depositional sedimentary environments: New York, Springer-Verlag, 439 p.

Pages 160-439 are devoted to summaries of many modern environments. Coverage is at the graduate student – professional sedimentologist level, but is patchy and rather uncritical. Vast reference lists are given, but it is hard to single out the very important papers from the trivial. The emphasis on modern environments is useful, but the book should not be used until one is at least somewhat familiar with specific environments.

Facies
Gressly, A., 1838, Observations geologiques sur le Jura Soleurois: Neue Denkschr. allg. schweiz, Ges. ges. Naturw., v. 2, p. 1-112.

Gressly's work first established the concept of facies in the geological literature.

Middleton, G.V., 1973, Johannes Walther's Law of the Correlation of Facies: Geol. Soc. Amer. Bull., v. 84, p. 979-988.

An excellent discussion of the use, misuse and implications of Walther's Law.

Moore, R.C., 1949, Meaning of Facies: in C.R. Longwell, ed., Sedimentary facies in geologic history: Geol. Soc. Amer., Memoir 39, p. 1-34.

This paper is from the first important North American volume on facies. It emphasizes the lateral variations of facies within a designated stratigraphic unit. Historically, an important paper, but now conceptually out of date (or out of fashion!).

Walther, J., 1893-4, Einleitung in die Geologie als historische Wissenschaft: Verlag von Gustav Fischer, Jena, 3 vols., 1055 p.

See Middleton, 1973, for a Commentary on the importance of Walther's work.

Facies Definition Using Statistical Methods
Harbaugh, J.W. and F. Demirmen, 1964, Application of factor analysis to petrologic variations of Americus Limestone (Lower Permian), Kansas and Oklahoma: Kansas Geol. Survey Sp. Dist. Publ. 15, 40 p.

A good example of factor analysis used to establish facies (termed "phases") in the Permian Americus Limestone (Kansas and Oklahoma). Maps show distribution of the phases, with interpretations of depositional environments.

Harbaugh, J.W. and D.F. Merriam, 1968, Computer applications in stratigraphic analysis: New York, Wiley, 282 p.

Chapter 7 is concerned with classification systems, and gives a good introduction to factor analysis and other techniques. Several useful examples are discussed.

Imbrie, J. and E.G. Purdy, 1962, Classification of modern Bahamian carbonate sediments: in W.E. Ham, ed., Classification of carbonate rocks: Amer. Assoc. Pet. Geol., Memoir 1, p. 253-279.

A good introduction to factor analysis, with an excellent example of how it can be used to define carbonate facies (with data from the Bahama Banks).

Klovan, J.E., 1964, Facies analysis of the Redwater Reef Complex, Alberta, Canada: Canadian Petrol. Geol. Bull., v. 12, p. 1-100.

Defines different types of carbonate particles, and uses a hierarchal representation technique to classify them.

The clusters of data so revealed were termed facies, and were used as the basis for an environmental interpretation of the Upper Devonian Redwater Reef (Alberta).

Facies, Facies Sequence, Facies Models

Cant, D.J., 1978, Development of a facies model for sandy braided river sedimentation: comparison of the South Saskatchewan River and the Battery Point Formation: *in* A.D. Miall ed., Fluvial sedimentology: Can. Soc. Petrol. Geol., Mem. 5, p. 627-639.
A detailed comparison of ancient sediments and recent sediments, emphasizing facies comparisons and the construction of a facies model.

Cant, D.J. and R.G. Walker, 1976, Development of a braided fluvial facies model for the Devonian Battery Point Sandstone, Quebec: Can. Jour. Earth Sci., v. 13, p. 102-119.
Selley's difference matrix is used to help define fluvial cycles in a sandy braided system. This paper is a type example of the construction of a local facies sequence.

De Raaf, J.F.M., H.G. Reading, and R.G. Walker, 1965, Cyclic sedimentation in the Lower Westphalian of North Devon, England: Sedimentology, v. 4, p. 1-52.
This paper gives the first published example of a facies relationship diagram, and uses the diagram to establish cyclicity in a series of prograding shoreline sediments. Cycles are defined by black mudstones resting on bioturbated sandstones.

Miall, A.D., 1973, Markov chain analysis applied to an ancient alluvial plain succession: Sedimentology, v. 20, p. 347-364.
A good introduction to Markov chain methodology with an example from the Devonian Peel Sound Formation of Prince of Wales Island, Arctic Canada.

Middleton, G.V., 1978, Facies: *in* R.W. Fairbridge, and J. Bourgeois, eds., Encyclopedia of Sedimentology: Stroudsberg, Pa., Dowden, Hutchinson and Ross, p. 323-325.
A concise statement of the facies concept, discussing the various ways in which the term has been used.

Selley, R.C.,1970, Studies of sequence in sediments using a simple mathematical device: Geol. Soc. London, Quart. Jour., v. 125, p. 557-581.
The first discussion of the difference matrix, and its possible use in describing and interpreting facies sequence. Contains written discussions of the paper by various authors, some of which are thought-provoking.

Walker, R.G. and J.C. Harms, 1971, The "Catskill Delta": a prograding muddy shoreline in central Pennsylvania: Jour. Geol., v. 79, p. 381-399.
Describes cyclic facies sequences that are defined by transgressive bioturbated sandstone horizons.

Williams, P.F. and B.R. Rust, 1969, The sedimentology of a braided river: Jour. Sediment. Petrol., v. 39, p. 649-679.
A good example of facies definition in a modern gravelly river (the Donjek, Y.T.), with definition of facies sequences and expression of a local "model" in terms of block diagrams.

MS received November 24, 1975. Revised April 1979. Reprinted from Geoscience Canada, Vol. 3, No. 1, p. 21-24.

Facies Models 2. Coarse Alluvial Deposits

Brian R. Rust
Department of Geology
University of Ottawa
Ottawa, Ontario K1N 6N5

Introduction
Coarse-grained alluvial deposits are a relatively minor component of the stratigraphic record, but have considerable importance because of their tectonic significance. They are good indicators of the sharp terrestrial relief produced by lithospheric uplift at continental margins, or by faulting within continental plates. In this respect they differ from glacial gravels (tills, the other main type of coarse terrestrial deposit), which do not necessarily indicate immediately adjacent relief. The principal differences between alluvial, glacial, beach and submarine fan gravels will be discussed later.

As noted by Walker (1976c), coarse-grained alluvial deposits are abundant in Canada, with both modern and ancient equivalents well represented. They also have economic importance, notable examples being the Witwatersrand placer gold and uranium ores of South Africa (Minter, 1978) and the similar uraniferous conglomerates of the Blind River-Elliot Lake area, Canada (Pienaar, 1963). Robertson et al. (1978) noted that uranium placer deposits are confined to rocks between 3.0 and 2.2 billion years old, because their formation ceased when the atmosphere became oxygenic. In other words they represent a clastic facies that is also subject to geochemical constraint.

The approach to facies models used here is essentially that advocated by Walker (1976a). The facies code for

Table I
Principal facies of alluvial gravels.

Major facies	**Gm:**	Horizontally bedded clast-supported gravel, commonly imbricate.
	Gms:	Muddy matrix-supported gravel; lacks imbrication and internal stratification
	Gt:	Trough cross-bedded clast-supported gravel
Minor facies	**Gp:**	Planar cross-bedded gravel, transitional from clast-supported through sand matrix-supported to sand (facies Sp)
	Sh:	Horizontally stratified sand
	St:	Trough cross-stratified sand
	Sp:	Planar cross-stratified sand
	Fm:	Massive fine sandy mud or mud
	Fl:	Laminated or cross-laminated very fine sand, silt or mud.
	P:	Concretionary carbonate (pedogenic)

alluvial deposits introduced by Miall (1977) will be adopted, as modified by Miall (1978a) and Rust (1978b). These facies (Table I) are the principal types found in alluvial gravels; the list is not exhaustive, but is a useful guide for constructing models based on the commonest facies inter-relationships.

Terminology
The term coarse-grained is used here for successions that contain 50 per cent or more gravels (particles greater than 2 mm in diameter). However, other authors have included as gravels deposits with only 25 per cent gravel-sized material (Pettijohn, 1975, Fig. 6-2). Another important distinction is that between clast-supported and matrix-supported gravel or conglomerate. Clast support indicates energetic aqueous transport that deposits gravel on the bed while sand is still carried in suspension; as flow velocity decreases, sand infiltrates into the spaces between the larger particles. Matrix-supported gravel is of two types: with stratified sand matrix, or unstratified matrix, usually of muddy sand [facies Gms of Miall (1978), Rust (1978b)]. The former case again indicates aqueous transport, but at a lower energy level in which sand and finer gravel particles are deposited together. The second type of matrix support points to deposition from mass flows, which in the alluvial context are debris flows (described later).

Alluvial gravels and their consolidated equivalents will be discussed here in terms of two principal environments: alluvial fans and rivers, the latter including alluvial plains (Walker, 1976a, Table I). Alluvial fans are morphological features: localised accumulations formed where a stream emerges from a confined valley onto a trunk river or a broad

Figure 1
Vertical air photograph (A15517-19) of upper reach of Slims River, Yukon (61°55'N, 138°38'W), showing marked contrast between tributary alluvial fans and trunk river. Note the entrenchment features of the lower left fan (see Bull, 1977, Fig. 20b). Original photo supplied by the Surveys and Mapping Branch, Department of Energy, Mines and Resources, Canada. Width of view about 7.5 km, north toward top of photograph.

alluvial plain (Holmes and Holmes, 1978, Fig. 18.43). In contrast, rivers and plains have lower gradients than their tributary fans, and they are much more extensive: rivers in one dimension, and alluvial plains in two. Most gravel-dominated rivers and alluvial plains have a braided channel pattern, but a few of the rivers meander. The morphological distinction between tributary fans and a braided

10

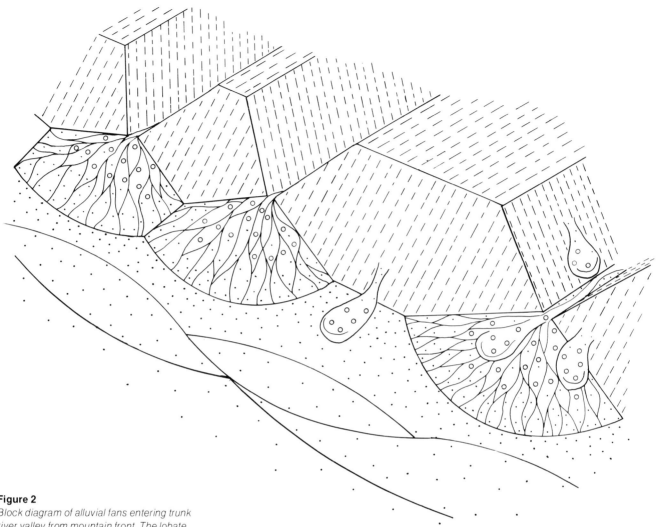

Figure 2
*Block diagram of alluvial fans entering trunk
river valley from mountain front. The lobate
features are debris flows.*

trunk river is clearly shown in Figures 1
and 2. As discussed later, however, the
recognition of fan and river deposits in
the ancient record is dependent on
identifying processes and the resulting
facies types, for morphological distinctions are rarely apparent.

Some authors have extended the term
alluvial fan to include what would be
regarded here as rivers or plains. For
example, Boothroyd and Nummedal
(1978) referred to coastal outwash
plains of Alaska and Iceland as humid
alluvial fans. The sediment is transported by meltwater issuing from glacial
termini, sources which migrate with time,
unlike the bedrock-controlled valley
mouth at the head of an alluvial fan. In
general the landforms are morphologi-
cally unlike fans, and their internal facies
are those of braided rivers and plains
(described below). Ancient alluvial

deposits have also been interpreted
rather broadly. For example, Pleistocene
sediments in the Netherlands which
closely resemble modern Icelandic
outwash plain sediments were termed
'classic' alluvial fan deposits by Reugg
(1977).

The term 'fan delta' has been applied
to alluvial fans deposited partly in
standing water in a lake or the sea
(Holmes and Holmes, 1978, p. 358-9).
However, apart from minor reworking by
waves, there is little evidence that
coarse-grained fans formed in this way
are significantly different from those of
completely subaerial origin (Fig. 3).
The deltaic connotation is misleading,
because the fans are dominated by
terrestrial processes, and do not show
the distinct separation at sea level
between subaerial and subaqueous
processes and deposits, which is so
characteristic of deltas. It therefore

seems preferable to call these features
coastal alluvial fans rather than fan
deltas, because in the ancient record
they would be identifiable only where the
alluvial deposits were interbedded with
strata containing marine fossils (Daily
et al., in press).

Alluvial Fans
Modern Examples. The classic descriptions of modern alluvial fans are mostly
from the mountainous semi-arid regions
of the south-western United States (Bull,
1963, 1964, 1972, 1977; Hooke 1967).
Fans of this type are uncommon in
Canada, but paraglacial fans (those
associated with retreat of valley glaciers) are relatively abundant (Ryder,
1971; Church and Ryder, 1972). In each
case the fans form adjacent to regions of
high relief, which are rapidly denuded to
provide the sediment which builds the

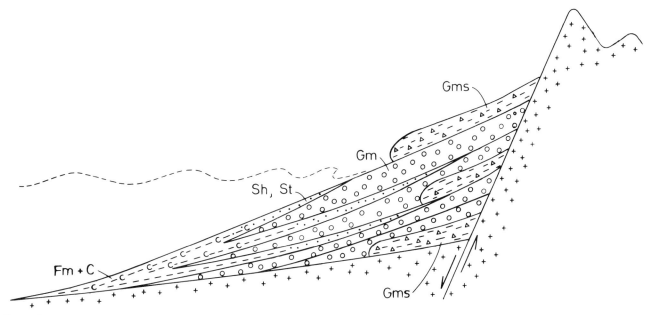

Figure 3

Diagrammatic cross-section of an alluvial fan, with the facies code explained in the text. For a terrestrial situation, disregard the dotted wavy line - but in the case of a coastal alluvial fan, the dotted line indicates sea or lake level. The internal facies distributions of the coastal fan are essentially the same as the terrestrial counterpart.

fans. In semi-arid environments the relief is commonly a faulted mountain front, and denudation is promoted by sparse vegetation and occasional intense rainfall. Paraglacial fans form where hanging tributary valleys enter a major glaciated valley (Fig. 4), in which case denudation is promoted by seasonal temperature fluctuations and the high spring runoff.

The characteristic shape of alluvial fans results mainly from increased frictional resistance as the relatively deep inter-montane stream changes to a braided system of broad shallow channels on the fan (Fig. 2). Bull (1977, p. 227) noted that there is commonly no change in slope from the canyon to the fan. There is, however, a marked slope contrast between tributary fans and related trunk rivers. For example, Spring Creek fan is a tributary to the Donjek River, Yukon (Fig. 4), with a slope of 0.019, while that of the trunk river at the same locality is 0.006. Spring Creek is the second largest tributary fan of the Donjek; smaller fans have correspondingly steeper slopes (Rust, 1972a, Table I).

Because of their topographic setting and hydraulic characteristics, alluvial fans are typified by coarse gravel in proximal reaches and rapidly decreasing grain size downfan (Fig. 3, and

Heward, 1978, Fig. 1). Most fans are dominated by water-laid deposits, which Bull (1972, p. 66-9) divided into sheet-flood, stream-channel and sieve deposits. The first two types can rarely be distinguished in ancient successions because channel dimensions commonly exceed those of outcrops, and sieve deposits are rare (Bull, 1972, p. 69). They form as lobes on fans which receive little sand or mud from their source areas. Transverse ribs are another minor but significant feature of alluvial fan channels, which permit estimation of paleo-depth, velocity and Froude Number (Koster, 1978). However, they have only recently been recognised in the stratigraphic record, by V.A. Gostin and the author (Koster, 1978, p. 186).

In terms of a facies model, it therefore seems best to avoid subdivisions of water-laid deposits other than facies defined on objective criteria. In proximal reaches the predominant facies is horizontally stratified, clast-supported coarse gravel, which is commonly imbricate (facies Gm of Miall, 1978). Imbrication commonly takes the form of upstream dip of the *ab* plane of clasts, with *a* transverse to flow (Rust, 1972b). Distally there is an increase in planar cross-stratal sets, with transitions from clast-supported fine-grained gravel,

through sand matrix-supported gravel to sand (facies Gp to Sp). This change reflects a gradual decrease in the particle size: water depth ratio as stream competence decreases downfan. Minor deposits of horizontally laminated sand (facies Sh) and laminated to massive mud (Fl, Fm) also increase downfan. Fans formed entirely of sand and finer sediment are rare, because they need a high-relief source of poorly consolidated sand or finer material, which is a short-lived feature of the landscape (Legget *et al.*, 1966).

Debris flow (or mudflow) deposits are the other principal component of most alluvial fan successions in both semi-arid and paraglacial environments (Figs. 2 and 3). According to Bull (1977, p. 236) debris flows are promoted by steep slopes, lack of vegetation, short periods of abundant water supply and a source providing debris with a muddy matrix. Johnson (1970) discussed debris flows, providing eye-witness accounts, as did Sharp and Nobles (1953), Curry (1966) and Wasson (in press). Middleton and Hampton (1976) pointed out that debris flows are one member of a continuous range of sediment gravity flows. In certain environments (especially subaqueous) it is difficult to distinguish debris flow from other types of mass flow deposits, but on alluvial fans they are usually quite distinct from waterlaid sediments. At their distal terminations debris flows are commonly lobate, and tend to concentrate the larger particles at the top and outer margin of the flow (Fig. 5).

Schumm (1977, p. 246) recognized two types of alluvial fans: "... dry or mudflow fans formed by ephemeral stream flow, and wet fans formed by perennial stream flow", the implication being that "wet" fans do not develop debris flows. It is true that consistent rainfall favours steady erosional processes rather than mass movements, but short team fluctuation in precipitation can undoubtedly produce debris flows in humid areas (Curry, 1966; Winder, 1965). The debris flow described by Wasson (in press) occurred on a fan fed by a perennial stream.

The example of a wet alluvial fan used by Schumm is the Kosi River, India, but facies types are not documented. The contours show that the Kosi lacks a fan morphology (Schumm, 1977, Fig. 7-5) and the size of the feature suggests that it is part of an alluvial plain rather than a fan. It is concluded that with few exceptions debris flow deposits characterize alluvial fans, and can be used (together with other criteria) to recognize ancient fan deposits, for fan-like morphology is rarely apparent.

Wasson (1977, p. 782-4) found that debris flow deposits in Quaternary fans in Tasmania lacked stratification and imbrication, and the larger clasts were dispersed in a poorly sorted finer matrix (facies Gms of Rust, 1978b; Miall, 1978a). Although they lack internal stratification, Bull (1963, p. 245) noted that more fluid (that is, proximal) debris flows may show graded bedding and subhorizontal orientation of flat megaclasts. However, more viscous (distal) debris flows tend to have larger clasts in vertical as well as other orientations (Fig. 6). In conclusion, debris flow deposits can normally be recognized as extensive units of massive, muddy matrix-supported gravel, with variable, non-imbricate internal fabric (facies Gms). This facies is particularly characteristic of alluvial fans, although not unique to them. Hooke (1967, p. 453) attributed wide variation in the proportion of debris flows on fans in similar topographic and climatic environments primarily to lithological control. The Trollheim, a modern fan with abundant debris flows studied by Hooke was used by Miall (1978, Table 2) to name his alluvial fan facies assemblage.

Figure 5
Leveed edge of debris flow lobe on west side of Donjek valley between Spring Creek and *Donjek Glacier. Pack (mid-ground) and figure (behind) give scale.*

Figure 4
Spring Creek alluvial fan, a tributary to the Donjek River, Yukon (see Koster, 1978, Fig. 1).

Figure 6
Rainwashed upper surface of debris flow (Fig. 5) showing variable but predominantly *subvertical megaclast fabric. Notebook 19 cm long.*

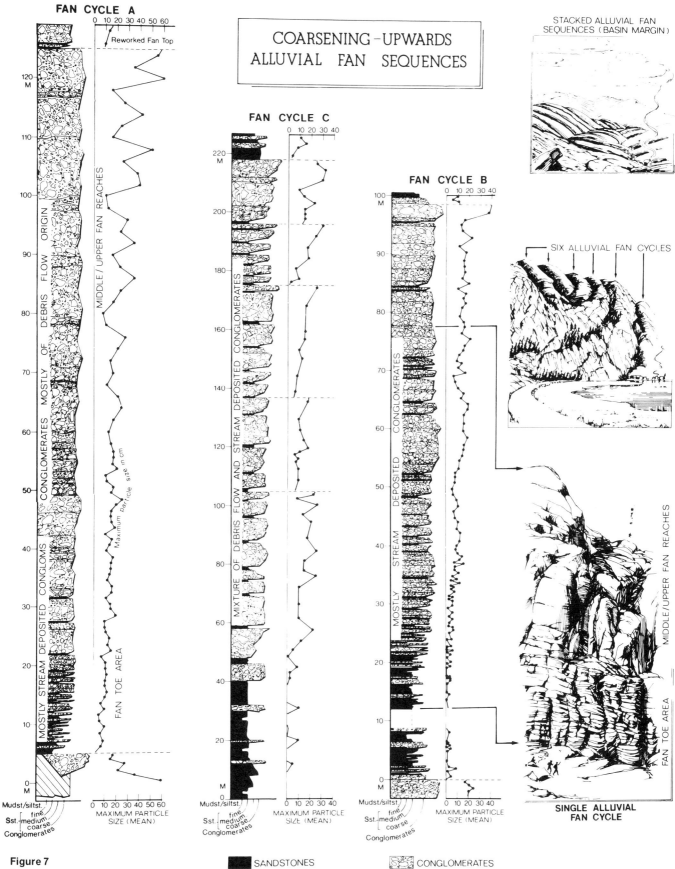

Figure 7
Devonian alluvial fan successions in the Hornelen Basin, Norway (from Steel et al., 1977, Fig. 3).

Ancient Examples. Some of the most impressive examples of ancient alluvial fan successions are those from the Devonian of Norway, discussed by Larsen and Steel (1978) and Steel *et al.* (1977). The latter authors described coarsening-upward cycles about 100 m thick, with subcycles in the 10 to 25 m thickness range (Fig. 7). The subcycles were attributed to rapid faulting of the basin margin; renewed relief stimulated fan progradation and the burial of fine distal deposits by coarse proximal detritus. A minority of the subcycles described by Steel *et al.* (1977) coarsen upward then fine upward, a sequence which would be expected if the sedimentary response to tectonic uplift had time to return to equilibrium (Fig. 8). Autocyclic mechanisms (those resulting from changes within a fan complex) can also influence fan successions, by switching of depositional centres on fans, and/or overbuilding of one fan by a neighbour. Heward (1978) also discussed cycles in fan deposits, applying the concepts to an ancient alluvial fan succession and its tectonic context.

Many authors have recognised debris flow deposits in ancient alluvial fan successions, for example Wasson (1977), Steel (1974) and Larsen and Steel (1978). Recognition is based on the internal characteristics discussed above, for the external morphology of debris flow lobes is rarely preserved. However, an example has been recognised in the Cambrian of South Australia by Daily *et al.* (in press).

McGowen and Groat (1971) interpreted the Ordovician or older sandstone and conglomerate succession of the Van Horn Sandstone, Texas as an alluvial fan deposit. They attributed the lack of debris flow deposits to low clay production in the source area, due to pre-vegetal origin and the nature of the source rocks. However, many of the source rocks cited produce clay on weathering, and there is no evidence that clay was less abundant before the appearance of terrestrial plants (Garrels and MacKenzie, 1971, Table 9.4). In view of its probable downslope extent (more than 30 km) and dominance by sandstone, an alternative explanation for the lack of debris flow deposits in the Van Horn Sandstone is that it formed on a relatively low-slope alluvial plain.

Most alluvial fan deposits in the stratigraphic record are relatively thick,

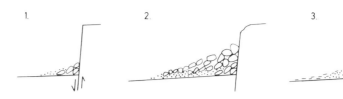

Figure 8
Diagram of sedimentary response on alluvial fan to faulting along adjacent mountain front. Progradation of fan (1) initially deposits fine sediments, subsequently followed by coarser material (2). As the uplifted source is worn down and the fan returns to equilibrium, finer sediment is again deposited (3). The result is an upward coarsening sequence followed by an upward fining sequence.

Figure 9
Proximal debris flow deposit with subhorizontal megaclast orientation: Lower Member, Cannes de Roches Formation, Gaspé, Canada (Rust, 1978b, Fig. 1). Tape 20 cm.

indicating formation in a tectonically-influenced setting. Red colouration and evaporitic paleosols (facies P, Table I) are also fairly common, and point to a semi-arid paleoclimate. The Devonian Peel Sound Formation of Arctic Canada contains red conglomerates and shows evidence of syntectonic activity (Miall, 1970). The fan conglomerates are transitional downslope into alluvial plain sandstones and thence to coastal deposits. Another Canadian example is the Carboniferous Cannes de Roche Formation of eastern Gaspé (Rust, in press). The lower and middle members of this formation are red beds with proximal and distal debris flow deposits (Figs. 9 and 10). Evaporitic paleosols are represented by nodular limestones which occur at the top of 2 to 5 m fining-upward cycles in the lower member and scattered throughout the finer middle member (Fig. 11), which is interpreted as a distal fan deposit.

Figure 10
Exposed top of distal debris flow deposit in the Lower Member, Cannes de Roche Formation (compare with Fig. 6). Rivière du Portage, Gaspé, Canada (Rust, in press). Tape 20 cm.

Braided Rivers and Alluvial Plains
Introduction. These two subenvironments are distinguished only on the basis of shape: braided rivers form one-dimensional valley fills, whereas braided alluvial plains (or braidplains: Allen, 1975) are two-dimensional. Braidplain deposits are therefore likely to be the

Figure 11
Vertical beds in Middle Member, Cannes de Roche Formation, Gaspé. Gm: horizontally stratified conglomerate, Fl/Fm: shale and mudstone, P: nodular limestone. The succession is interpreted as a distal alluvial fan deposit (Rust, in press). Tape 50 cm.

Figure 12
Horizontally stratified conglomerate (facies Gm) in the Devonian Malbaie Formation near *Pointe Verte, Gaspé, Canada (Rust, 1976, Fig. 1). Notebook 19 cm long.*

Figure 13
Bedding plane of imbricate conglomerate (facies Gm) in the Malbaie Formation near *Belle Anse, Gaspé (Rust, 1976, Fig. 1). Notebook 19 cm long.*

more abundant type in the stratigraphic record, and may be distinguished on the basis of extent perpendicular to the paleoslope. Another distinction is the directional relationship with tributary fan successsions. Alluvial fans which occur as complexes along mountain fronts lose their topographic individuality downstream, merging into braidplains (or bajadas) with the same mean flow direction. In contrast, fans tributary to braided rivers enter them laterally, with distinct topographic contrast (Figs. 1 and 2). However, the internal facies are identical in both subenvironments, and will be described together here. They differ from alluvial fans principally by a lack of debris flow deposits, and much more gradual facies changes downstream.

Proximal. The most abundant facies of coarse-grained proximal braided rivers is horizontally bedded, imbricate gravel, which may appear massive where bedding is thick and texture uniform (Figs. 12 and 13). This facies dominates proximal braided rivers of paraglacial environments (Boothroyd and Ashley, 1975; Church and Gilbert, 1975; Rust 1972a, 1975) as well as braided gravels not influenced by glacial melting (Ore, 1964, p. 9; Smith, 1970, p. 2999). The dominance of facies Gm reflects the low ratio of mean particle size to water depth, in turn a function of the relatively low relief of bars and channels in proximal reaches. The bars are mostly longitudinal: elongate parallel to flow, with gentle slopes into surrounding low sinuosity channels; diagonal bars are similar, but

oblique to flow (Smith, 1974, p. 210). Leopold and Wolman (1957) proposed that longitudinal bars start as a nucleus of the coarsest bedload fractions deposited in mid-channel as flow diminishes, and grow by addition of finer sediment mainly downstream from the nucleus. Smith (1974) observed similar processes during diurnal stage fluctuations in the Kicking Horse River, British Columbia: lateral and downstream growth of 'unit bars' with predominantly

depositional morphology. In the same river Hein and Walker (1977) observed an initial stage of bar formation as 'diffuse gravel sheets' a few pebble diameters thick; they postulated that the sheets evolve into longitudinal or diagonal bars with horizontal stratification, or transverse bars with cross-strata. The latter, however, are rare in gravel-bed braided streams (Smith, 1974, p. 218).

It is clear that falling-stage modifications of gravel bars occur by deposition-

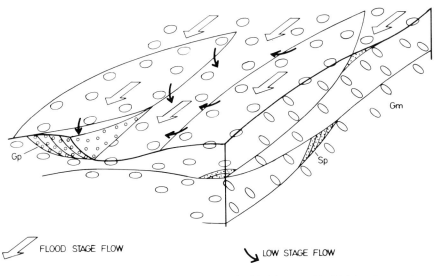

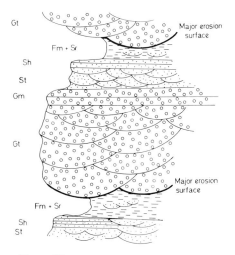

FLOOD STAGE FLOW

LOW STAGE FLOW

Figure 15
*Diagram of a fining-upward sequence in the
Upper Member of the Cannes de Roche
Formation.*

Figure 14
Depositional model for gravels of proximal

*braided rivers and braidplains. See text for
explanation of facies codes.*

al and erosional processes, but observations during flood stage are hampered by cloudy water and the impossibility of traversing bars, let alone channels. Remote sensing is also impracticable under these circumstances. Rust (1978b, p. 614-5) suggested that longitudinal bars are stable bedforms at flood stage, when *all* the bedload is in motion. An indication that this is so comes from the giant longitudinal bars (1.5 to 2.5 km long, 15 to 45 m high) of catastrophic Pleistocene floods (Bretz *et al.,* 1956; Malde, 1968). Giant cross-bed sets with boulders up to 3 m in diameter formed in estimated water depths up to 100 m (Malde, 1968), but longitudinal bar forms were preserved. Preservation of such large bars implies stability under the conditions prevailing. A possible analogy may be with the apparent bar forms in channels on Mars (Baker, 1978).

Debris flows occasionally travel beyond the margins of tributary alluvial fans, but they are rarely preserved in braided river deposits, because the consistent aqueous flow is likely to rework them before consolidation. Sand facies (Sp, Sh of Miall, 1978) are uncommon in proximal braided gravels (Fig. 14), and mud deposits are very rarely preserved. All these facies increase in abundance downstream, but unlike alluvial fan deposits the change is very gradual. For example, clast-supported gravel is the principal lithotype of the Donjek River 50 km from its glacial source (Area 2, Rust 1972a).

An ancient example of proximal braidplain gravels is present as conglo-merate units in the Devonian Malbaie Formation of eastern Gaspé (Rust, 1978b, p. 615-6). Horizontally bedded conglomerate is the dominant facies (Fig. 12), with remarkably consistent directions of clast imbrication (Fig. 13, and Rust, 1978b, Fig. 8). Planar cross-stratified conglomerate (facies Gp) makes up about 20 per cent of the conglomerate, a much higher proportion than in the modern equivalents described above. A possible explanation is that in Devonian times, before terrestrial plants had become established, gravel-dominated braidplains formed in all types of climatic environment, rather than in glacially-influenced and semi-arid environments, as is largely the case today. Sediment transport and accumulation rates would have been highest under humid paleoclimates, so the pre-late Paleozoic record of braidplain deposits is probably biased towards high rainfall areas. Given higher and more consistent rainfall, channel depth and bar relief would have been greater; hence we can expect a higher proportion of facies Gp formed by lateral outbuilding from longitudinal gravel bars into adjacent deep channels (Fig. 14).

The Lower Cretaceous Cadomin Formation contains further examples of proximal braidplain deposits (McLean, 1977). The dominant facies in conglo-merate units is horizontally stratified, clast-supported conglomerates (McLean, 1977, Plates 3 and 4A). The interpretation made here that the Cadomin is largely a braidplain rather than an alluvial fan deposit is based on

the extent of the formation, and on the lack of debris flow deposits (McLean, 1977, p. 809).

Distal. The facies model for distal braided gravels discussed here is not as well established as that for proximal equivalents; it is offered as a guide for further investigation. Miall (1977, 1978) chose the middle reaches of the Donjek River (Area 2 of Rust, 1972a) as the type example of his facies assemblage for distal gravelly braided rivers. Clast-supported gravel is the dominant lithotype in this reach of the river, and is abundant in the large, active channels. In a similar setting in the Knik River, Alaska, Fahnestock and Bradley (1973, p. 241 identified dunes of fine gravel by echo sounding. They probably resemble the crescentic gravel bedforms observed in the North Saskatchewan River by Galay and Neill (1967), and would generate sets of trough cross-strata (facies Gt) by migration in flood. An alternative possibility in shallower rivers is the formation of transverse gravel bars (Hein and Walker, 1977, Fig. 3) which would generate principally planar cross-strata (facies Gp) on migration.

The mid reaches of the Donjek are also characterised by inactive tracts on levels or terraces slightly above the active tract (Williams and Rust, 1969). The inactive tract is primarily subject to vertical accretion of fine sediment, and supports abundant vegetation. In flood, however, minor channels transport sand and fine gravel across the inactive tract. In time, migration of the active and inactive tracts can be expected to deposit a sequence which fines upward from trough cross-stratified clast-supported gravel through sand to mixtures containing mud and plant material (Fig. 16).

An ancient equivalent of the middle reach of the Donjek is the Upper Member of the Carboniferous Cannes de Roche Formation of eastern Québec (Rust, in press). Trough cross-stratified clast-supported conglomerate occurs in multiple sets above a sharp erosional base (Fig. 15). This facies (Gt) fines upwards to trough cross-stratified sandstone (Sh), commonly through intermediate units of horizontally bedded conglomerate (Gm). This succession is interpreted as a response to shallowing of water over bars and channels as they accrete, accompanied by, or in response to migration of the active tract. The sequence ends with mudstone and organic material deposited as the tract became inactive and started to support vegetation (Fig. 16).

Meandering Rivers

The definition of a meandering river used here is that of Rust (1978a): a single channel (braiding parameter less than 1) with sinuosity greater than 1.5. This excludes from consideration coarse-grained deposits of low sinuosity rivers such as the Nueces, Texas (Gustavson, 1978).

Several meandering rivers with gravelly beds have been described, but not all are coarse-grained by the definition given earlier in this paper. For example, the grain size data for point bars of the Amite River, Louisiana and the Colorado (Texas) show only one sample with a mean significantly coarser than −1.0 phi, the gravel/sand boundary (McGowen and Garner, 1970, Figs. 22 and 23). The sections given by these authors show that in general the sediments are sands or pebbly sands, with pebbles locally concentrated into lag gravels of limited vertical extent. The development of a lag or armour from mixed sediment is a surface phenomenon, caused mainly by selective transport of the finer fractions

at low stage (Day, 1976). The Endrick River, Scotland is similar in that its bar deposits contain pebbly sand with sheets of clast-supported gravel formed as lag concentrates (Bluck, 1971, p. 100). Trenches and cutbank sections of the Lower Wabash River, Illinois and Indiana show similar gravel concentrations with mean grain size about −1.0 phi at various horizons (Jackson, 1976, Figs. 6 to 8; Fig. 17 of this paper).

The Lower Wabash, Endrick, Amite and Colorado Rivers fall within category 4 of Jackson (1978): graveliferous sand bed streams, none of which rate the designation coarse-grained, as defined here. Jackson's fifth category of meandering river is exemplified by the Little Wind River, Wyoming, and is certainly coarse-grained (Jackson, 1978, p. 560-1). Jackson (pers. commun., 1979) reports difficulty in coring or trenching gravel in this river, a necessary procedure to establish the nature and preservability of gravel in the stratigraphic record. The data from a single trench show that clast-supported

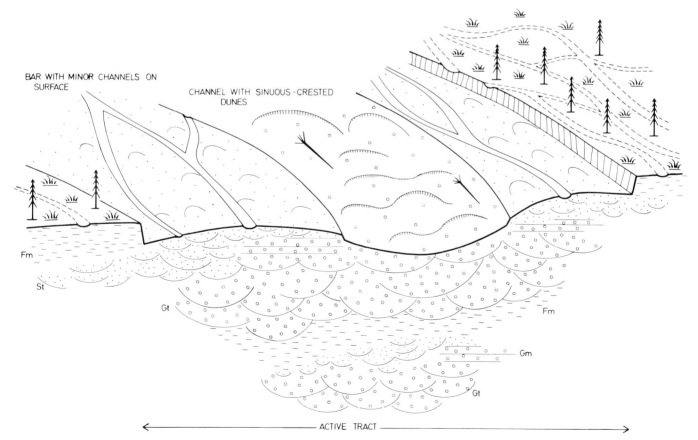

Figure 16
Depositional model for distal coarse-grained braided alluvium based on mid reaches of the Donjek River, Yukon and the Upper Member of the Cannes de Roche Formation. The Cannes de Roche cycle would be generated by the migration of the vegetated terrace and sandy bar with minor channels over the main gravelly channel with sinuous crested dunes.

gravel is present, but there is little evidence that it is the dominant facies.

Rust (1978b) proposed that the prime distinguishing characteristic of gravelly braided alluvium is the dominance of clast-supported gravel. This feature is easily determined in ancient successions, and is notably absent from sections of recent gravelly meandering rivers documented so far. Jackson (1978, p. 568-9) suggested that mean size above 0.5 m was a suitable criterion for distinguishing coarse braided alluvium from gravel deposits of meandering rivers, but such a texture is uncommon even for proximal alluvial fan deposits. Some of Jackson's other criteria for distinguishing braided and meandering gravelly alluvium require further amplifi-

cation. Low paleocurrent variance of imbricate gravel is a characteristic feature *only* of horizontally stratified gravel (facies Gm) in braided alluvium. Gravel fabrics measured by the author in planar cross-stratified conglomerate in the Malbaie Formation are highly variable. Another probable indicator of braided origin is low paleocurrent variance of trough cross-stratified clast-supported gravel (facies Gt) formed in the main channels of distal braided reaches (Fig. 16). It is very difficult to determine the orientation of equivalent bedforms in modern rivers, but by analogy with sinuous-crested dunes in braided channels of the sandy South Saskatchewan River (Cant, 1978), gravel dunes should also have low directional variance.

Alluvial and Other Gravels Compared

Glacial. Gravels deposited by glaciers (tills) differ from alluvial gravels in a number of respects. They are poorly sorted, lack stratification, and the more extensive (basal) till sheets commonly have a preferred fabric in which elongate clasts dip up-glacier (Price, 1973, p. 68-70; Pettijohn, 1975, p. 69). The only alluvial gravels with which they could be confused are debris flow deposits, but the lateral extent and fabric of basal tills would normally be distinctive.

Flow tills (Boulton, 1972) are small bodies of till that slide off the glacier or other topographic highs. They are a form of debris flow, but successions in which they occur can usually be distinguished from alluvial fan deposits by the abundance of ice-contact deformation and striated clasts. Winterer and von der Borch (1968) observed striated clasts in a recent debris flow deposit at Hookina Creek, South Australia. V.A. Gostin and the author re-examined this deposit, and found that the striations are fine, and are restricted to softer clasts in a mixed suite also containing much harder lithologies. We also found similar striated clasts in river gravels of varying hardness. Thus striated clasts are not unique to glacial gravels; the distinguishing feature seems to be that glacial striae are deeper, and are present on hard as well as soft clasts within a mixed suite.

Other Gravels. Beach gravels are formed by high-energy wave action, which achieves a high degree of sorting and rounding relatively close to the source. For example, Figure 18 shows a beach gravel at Nash Point, South Wales, derived from the vertical 30 m cliff in the background. The gravel is well sorted and rounded, and shows another typical feature of beach gravels: low-angle imbrication dipping seaward (Cailleux, 1945).

Clifton (1973) compared sorting and bedding characteristics of beach and alluvial gravels. He found that wave-worked gravels tend to be better segregated into discrete beds (with sand beds intervening) of greater lateral extent than those deposited by alluvial processes. In most cases the presence of marine fossils and association with offshore marine strata would also distinguish beach and alluvial deposits. However, as discussed above, alluvial fans may prograde into the sea, and hence gravels formed by alluvial processes

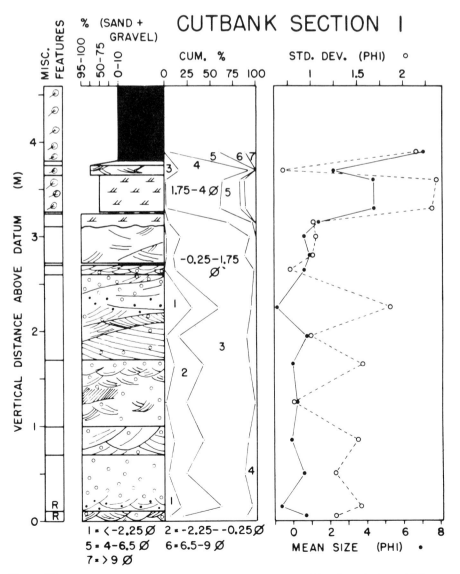

Figure 17
Cutbank Section 1 (of Jackson) from the Lower Wabash River (from Jackson, 1976, Fig. 10).

Figure 18
Gravel on storm beach, Nash Point, South Wales. The gravel is derived from thin- *bedded Liassic limestones in the cliff behind. Figures in background give scale.*

and with alluvial characteristics may contain marine fossils (Daily *et al.*, in press). Modern examples of deposits formed in this way occur in the Gulf of California (Walker, 1967; Meckel, 1975) and the Gulf of Elat (Gvirtzman and Buchbinder, 1978).

Submarine fan gravels and their consolidated equivalents have been discussed extensively by Walker (1976b, 1977, 1978), and require but brief mention here. The main criteria for distinguishing them from alluvial gravels are the association of submarine fan gravels with turbidites, including "classical" Bouma types, their resedimented marine faunas and fabric. The typical fabric has clast long (*a*) axes dipping upcurrent, in contrast to the common alluvial fabric in which the *ab* plane dips upstream, and *a* is transverse to flow (Rust, 1972b; Walker, 1976b, p. 30-31).

Summary
Models for coarse-grained alluvial deposition can be summarised for various environments as follows:

Alluvial Fans. Normally characterised by abundant (and commonly coarse) gravel proximally, alluvial fans show rapid fining downstream. In contrast to proximal braided rivers and braidplains, they show greater lithological diversity at any point in the fan, and commonly show grading (fining upwards, coarsening

upwards, or coarsening then fining) on a scale of several metres related to tectonic movements of the adjacent bedrock source. A characteristic feature is the presence of debris flow deposits (facies Gms), mostly in proximal regions.

Proximal Braided Rivers and Braidplains. These two subenvironments are only distinguishable in the ancient record on the basis of extent, and paleocurrent relationships to tributary alluvial fan deposits. Otherwise, they are identical, and are characterised by an abundance of facies Gm: horizontally bedded, clast-supported, imbricate gravel. Devonian and older deposits may have appreciably higher contents of cross-stratified conglomerate (facies Gp), due to formation in humid climates before the advent of terrestrial vegetation. Other facies are comparatively rare.

Distal Braided Rivers and Braidplains. Gravel deposits in these environments are characterised by fining-upward cycles dominated by clast-supported, trough cross-stratified gravel (facies Gt). This passes upward to horizontally stratified gravel (Gm), or to trough cross-stratified and horizontally stratified sand (facies St and Sh). Each cyclic sequence ends with massive or laminated mud (Fm, Fl) containing plant material, and the next cycle commences with an irregular erosion surface.

Meandering Rivers. Gravelly deposits of meandering rivers can be distinguished from coarse braided alluvium by the lack of clast-supported gravel as a dominant facies. In most cases, clast-supported gravel forms as thin lag sheets by reworking of pebbly sand at low stage. A coarser type, exemplified by the Little Wind River (Jackson, 1978) requires further investigation.

Acknowledgements
The work on which this paper is based was supported by grant A 2672 from the National Science and Engineering Research Council of Canada, which is gratefully acknowledged. I would also like to thank Andrew Miall, Roger Walker and Roscoe Jackson for comments on the manuscript, Roscoe Jackson for unpublished field data from the Little Wind River, Edward Hearn for drafting and photography and S.L. Meunier and H.J. De Gouffe for typing.

References

Basic References

Walker, R.G., 1975, Conglomerate: Sedimentary structures and facies models, *in* J.C. Harms, J.B. Southard, D.R. Spearing, and R.G. Walker, Depositional environments as interpreted from primary sedimentary structures and stratification sequences: Soc. Econ. Paleont. Min., Short Course 2, p. 133-161.
A good introduction to the use of sedimentary concepts to build facies models for conglomerate deposition.

Collinson, J.D., 1978, Alluvial sediments. *in* H.G. Reading, ed., Sedimentary environments and facies: Oxford, Blackwell Scientific Publications, p. 15-60.
An excellent discussion of the whole spectrum of alluvial deposits, and a good paper to review after reading Papers 2 and 3 of this reprint volume.

Bull, W.B., 1977, The alluvial fan environment: Progr. Phys. Geog., v. 1, p. 222-270.
The latest review by an author who has contributed much to our understanding of alluvial fans. Mainly deals with morphology and deposits of modern fans, but ancient equivalents are also discussed.

Steel, R.J., S. Maehle, H. Nilsen, S.L. Røe, and Å. Spinnangr, 1977, Coarsening-upward cycles in the alluvium of Hornelen Basin (Devonian), Norway: Sedimentary response to tectonic events: Geol. Soc. Amer. Bull., v. 88, p. 1124-1134.
A well illustrated account of a remarkable succession of ancient alluvial fan conglomerates and their tectonic setting.

Miall, A.D., 1977, A review of the braided-river depositional environment: Earth Sci. Revs., v. 13, p. 1-62.

A review of modern and ancient braided alluvial deposits, which introduces the facies code used in this paper.

Jackson, R.G., 1976, Depositional model of point bars in the lower Wabash River: Jour. Sed. Pet., v. 46, p. 579-594.
An account of gravelly sands formed by point bar migration in a meandering river.

Miall, A.D., ed., 1978, Fluvial Sedimentology: Can. Soc. Petrol. Geol., Mem. 5, 859 p.
Last but not least: the latest word on just about all aspects of fluvial sedimentology. Beginners beware of indigestion!

General

Boulton, G.S., 1972, Modern arctic glaciers as depositional models for former ice sheets: Jour. Geol. Soc., v. 128, p. 361-393.

Cailleux, A., 1945, Distinctions des galets marins et fluviatiles: Bull. Soc. Géol. France, v. 15, p. 375-404.

Clifton, H.E., 1973, Pebble segregation and bed lenticularity in wave-worked versus alluvial gravel: Sedimentology, v. 20, p. 173-187.

Day, T.J., 1976, Preliminary results of flume studies into the armouring of a coarse sediment mixture: Geol. Survey Canada, Paper 76-1C, p. 277-287.

Garrels, R.M. and MacKenzie, F.T., 1971, Evolution of sedimentary rocks: New York, Norton, 397 p.

Miall, A.D., 1978a, Lithofacies types and vertical profile models in braided river deposits: a summary, in A.D. Miall, ed., Fluvial Sedimentology: Can. Soc. Petrol. Geol. Mem. 5, p. 597-604.

Minter, W.E.L., 1978, A sedimentological synthesis of placer gold, uranium and pyrite concentrations in Proterozoic Witwatersrand sediments, in A.D. Miall, ed., Fluvial Sedimentology: Can. Soc. Petrol. Geol. Mem. 5, p. 801-829.

Pettijohn, F.J., 1975, Sedimentary rocks: New York, Harper and Row, Third edition, 628 p.

Pienaar, P.J., 1963, Stratigraphy, petrology and genesis of the Elliot Group, Blind River, Ontario, including the uraniferous conglomerate: Geol. Survey Canada, Bull. 83, 140 p.

Price, R.J., 1973, Glacial and fluvioglacial landforms: London, Longman, 242 p.

Robertson, D.S., J.E. Tilsley, and G.M. Hogg, 1978, The time-bound character of uranium deposits: Econ. Geol. v. 73, p. 1409-1419.

Rust, B.R., 1976, Stratigraphic relationships of the Malbaie Formation (Devonian), Gaspé, Quebec: Can. Jour. Earth Sci., v. 13, p. 1556-1559.

Rust, B.R., 1978a, A classification of alluvial channel systems, in A.D. Miall, ed., Fluvial Sedimentology: Can. Soc. Petrol. Geol. Mem. 5, p. 187-198.

Schumm, S.A., 1977, The Fluvial System: New York, Wiley-Interscience, 335 p.

Walker, R.G., 1976a, Facies models 1, General introduction: Geosci. Canada, v. 3, p. 21-24.

Walker, R.G., 1976b, Facies models 2, Turbidites and associated coarse clastic deposits: Geosci. Canada, v. 3, p. 25-36.

Walker, R.G., 1977, Deposition of upper Mesozoic resedimented conglomerates and associated turbidites in southwest Oregon: Geol. Soc. Amer. Bull., v. 88, p. 273-285.

Walker, R.G., 1978, Deep-water sandstone facies and ancient submarine fans: Models for exploration for stratigraphic traps: Amer. Assoc. Petrol. Geol. Bull., v. 62, p. 932-966.

Alluvial Fans

Bull, W.B., 1963, Alluvial fan deposits in western Fresno County, California: Jour. Geol., v. 71, p. 243-251.

Bull, W.B., 1964, Alluvial fans and near-surface subsidence in western Fresno County, California: U.S. Geol. Survey Prof. Paper 437-A, 71 p.

Bull, W.B., 1972, Recognition of alluvial-fan deposits in the stratigraphic record, in W.K. Hamblin and J.K. Rigby, eds., Recognition of ancient sedimentary environments: Soc. Econ. Paleont. Mineral. Spec. Pub. 16, p. 63-83.

Church, M. and J.M. Ryder, 1972, Paraglacial sedimentation: a consideration of fluvial processes conditioned by glaciation: Geol. Soc. Amer. Bull., v. 83, p. 3059-3072.

Curry, R.C., 1966, Observation of Alpine mudflows in the Tenmile Range, Central Colorado: Geol. Soc. Amer. Bull., v. 77, p. 771-776.

Daily, B., P.S. Moore, and B.R. Rust, in press, Terrestrial-marine transition in the Cambrian rocks of Kangaroo Island, South Australia: Sedimentology.

Gvirtzman, G. and B. Buchbinder, 1978, Recent and Pleistocene Coral Reefs and Coastal Sediments of the Gulf of Elat: Postcongress Guidebook, Internatl. Cong. Sedim., Jerusalem, p. 161-191.

Heward, A.P., 1978, Alluvial fan sequence and megasequence models: with examples from Westphalian D - Stephanian B coalfields, Northern Spain, in A.D. Miall, ed., Fluvial Sedimentology: Can. Soc. Petrol. Geol. Mem. 5, p. 669-702.

Holmes, A. and D.L. Holmes, 1978, Principles of Physical Geology: Sunbury-on-Thames, Nelson, Third edition, 730 p.

Hooke, R.L.B., 1967, Processes on arid-region alluvial fans: Jour. Geol., v. 75, p. 438-460.

Johnson, A.M., 1970, Physical Processes in Geology: San Francisco, Freeman, 575 p.

Koster, E.H., 1978, Transverse ribs: their characteristics, origin and paleohydraulic significance, in A.D. Miall, ed., Fluvial Sedimentology: Can. Soc. Petrol. Geol. Mem. 5, p. 161-186.

Larsen, V. and R.J. Steel, 1978, The sedimentary history of a debris-flow dominated, Devonian alluvial fan - a study of textural inversion: Sedimentology, v. 25, p. 37-59.

Legget, R.F., R.J.E. Brown, and G.H. Johnson, 1966, Alluvial fan formation near Aklavik, Northwest Territories, Canada: Geol. Soc. Amer. Bull., v. 77, p. 15-30.

McGowen, J.H. and C.G. Groat, 1971, Van Horn Sandstone, West Texas: An alluvial fan model for mineral exploration: Bur. Econ. Geol., Univ. Tex. Rept. Invest. 72, 57 p.

Meckel, L.D., 1975, Holocene sand bodies in the Colorado delta area, northern Gulf of California, in M.L. Broussard, ed., Deltas, Models for Exploration: Houston Geol. Soc., p. 239-265.

Miall, A.D., 1970, Devonian alluvial fans, Prince of Wales Island, Arctic Canada: Jour. Sed. Pet., v. 40, p. 556-571.

Middleton, G.V. and M.A. Hampton, 1976, Subaqueous sediment transport and deposition by sediment gravity flows, in D.J. Stanley and D.J.P. Swift, eds., Marine sediment transport and environmental management: New York, Wiley, p. 197-218.

Rust, B.R., in press, The Cannes de Roche Formation: Carboniferous alluvial deposits in eastern Gaspé, Canada: Comptes Rendus, Internatl. Carb. Congr., Urbana.

Ryder, J.M., 1971, The stratigraphy and morphology of paraglacial alluvial fans in south-central British Columbia: Can. Jour. Earth Sci., v. 8, p. 279-298.

Sharp, R.P. and L.H. Nobles, 1953, Mudflows of 1941 at Wrightwood, Southern California: Geol. Soc. Amer. Bull., v. 64, p. 547-560.

Steel, R.J., 1974, New Red Sandstone floodplain and piedmont sedimentation in the Hebridean Province, Scotland: Jour. Sed. Pet., v. 44, p. 336-357.

Walker, T.R., 1967, Formation of red beds in modern and ancient deserts: Geol. Soc. Amer. Bull., v. 78, p. 353-368.

Wasson, R.J., 1977, Last-glacial alluvial fan sedimentation in the Lower Derwent Valley, Tasmania: Sedimentology, v. 24, p. 781-799.

Wasson, R.J., in press, A debris flow at Reshun, Pakistan Hindu Kush: Geograf. Annal.

Winder, C.G., 1965, Alluvial cone construction by alpine mudflow in a humid temperate region: Can. Jour. Earth Sci., v. 2, p. 270-277.

Winterer, E.L. and C.C. von der Borch, 1968, Striated pebbles in a mudflow deposit, South Australia: Paleogeog., Paleoclim., Paleoecol., v. 5, p. 205-211.

Rivers

Allen, P., 1975, Wealden of the Weald: a new model: Proc. Geol. Assoc., v. 86, p. 389-437.

Baker, V.R., 1978, The Spokane flood controversy and the Martian outflow channels: Science, v. 202, p. 1249-1256.

Bluck, B.J., 1971, Sedimentation in the meandering River Endrick: Scot. Jour. Geol., v. 7, p. 93-138.

Boothroyd, J.C. and G.M. Ashley, 1975, Processes, bar morphology, and sedimentary structures on braided outwash fans, northeastern Gulf of Alaska, *in* A.V. Jopling and B.C. McDonald, eds., Glaciofluvial and Glaciolacustrine Sedimentation: Soc. Econ. Paleont. Mineral. Spec. Publ. 23, p. 193-222.

Boothroyd, J.C. and D. Nummedal, 1978, Proglacial braided outwash: a model for humid alluvial fan deposits, *in* A.D. Miall, ed., Fluvial Sedimentology: Can. Soc. Pet. Geol. Mem. 5, p. 641-668.

Bretz, J.H., H.T.U. Smith, and G.E. Neff, 1956, Channelled scabland of Washington: new data and interpretations: Geol. Soc. Amer. Bull., v. 67, p. 957-1049.

Cant, D.J., 1978, Development of a facies model for sandy braided river sedimentation: Comparison of the South Saskatchewan River and the Battery Point Formation, *in* A.D. Miall, ed., Fluvial Sedimentology: Can. Soc. Petrol. Geol. Mem. 5, p. 627-639.

Church, M. and R. Gilbert, 1975, Proglacial fluvial and lacustrine environments, *in* A.V. Jopling and B.C. McDonald, eds., Glaciofluvial and Glaciolacustrine Sedimentation: Soc. Econ. Paleont. Mineral. Spec. Publ. 23, p. 22-100.

Fahnestock, R.K. and W.C. Bradley, 1973, Knik and Matanuska Rivers, Alaska: a contrast in braiding, *in* M. Morisawa, ed., Fluvial Geomorphology: State Univ. New York, Binghampton, Proc. 4th Geomorph. Symp., p. 220-250.

Galay, V.J. and C.R. Neill, 1967, Discussion of "Nomenclature for bed forms in alluvial channels": Jour. Hydr. Div., Amer. Soc. Civ. Eng., v. 93, p. 130-133.

Gustavson, T.C., 1978, Bed forms and stratification types of modern gravel meander lobes, Nueces River, Texas: Sedimentology, v. 25, p. 401-426.

Hein, F.J. and R.G. Walker, 1977, Bar evolution and development of stratification in the gravelly, braided, Kicking Horse River, British Columbia: Can. Jour. Earth Sci., v. 14, p. 562-570.

Jackson, R.G., 1978, Preliminary evaluation of lithofacies models for meandering alluvial streams, *in* A.D. Miall, ed., Fluvial Sedimentology: Can. Soc. Petrol. Geol. Mem. 5, p. 543-576.

Leopold, L.B. and M.G. Wolman, 1957, River channel patterns: Straight, meandering and braided: U.S. Geol. Survey Prof. Paper 282-B, p. 39-85.

Malde, H.E., 1968, The catastrophic Late Pleistocene Bonneville Flood in the Snake River Plain, Idaho: U.S. Geol. Survey Prof. Paper 596.

McGowen, J.H. and L.E. Garner, 1970, Physiographic features and stratification types of coarse-grained point bars: Modern and ancient examples: Sedimentology, v. 14, p. 77-111.

McLean, J.R., 1977, The Cadomin Formation: Stratigraphy, sedimentology and tectonic implications: Can. Petrol. Geol. Bull., v. 25, p. 792-827.

Ore, H.T., 1964, Some criteria for recognition of braided stream deposits: Univ. Wyoming Contr. Geol., v. 3, p. 1-14.

Ruegg, G.H.J., 1977, Features of Middle Pleistocene sandur deposits in the Netherlands: Geol. Mijnb., v. 56, p. 5-24.

Rust, B.R., 1972a, Structure and process in a braided river: Sedimentology, v. 18, p. 221-245.

Rust, B.R., 1972b, Pebble orientation in fluvial sediments. Jour. Sed. Petrol., v. 42, p. 384-388.

Rust, B.R., 1975, Fabric and structure in glaciofluvial gravels, *in* A.V. Jopling and B.C. McDonald, eds., Glaciofluvial and Glaciolacustrine Sedimentation: Soc. Econ. Paleont. Min. Spec. Publ. 23, p. 238-248.

Rust, B.R., 1978b, Depositional models for braided alluvium: *in* A.D. Miall, ed., Fluvial Sedimentology: Can. Soc. Pet. Geol. Mem. 5, p. 605-625.

Smith, N.D., 1970, The braided stream depositional environment: Comparison of the Platte River with some Silurian clastic rocks, north-central Appalachians: Geol. Soc. Amer. Bull., v. 81, p. 2993-3014.

Smith, N.D., 1974, Sedimentology and bar formation in the upper Kicking Horse River, a braided outwash stream: Jour. Geol., v. 82, p. 205-223.

Walker, R.G., 1976c, Facies models 3: Sandy fluvial systems: Geosci. Canada, v. 3, p. 101-109.

Williams, P.F. and B.R. Rust, 1969, The sedimentology of a braided river: Jour. Sed. Pet., v. 39, p. 649-679.

MS received May 29, 1979.

Facies Models 3. Sandy Fluvial Systems

Roger G. Walker and Douglas J. Cant
Department of Geology
McMaster University
Hamilton, Ontario L8S 4M1

Introduction

We will describe in general terms the depositional environments of modern fluvial systems, and attempt to formulate a general model. We will then show how the model acts, 1) as a *norm* for purposes of comparison, 2) as a *framework* and *guide* for future observations, 3) as a *predictor* in new geological situations, and 4) as a *basis for hydrodynamic interpretation*.

For reasons of space, we will consider here only the sandy systems. Gravelly and bouldery systems are less well understood, although their deposits are abundant in Canada and locally are economically important (e.g., the uraniferous conglomerates in the Elliot Lake area are generally considered to be fluvial).

Sandy rivers can be straight, meandering or braided, although natural straight rivers are very uncommon. There would appear to be a spectrum of types from meandering to braided; meandering systems are fairly well understood, and sandy braided systems less well understood. At the moment it seems best to consider them separately, and use two facies models. Comparison of new situations with the meandering norm and the braided norm should help to establish the range of variation between the two types of system.

Sandy fluvial deposits are known from all geological systems, Archean to recent. They form important

hydrocarbon reservoirs, the meandering systems depositing elongate shoestring sands stratigraphically bounded by shales, and the sandy braided systems forming thicker and laterally more extensive sand bodies.

Meandering Systems

The main elements of a modern meandering system [exemplified by the Mississippi or Brazos (Texas) Rivers] are shown in Figure 1. Sandy deposition is normally restricted to the main channel, or to partially or completely abandoned meander loops; deposition of fines (silt and clay) occurs on levees and in flood basins. It is surprising that there are so few integrated studies of modern meandering systems in the literature of the last twenty years. The most important papers include those of Sundborg (1956; River Klaralven), Harms *et al.* (1963; Red River), Bernard *et al.* (1970; Brazos River), McGowan and Garner [1970; Colorado (Texas) and Amite Rivers] and Jackson (1976; Wabash River). Although the meandering river model is well established, its basis is somewhat dated. Further integrated work (not just on one point bar) would seem important.

a) The Main Channel. Meandering in the channel is maintained by erosion on the outer banks of meander loops, and deposition on the inner parts of the loops. The main depositional environment is the *point bar,* which

builds laterally and downstream across the flood plain.

The channel floor commonly has a coarse "lag" deposit of material that the river can only move at peak flood time. This material would include the gravelly component of the clastic load, together with water-logged plant material and partly consolidated blocks of mud eroded locally from the channel wall. Above the lag, sand is transported through the system as bedload. During average discharge, the typical bedform on the channel floor consists of sinuous-crested dunes (Fig. 1) ranging in height from about 30 cm to one metre. Preservation of these dunes results in trough cross-stratification. In shallower parts of the flow, higher on the point bar, the bedform is commonly ripples (preserved as trough cross lamination; Fig. 1). As a broad generalization, we may propose that the preserved deposits of the active channel will pass from trough cross-bedded coarse sands to small scale, trough cross-laminated fine sands upward (Fig. 1).

The development of a plane bed (without ripples or dunes) is favoured by higher velocities and shallower depths, and deposition on the plane bed results in horizontal lamination. The particular combinations of depth and velocity required to produce plane bed can occur at various river stages, and hence parallel lamination can be formed both low and high on the point bar. It can therefore be preserved interbedded with

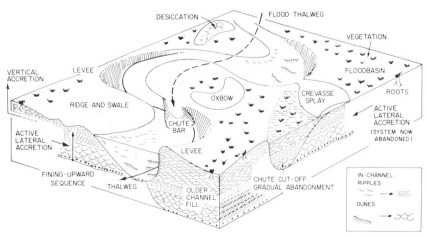

Figure 1
Block diagram showing major morphological elements of a meandering river system. Erosion on the outside bend of meander loops leads to lateral accretion on the opposite

point bar - the dunes and ripples in the channel give rise to trough cross-bedding and ripple cross-lamination, respectively (inset, lower right), which are preserved in a fining-upward sequence. See text for details.

trough cross-bedding, or small scale trough cross-lamination (Figs. 2,3).

The preservation of all of these features is basically due to the lateral and downstream accretion of the point bars. Channel-floor dunes may be driven slightly diagonally onto the lower parts of the point bar, thus helping the point bar to accrete laterally on top of the channel-floor lag. The upper part of the point bar may be composed of a series of ridges and swales (Fig. 1), the swales acting to funnel flood waters across the point bar with a much straighter thalweg than the normal low-stage meandering thalweg. If coarse bed load is funnelled through the swales at high flow, it may be deposited at the downstream end in the form of "chute bars" (Fig. 1; McGowan and Garner, 1970). These chute bars act to lengthen the point bar in the downstream direction, hence constricting the flow and causing increased erosion immediately downstream.

Pure meandering streams rarely have exposed bars in the middle of their channels, and hence the sandy active-channel deposits can all be termed LATERAL ACCRETION deposits, related to lateral migration of point bars (Fig. 1).

b) Channel Abandonment. Meander loops can be abandoned gradually (chute cut-off) or suddenly (neck cut-off) (Allen, 1965, p. 118-9, 156). During chute cut-off, the river gradually re-occupies an old swale, and simultaneously flow gradually decreases in the main channel. Gradual abandonment thus results in gradual flow decrease, and this could be reflected in the sediments by the development of a thick sequence of low-flow sedimentary structures - essentially ripple cross-lamination (Fig. 4). After complete abandonment, forming an ox-bow lake, sedimentation would be restricted to fines (silt, mud) introduced into the ox-bow during overbank flooding from the main stream (Fig. 1).

Neck cut-off involves the breaching of a neck between two meanders, and the sudden cut-off of an entire meander loop. Both the entrance to and exit from the loop tend to be rapidly plugged with sand. Flow diminishes to zero very quickly and the resulting sequence of deposits is dominated by later, flood-introduced silts and muds (Fig. 4).

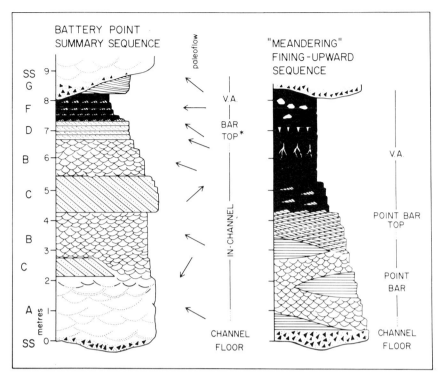

Figure 2

Comparison of a summary sequence of facies for the Devonian Battery Point Sandstone (Gaspé, Quebec: Cant and Walker, 1976) with the meandering river model. The model is redrawn to correct scale

from data in Allen, 1970. Implications of the comparison are discussed in the text. Letters beside the Battery Point sequence refer to the facies of Cant and Walker (1976). V.A. indicates vertical accretion.

c) Vertical Accretion Deposits. Outside the main channel, deposition in the flood basins, ox-bows and levees takes place by addition of sediment during flood stage when the river overtops its banks (Fig. 1). In contrast to the lateral accretion within the main channel, overbank deposition causes upbuilding of the flood plain, hence the term VERTICAL ACCRETION. Near to the main channel, where the flood waters sweep along as a stream, the vertical accretion deposits tend to be silty, and are commonly cross-laminated. Farther from the river, flood waters may stagnate and only mud is deposited. After retreat of the flood, the mud and silt commonly dry out, and dessication cracks are formed. The flood basins and levees of most river systems (post-Silurian) tend to be abundantly vegetated, and hence the deposits contain root traces (Fig. 5). In some climatic regimes, the vegetation may grow sufficiently abundantly to form coal seams. In semi-arid or arid environments, the fluctuating water table and drying at the surface favour the formation of caliche-like nodules within the vertical accretion deposits.

During rising flood stage, the levees can be breached, causing the formation of a "crevasse-splay" (Fig. 1) - a wedge of sediment suddenly washed into the flood basin, and commonly containing some of the coarse bedload portion of the river sediment. The crevasse splay deposit may resemble a classical turbidite in having a sharp base, overall graded bedding, and a sequence of sedimentary structures indicative of decreasing flow during deposition.

The only other deposits that may rarely be preserved as part of the vertical accretion sequence are windblown, and may be either loess, or coarser sandy deposits blown in as large dunes.

d) Meandering River Facies Sequence. The distillation of observations from a large number of modern meandering streams, and from many ancient formations interpreted as meandering-fluvial, allows a general facies sequence to be formulated. One version of this sequence is shown in Figure 2; it was distilled by Allen (1970) and is redrawn to scale here. In its simplest form, the sequence is FINING-UPWARD and consists of in-channel deposits (lateral accretion), followed by overbank fines (vertical accretion) (Figs. 6, 7).

In this particular sequence, the facies relationships were determined statistically for a large number of Devonian outcrops in Britain and North America, but application of the model has demonstrated that it can be used appropriately in many other areas. The lag deposits are overlain by trough cross-bedding, which is in turn overlain by small scale trough cross-lamination. Horizontal lamination can occur at several places within this sequence (Fig. 2), depending on the river stage at the time when the depth/velocity criteria for plane bed were met.

After the channel migrated away laterally, the facies sequence continued with vertical accretion deposits introduced at flood stage. The diagram (Fig. 2) shows root traces, dessication cracks and caliche-like concretions. Using the data presented by Allen (1970, Table 9), it can be seen that the vertical and lateral accretion deposits in the meandering model are on average roughly equal in thickness.

Allen's model serves excellently as a norm with which to compare other fining-upward sequences (see particularly Allen, 1964). In its construction, Allen apparently used a

Figure 3
Fining-upward sequence, Cretaceous Belly River Formation, on Trans-Canada Highway between Calgary and Banff. Note sharp base *to sand body, and cross bedding (by note-book) in lower part. Upper part of sand body (by geologist) is ripple cross laminated and overlain by fines.*

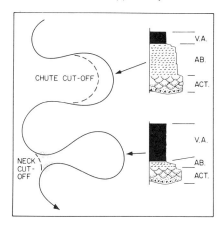

Figure 4
Meander loops can be abandoned by chute or neck cut-off. Old channel shown solid, new channels dashed. Chute cut-off involves reoccupation of an old swale and gradual abandonment of the main channel. The stratigraphic sequence will consist of some trough cross-bedded deposits of the active river (ACT) and a thick sequence of ripple cross-laminated fine sands representing gradual abandonment (AB). After cut-off, the sequence is completed by vertical-accretion (V.A.) deposits. By contrast, after neck cut-off, the meander loop is suddenly abandoned and sealed off by deposition of sand plugs (stipple). After the active deposits, the ripple cross-laminated fine sands representing low flow during abandonment (AB) are very thin, and the bulk of the sequence consists of vertical-accretion (V.A.) deposits washed into the abandoned loop at flood time. Compare with the active lateral-accretion sequence (Fig. 2).

Figure 5
Large root system in fluvial overbank deposits, part of the U. Devonian (Catskill) clastic wedge of the Appalachians. Photo from River Road, West Virginia (near Hancock, Md.).

Figure 6
A complete fining-upward sequence form the Maringouin Formation, Nova Scotia. Note sharp base, interbedding of sandstones and shales toward top of sand body, and overbank fines before incoming of next sand body (top left). Directly above geologist's head is a small mud-filled channel that cuts out the interbedded sandstones and shales – it may represent an abandoned swale on a point bar.

Figure 7
A complete fining-upward sequence from the upper part of the Battery Point Formation (L. Devonian) at Penouille (near Gaspé), Quebec. Base (B) rests on massive red mudstones (the vertical-accretion deposits of the underlying sequence). The coarse member is cross-bedded (above figure) and passes up into thick vertical-accretion red siltstones. A new fining-upward sequence begins on the left of the photo.

wide variety of fining-upward sequences. Comparison of sequences such as those in Figure 4 with Allen's norm immediately shows that the trough cross-bedding is very reduced in thickness, that one of the sequences contains an abnormal thickness of ripple trough cross-lamination, and that both contain unusual thicknesses of vertical accretion deposits. The comparison with Allen's model suggests the interpretations shown in Figure 4; without the model, we would not be so conscious that the sequences in Figure 4 differed significantly from the sequence developed by lateral accretion in an active channel.

e) Sand Body Geometry and Flood-Plain Aggradation. One of the essential components of a meandering model is the fact that meander loops are cut off, abandoned, and ultimately filled with fines – silt and clay. Through time, these clay plugs, along with thick back-swamp clays, may become abundant because the plugs are relatively hard to erode. Once confined, the entire meander belt may become raised above the general level of the flood plain by vertical accretion (Fig. 8A). This situation can persist until one catastrophic levee break results in the sudden switch of the entire river to a lower part of the floodplain ("avulsion", Fig. 8A). Thus the sand body geometry of a highly sinuous meandering stream will be essentially elongate ("shoestring"), bounded below and on both sides by flood-basin fines. The shoestring will also stand a good chance of being covered by overbank fines from the active river in its new position. Thus the *high sinuosity meandering model* predicts that, given continuing supply and basin subsidence, a series of sand lenticles interbedded with shales should be developed. The internal structure of the sand lenticles themselves should conform roughly to the pattern shown in Figure 1. A single-sequence sand body should be about as thick as the river was deep; however, stacking of sand bodies can occur with younger channel deposits cutting into older ones. The vertical scale in Figure 8A is considerably exaggerated, and individual shoestrings will probably be many times wider than they are thick.

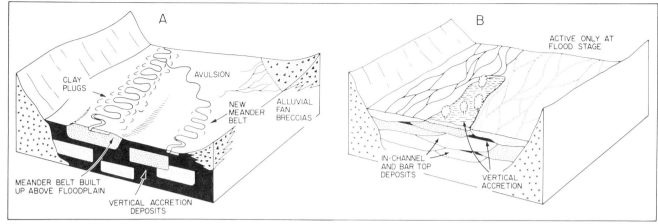

Figure 8
*A, Block diagram of flood-plain aggradation
with very sinuous rivers. Shoestring sands are
preserved, and are surrounded by vertical*

*accretion siltstones and mudstones. Vertical
scale is highly exaggerated. Compare with B,
block diagram of a braided sandy system with
low sinuosity channels. Vertical accretion*

*can occur during flood stage, for example on
the vegetated island, but deposits are rarely
preserved. Diagrams modified from those in
Allen (1965).*

Sandy Braided Fluvial Systems

In contrast to meandering rivers, sandy
braided systems have received
relatively little study. The best known
rivers include the Durance and Ardèche
(Doeglas, 1962), Brahmaputra (Coleman, 1969), Platte (Smith, 1970), Tana
(Collinson, 1970) and South Saskatchewan (Cant and Walker, 1978). The
morphological elements of these rivers
(Fig. 9) are complex, and include (in
increasing scale) individual bedforms,
small "unit" bars, bar complexes (or
sandflats), and mature vegetated
islands. The river itself flows over and
between these sand accumulations in a
constantly branching and rejoining
braided pattern. The finer material (silt
and clay) tends to be transported
through the system without
accumulation; vertical accretion
deposits are rarely preserved, and
deposition in flood basins is not such an
important process as it is in meandering
systems.

a) Braiding vs Meandering – Controls.
The fundamental processes that control
whether a river has a braided or
meandering pattern are not completely
understood, but we do know that
braiding is favoured by rapid discharge
fluctuations, of a greater absolute
magnitude than in meandering rivers.
Braided rivers also tend to have higher
slopes, a coarser load, and more easily
erodable banks. In combination, these
features would suggest that braiding is
more characteristic of the upstream
reaches of a river, with meandering

becoming more common downstream
as the slope and coarseness of load
decrease.
*b) Braided Channels and Sand
Accumulations.* The channels tend to be
very variable in depth and width, and do
not conform to the simple pattern shown
by meandering rivers. The channel floor
commonly has a lag deposit, and above
the lag, sand is transported through the
system as bedload. Bedforms in the
deeper channels (3 m or deeper) tend to
be sinuous crested dunes that give rise
to trough cross-bedding. Deposition
within channels during waning flood
stage can cause channel beds to

aggrade, preserving flood stage sedimentary structures. In shallower channels, and on bar tops when they are
submerged at flood stage, the bedforms
now known as sandwaves are common
(see Harms *et al.*, 1975, p. 24, 47-49).
Sandwaves tend to be straight crested,
and have a very long wavelength (many
metres) compared with their height (20-
50 cm); they give rise to planar tabular
sets of cross bedding when preserved
(Fig. 9, number 5).

In contrast to meandering systems,
point bars are very uncommon in

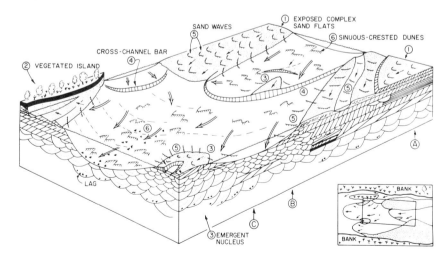

Figure 9
*Block diagram showing elements (num-
bered) of a braided river (based on the South
Saskatchewan). Stippled areas exposed, all
other features underwater. Bar in left corner is
being driven laterally against a vegetated
island, and is forming a slough in which mud is*

*being deposited. Large sandflats (e.g., right
hand corner) may develop by growth from an
emergent nucleus on a major cross-channel
bar (see Fig. 10). Vertical fining-upward
sequences A, B and C are shown in Figure 12,
and include in-channel and bar top* deposits.
See Figure 2 and text for details.*

braided systems. Instead, bars tend to occur within channels, and if the bar is single and represents only one episode of growth, it can be termed a "unit" bar (Smith, 1970). Unit bars can be longitudinal (elongated parallel to flow) or transverse; in the South Saskatchewan transverse bars can extend across the entire width of channels and have been termed cross-channel bars by Cant and Walker (1978); Fig. 10 of this article). In the Platte, Tana and Brahmaputra, many of the transverse bars are linguoid (tongue) shaped, and there is a wide spectrum of intermediate shapes. The term diagonal bar has been given to those which have a foreset trending at an angle to the main direction of the channel.

The cross-channel bars shown in Figure 9 are forms of transverse bars. In the South Saskatchewan, many cross-channel bars have a "nucleus" that is emergent at low stages (Figs. 9, 10; Cant and Walker, 1978). The nucleus grows by lengthening downstream as sand is swept around in two "horns" (Fig. 10), and it also grows in the upstream direction as dunes and sand waves are driven up from the channel floor. As the nucleus grows, possibly with other bars coalescing onto it, the original unit bar expands into a large sandflat (Cant and Walker, 1978; Fig. 11). The South Saskatchewan sandflats are complex, and their original shape has been obscured by dissection and redeposition during changing river stage (Fig. 9). They are one to two km long in the South Saskatchewan, three km in the Tana (Collinson, 1970) and up to 10 km in the Brahmaputra (Coleman, 1969). In the South Saskatchewan, they remain constant for at least five to six years, and because of their size, they would seem likely parts of the braided system to be preserved in the stratigraphic record.

In the South Saskatchewan, one other type of bar is commonly developed. The bar is elongated parallel to the channel trend, but flow is diagonally across the bar, resulting in a foreset slope dipping almost perpendicularly to the channel trend (bar in left hand corner, Fig. 9). They form in areas of flow expansion and commonly isolate a quiet slough between the bar and an adjacent bank or vegetated island. The slough is an area

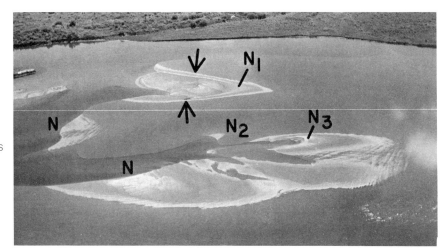

Figure 10
Cross channel bar with several nucleii (N). South Saskatchewan R., near Outlook, Sask., flow to left. Nucleus N₁ has the welldeveloped "horns" that have grown downstream. Fore-set avalanche faces dip in direction of arrows. Nucleus N₂ is in its earliest and simplest state, and N₃ has grown horns downstream but has also aggraded considerably on the upstream side.

Figure 11
Compound sandflat in South Saskatchewan R., at Outlook, Sask., flow to left. In foreground, nucleus N has extensive "horns" growing downstream. Upstream from nucleus large sandwaves can just be seen through the water. As they are driven onto the nucleus, it will aggrade and grow in the upstream direction.

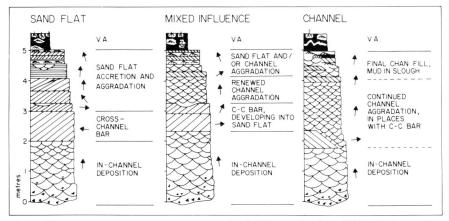

Figure 12
Three proposed sequences of sedimentary structures, based on the South Saskatche-wan River. "Sand Flat" corresponds to A on Fig. 9, "Channel" to C, and "Mixed Influence" to B. Arrows indicate generalized paleoflow directions, and sequences are explained in the text.

of mud deposition at low stage, or may be a passage for sandwaves at high stage.

From our understanding of the South Saskatchewan (Cant and Walker, 1978), we propose a series of stratification sequences (Fig. 12) that might characterize the deposits of this type of river. The channel sequence (Fig. 12) would consist of a lag, overlain by trough cross stratification – in our sketch, we also show partial preservation of planar-tabular cross bedding formed by a cross-channel bar. The sand flat sequence also begins with a lag and trough cross bedded sand sequence, but sand flat development is initiated by formation and preservation of a cross channel bar. A spectrum of stratification sequences probably exists between these two end members – we show one (Fig. 12) that begins as a channel sequence, and is followed by the initiation of a sand flat and the trough cross stratification suggests channel aggradation.

The sandy tops of all of these sequences are composed of small planar and trough cross beds, and rippled sands, making up the feature termed bar top* in Fig. 9. The bar top* (with asterisk) implies that deposition and modification are not restricted to the exposed bar tops, but may also take place in shallow dissection channels. The terminology of in-channel and bar top* was first used for ancient rocks (Cant and Walker, 1976; Fig. 2 of this article); it is important that the same terms be used for ancient and recent sediments where possible.

c) *Vertical Accretion Deposits.* In contrast to meandering streams, the vertical accretion deposits of braided streams are less commonly deposited and only rarely preserved. At low stage, the river may only occupy one or two of the available channels on the flood plain (Fig. 8B). It adjusts to flood stage by re-using the empty or abandoned channels, and only during major floods does the river spill from its main channel system onto the surrounding flood plain. In the South Saskatchewan between the Gardiner Dam and Saskatoon, the flood plain is very narrow and the braided portion is essentially confined between Pleistocene bluffs. Consequently, the narrow flood plain and the vegetated islands can relatively easily be submerged and receive vertical accretion deposits.

The Brahmaputra spills into its flood basins every year, but the clays settling from the flood waters are deposited slowly, with thickness of two cm or less per annum. However, vegetation is abundant in these flood basins, and peat deposits one to four m in thickness are forming (Coleman, 1969, p. 232-3). These various sub-environments of the braided sandy system are sketched in Figure 9, but there is certainly more complexity in the deposits than is indicated in the diagram.

d) *Ancient Sandy Braided Fluvial Deposits.* Very few ancient sandy systems have been positively identified as braided (or low sinuosity) rivers. The best studies include those of Moody-Stuart (1966; Devonian of Spitsbergen), Kelling (1968; Coal Measures, South Wales) and Cant and Walker (1976; Devonian Battery Point Sandstones of Quebec Appalachians). We will briefly comment on the Battery Point Sandstones because we are familiar with them.

In the field area near Gaspé, Quebec, we found a 110 m sequence that could be divided into at least 10 generally fining-upward sequences. Within the section, we were also able to define eight distinct facies, characterized by their various scales and combination of sedimentary structures. The sequence of facies was "distilled" (Walker, first paper in this volume) in order to look for a general facies sequence that could act as a basis for interpretation – see also Miall (1973), Harms *et al.,* (1975, p. 68-73) and Cant and Walker (1976, p. 111-114). The end result of the Battery Point distillation was the sequence shown here in Figure 2. It is *not* a model – it is only a summary of a local example that could, in the future, be re-distilled with local examples from other areas to produce a general facies model. In the Battery Point summary sequence, we identified a channel-floor lag overlain by poorly defined trough cross-bedding (Facies A, Fig. 2). The in-channel deposits consisted of well-defined trough cross-bedding (B) and large sets of planar-tabular cross-bedding (C) that commonly showed a large paleocurrent divergence form the trough cross-bedding (Figs. 2, 9; Cant and Walker, 1976, Fig. 7). The bar-top* deposits consisted mainly of small sets of planar-tabular cross-bedding (D), and the thin record of vertical accretion included

cross-laminated siltstones interbedded with mudstones (F), and some enigmatic low-angle cross-stratified sandstones (G).

Upon developing this summary sequence, our first reaction was to compare it to the existing fluvial (meandering) norm (Fig. 2). Although both sequences showed channelled bases, followed by fining-upward sequences, there appeared to be sufficient differences that the norm would *not* act reliably as a basis for interpretation (Walker, first paper in this volume). In other words, the meandering model seemed inappropriate for the Battery Point Sandstone.

Comparison with the norm nevertheless highlighted the major differences, and this gave us added understanding of the Battery Point. Similar comparisons of other systems with the two sequences in Figure 2 should also give added understanding. For example, the vertical-accretion deposits in the Battery Point are very thin compared with the meandering norm, both in absolute terms, and in proportion to the amount of in-channel sandstone. The in-channel sandstones do not contain parallel lamination, but planar-tabular sets of cross-bedding are common, and show high paleocurrent divergences from the main channel trend. All of these points of comparison aided in making our "braided" interpretation (Cant and Walker, 1976, p. 115-8). Refer also to annotated comments about Campbell, 1976, in references.

e) *Sand Body Geometry and Flood-Plain Aggradation.* One major point of contrast with the meandering system is that braided rivers tend to have easily erodible banks, and no clay plugs. The area occupied by the braided river may therefore be very wide (see Campbell, 1976), and coalescing bars and sand-flats will result in a laterally continuous and extensive sand sheet, unconfined by shales (Fig. 8B). Vertical accretion deposits (if formed) will tend to be quickly eroded because of the comparatively rapid lateral migration of channels. Consequently, any shales preserved in the section will tend to be patchy, laterally discontinuous, and relatively ineffective barriers to vertical hydrocarbon migration. This will not be the case for meandering systems.

Sandy Fluvial Facies Models

The meandering model effectively does all of the things a facies model should. It is a well-established *norm,* and serves as a *guide* for future observations. It has been used as a *basis* for hydrodynamic interpretation (Allen, 1970; Cotter, 1971), and has served many oil companies as a *predictor* in new situations. The braided "model", in as much as one exists, does none of these things very well.

Comparisons of new braided examples with the existing meandering norm will be instructive, but will not add to the braided "norm". What is needed is a well-established braided norm, so that more systematic work on the variability of modern braided rivers and their ancient deposits. In this way the range of variability will become better known, and the implications of a braided river "norm" will be better understood.

Basic References on Sandy Fluvial Sedimentation

Reineck, H. E. and I. B. Singh, 1973, Depositional Sedimentary Environments: New York, Springer-Verlag, 439 p.

A basic summary of all fluvial types of sedimentation can be found in the section on Fluvial environment, p. 225-263. This is a good, well-illustrated review for readers unfamiliar with fluvial sedimentation, and is recommended over Allen, 1965 (below), which is too long and a little out-of-date.

Harms, J. C. and R. K. Fahnestock, 1965, Stratification, bed forms and flow phenomena (with an example from the Rio Grande), *in* G. V. Middleton, ed., Primary sedimentary structures and their hydrodynamic interpretation: Soc. Econ. Paleontol. Mineral., Spec. Publ. 12, p. 84-115.

This paper is a basic source of information on bed forms, stratification and flow regimes in rivers. Some of the more recent ideas can also be found in chapters 2, 3 and 4 of Harms *et al.,* 1975 (below).

McGowan, J. H. and L. E. Garner, 1970, Physiographic features and stratification types of coarse-grained point bars: modern and ancient examples: Sedimentology, v. 14, p. 77-111.

This is one of the most recent integrated descriptions of meandering systems, and the comparison of ancient and modern sediments is particularly useful.

Smith, N. D., 1970, The braided stream depositional environment: comparison of the Platte River with some Silurian clastic rocks, north-central Appalachians: Geol. Soc. Amer. Bull., v. 81, p. 2993-3014.

Good description of downcurrent changes in bar type, on a regional scale, and comparison with ancient rocks.

Cant, D. J. and R. G. Walker, 1976, Development of a braided-fluvial facies model for the Devonian Battery Point Sandstone, Quebec: Can. Jour. Earth Sci., v. 13, p. 102-119.

Distillation of a local summary facies sequence from an ancient sandstone, emphasizing facies descriptions, analytical methods, and comparisons with a meandering norm.

Miall, A.D., 1977, A review of the braided-river depositional environment: Earth-Sci. Review, v. 13, p. 1-62.

Miall reviews gravelly and sandy systems, and suggest that all fluvial deposits can be described using a standard set of facies (3 gravelly, 5 sandy, 2 fine-grained). Following a "distillation" of these facies in many examples, he suggests four braided-river sequences – the Scott, Donjek, Platte and Bijou Creek types (the S. Saskatchewan type has since been added, Miall, 1978).

Miall, A.D., 1978, ed., Fluvial Sedimentology: Calgary, Canadian Soc. Petrol. Geol., Memoir 5, 859 p.

This volume stems from the first international congress on Fluvial Sedimentology, October, 1977 in Calgary. It contains 41 papers covering all aspects of fluvial sedimentation, and is now *the* most important source book for ideas, data and references. Extensive and important discussions of facies models are given by Jackson, Collinson, Leeder, Miall, Rust and Cant.

Collinson, J.D., 1978, Alluvial sediments: *in* H.G. Reading, ed., Sedimentary environments and facies. Oxford, Black-well, 1978, p. 15-60.

This chapter is now the most useful modern general review of alluvial systems. For readers unfamiliar with rivers, it is the next thing to read after our chapter in this reprint volume.

Horne, J.C. *et al.,* 1978, Depositional models in coal exploration and mine planning in Appalachian region: Amer. Assoc. Petrol. Geol., Bull., v. 62, p. 2379-2411.

An excellent, well-illustrated study of how fluvial (and marginal marine) models can be used in exploration.

Other References Cited in Text

Allen, J. R. L., 1964, Studies in fluviatile sedimentation: six cyclothems from the Lower Old Red Sandstone, Anglo-Welsh Basin: Sedimentology, v. 3, p. 163-198.

Contrasts the features of six fining-upward sequences, and interprets their origin in terms of river type, behaviour, and sub-environment. An interesting and well illustrated detailed comparison.

Allen, J. R. L., 1965, A review of the origin and characteristics of Recent alluvial sediments: Sedimentology, v. 5, p. 89-191.

An excellent review, with abundant citations, of fluvial systems. Because of its length, I recommend readers begin with Reineck and Singh (quoted above) before working through Allen.

Allen, J. R. L., 1970, Studies in fluviatile sedimentation: a comparison of fining-upward cyclothems, with special reference to coarse-member composition and interpretation: Jour. Sediment. Petrol., v. 40, p. 298-323.

Good technical discussion of the internal structures of the coarse member, and their hydrodynamic interpretation.

Bernard, H. A., C. F. Major, Jr., B. S. Parrott, and R. J. LeBlanc, Sr., 1970, Recent Sediments of Southeast Texas: Bur. Econ. Geol. Texas, Guidebook No. 11.

The first part of the guidebook contains a very brief description but abundant illustrations of the Brazos River.

Campbell, C.V., 1976, Reservoir geometry of a fluvial sheet sandstone: Amer. Assoc. Petrol. Geol., Bull., v. 60, p. 1009-1020.

The Morrison Formation in northwestern New Mexico consists of trough cross beds. Sand bodies are typically channelized and cut into one another, forming a sand sheet that can be traced at least 85 km in an E-W direction, with a minimum area of 2600 km^2. The block diagram showing a sequence of sedimentary structures and textures differs very considerably from our Figure 2 – the overall abundance of *trough* cross bedding in ancient braided sandy systems is not known.

Cant, D.J. and R.G. Walker, 1978, Fluvial processes and facies sequences in the sandy braided South Saskatchewan River, Canada: Sedimentology, v. 25, p. 625-648.

Description of bar and sandflat development, emphasizing the development of a summary block diagram (Fig. 9 of our paper in this reprint volume).

Coleman, J. M., 1969, Brahmaputra River: channel processes and sedimentation: Sedimentary Geol., v. 3, p. 129-239.

Excellent description of the Brahmaputra system. The best, most detailed (but longest) description of a braided river yet published.

Collinson, J. D., 1970, Bedforms of the Tana River, Norway: Geograf. Ann., v. 52A, p. 31-56.

Good description of bedforms and their interaction with changing river stage.

Cotter, E., 1971, Paleoflow characteristics of a late Cretaceous river in Utah from analysis of sedimentary structures in the Ferron Sandstone: Jour. Sediment. Petrol., v. 41, p. 129-138.

An example of the type of hydrodynamic predictions that can be made from ancient fluvial sediments.

Doeglas, D. J., 1962, The structure of sedimentary deposits of braided rivers: Sedimentology, v. 1, p. 167-190.

A pioneering paper on fluvial stratification in braided systems; contains excellent descriptions and illustrations that are still useful.

Harms, J. C., D. B. MacKenzie, and D. G. McCubbin, 1963, Stratification in modern sands of the Red River, Louisiana: Jour. Geol., v. 71, p. 566-580.

One of the first full descriptions of the internal structures of point bars, and still one of the best.

Harms, J. C., J. B. Southard, D. R. Spearing and R. G. Walker, 1975, Depositional environments as interpreted from primary sedimentary structures and stratification sequences: Soc. Econ. Paleontol. Mineral., Short Course 2 (Dallas, 1975), 161 p.

Summary of modern ideas on bedforms and stratification in Chapters 2 and 3.

Jackson, R.G., II., 1976, Largescale ripples of the lower Wabash River: Sedimentology, v. 23, p. 593-632.

The paper emphasizes bedforms, their hydrodynamic conditions of stability, and the resulting stratification.

Kelling, G., 1968, Patterns of sedimentation in Rhondda Beds of South Wales: Amer. Assoc. Petrol. Geol., Bull., v. 52, p. 2369-2386.

Emphasizes fining-upward sequences and paleocurrent variations in late Pennsylvanian (Coal Measure) braided stream deposits.

Miall, A. D., 1973, Markov chain analysis applied to an ancient alluvial plain succession: Sedimentology, v. 20, p. 347-364.

Discussion of analytical methods of sequence analysis, applied to Devonian rocks in Arctic Canada.

Moody-Stuart, M., 1966, High and low sinuosity stream deposits with examples from the Devonian of Spitsbergen: Jour. Sediment. Petrol., v. 36, p. 1102-1117.

Description of rocks in Spitsbergen, with a general discussion of low and high sinuosity models.

Sundborg, A., 1956, The river Klaralven: a study of fluvial processes: Geog. Ann., v. 38, p. 127-316.

Walker, R. G., 1976, Facies models - 1. General Introduction: Geosci. Can., v. 3, p. 21-24.

This paper has been extensively rewritten and now appears as paper 1 in this volume.

Canadian Examples

There is an astonishing absence in the Canadian literature of detailed interpretations of ancient sandy fluvial depositional environments. A quick review of the total contents of the *Canadian Journal of Earth Sciences* and *Bulletin of Canadian Petroleum Geology* gave no references at all. There are obviously great thicknesses of terrestrial deposits in the clastic wedges of the Cordilleran, Arctic Island, and Appalachian foldbelts, but if detailed interpretations of the sandy fluvial systems exist, they are in government reports. There are, of course, many examples described as fluvial, with some petrographic and paleocurrent information. However, none contains the necessary data on sedimentary structures and their sequence, integrated with paleoflow data in such a way that they contribute to sandy fluvial facies models.

New references to fluvial deposits, Canadian and elsewhere, appear in Miall, 1978 (above).

Appalachian Area

Belt, E. S., 1968, Carboniferous continental sedimentation, Atlantic Provinces, Canada, *in* G. DeV. Klein, ed., Late Paleozoic and Mesozoic continental sedimentation, northeastern North America: Geol. Soc. Amer., Spec. Paper 106, p. 127-176.

Discusses four facies – fanglomerate, fluvial, lacustrine and mixed fluvial/lacustrine. Many formations discussed and citations given.

Cant, D.J. and R.G. Walker, 1976 (see above).

MS received, February 27, 1976. Revised version for reprint volume, April, 1979. Reprinted from Geoscience Canada, v. 3, p. 101-109.

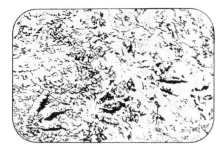

Facies Models 4. Eolian Sands

Roger G. Walker
and Gerard V. Middleton
Department of Geology,
McMaster University,
Hamilton, Ontario L8S 4M1

Summary

Facies models for extensive eolian sand deposits must be based on the characteristics of modern ergs ("sand seas"). Peripheral areas of ergs, where supply of sand is sparse, are characterized by barchans and longitudinal dunes: neither of these seems very likely to be preserved in the geological record. The central parts of ergs, where accumulated sand thickness reaches several hundred metres, are composed of complex pyramidal dunes (draas) with superimposed complex transverse dune types. Almost nothing is known by direct observation of the cross-bedding formed in such dunes, but it is possible that it is not as variable as the complex external morphology might suggest.

Identification of ancient sandstone formations as eolian is based mainly upon one major criterion: the very large scale and relatively high angle of the cross-bedding. In many cases minor indicators, such as eolian ripples (orientated at large angles to the trend of the major foresets), avalanch scours, animal tracks, soft-sand faults, raindrop impressions, etc., are also present. Negative evidence is equally important: the only known alternative model, that of a submarine tidal sand-wave field, is *not* known to produce large scale, high angle cross-bedding: it probably produces medium-scale cross-bedding resulting mainly from migration of megaripples or sand waves of relatively small scale, superimposed on the large

scale features. The cross-bedding probably shows at least some bipolar orientation of cross-bed directions, a feature absent from all classic ancient eolian sandstones. Furthermore, in the tidal sand wave model, sand deposits are associated with muddy sediments bearing a marine fauna, and this is not the case in the classic ancient eolian sandstones of the western U.S.A. and elsewhere.

Introduction

Eolian sandstones have fascinated sedimentologists, mainly because of the immense scale (up to 35 m) of the cross-bedding associated with dune migration. Recently, however, some of the "classic" eolian units, particularly the Navajo Sandstone, have been reinterpreted as shallow marine. These reinterpretations have been based upon a comparison with tidal ridges and sand waves on the present Continental Shelves. It therefore seemed appropriate in this facies model series to examine eolian sands, and to see what (if any) foundation the reinterpretations may have. A second purpose is to point out that there is no well documented, published account of an ancient Canadian eolian sandstone – perhaps an examination of eolian features will direct attention to possible Canadian examples.

In this paper, we will describe modern deserts and dunes, and then assemble information from the classic ancient examples, mostly from the Southwestern U.S.A. We will also describe briefly the large sand waves and tidal ridges on modern Continental Shelves, and review the controversial interpretations.

Modern Eolian Sands

Modern eolian sands are found in two main settings: sandy deserts and coastal dunes directly associated with sandy beaches. Of these the deposits of sand in deserts are by far the most extensive. Arid and semi-arid regions occupy about one third of the present land surface and comprise three main sedimentary environments: alluvial fans and ephemeral streams, inland sebkhas or playas, and sandy deserts, also called "sand seas" or ergs. Much of the area of modern deserts is composed of eroding mountains (40%) and stony areas (10 to 20%) where erosion, rather than deposition, is taking place, but on the

average sandy deserts form about 20 per cent of the area of modern deserts (Cooke and Warren, 1973, p. 52-53). As our ideas about modern environments tend to be strongly influenced by the examples with which we are personally familiar, it is worth noting that the desert areas of North America are not at all typical of those of the rest of the world: in America alluvial fans are much more important (30%) and sandy deserts much less important (less than 1%) than in other major deserts.

The largest desert in the world is the Sahara (7 million km^2). Traditionally it has been explored by the French, who have recently made good use of both conventional aerial and also satellite photography (e.g., Mainguet and Callot, 1974, Mainguet, 1976; also Wilson, 1971, 1972, 1973). The Sahara includes several major ergs, arranged in three main belts. Individual ergs cover areas as large as 500,000 km^2 (twice the area of Nevada): they are generally located in basins (both physiographic and structural) with a long history of sedimentary accumulation, including extensive fluvial deposition in Tertiary and Pleistocene times. The modern eolian deposits, however, are rarely more than 100 m in thickness.

Wilson (1971) found that ergs do not generally develop in areas of confluence of wind patterns: instead the predominant winds cross the ergs in more or less parallel lines. The main reason for the accumulation of sand in ergs seems to be the presence of a topographic depression, and the trapping action that the dunes themselves have on the movement of sand, once accumulation is initiated. Mainguet and Callot (1974) made a study of the Fachi-Bilman Erg, based on extensive air and satellite photography. The erg is situated in the southern Sahara, in a region of easterly trade winds. The trade winds are deflected by the Tibesti massif, and the erg is situated on the southwestern side of the massif, where two wind streams converge (Fig. 1). Within the erg, there is a definite spatial zonation of dune types. Barchans are found on all sides of the erg, forming an outer zone of relatively small, mobile bed forms in areas not yet fully "saturated" with sand. Inward, the barchans coalesce and are transformed into larger, less mobile forms: first sinuous seif dunes and then fully developed longitudinal dunes (called silks). In the upwind, central part of the erg is a zone of large pyramidal dunes (oghurds or draas) of considerable height (more than 100m) and width (0.5 to 1.5 km). Small seifs radiate in several directions from each of these large sand masses, which may nevertheless themselves be arranged in regular geometric patterns or rows. The main zone of longitudinal dunes (silks) is on the lee side of the central oghurds.

This particular erg appears to be fairly typical. As Wilson (1972) has pointed out, there are three main scales of eolian bed forms: small ripples (which are flatter than ripples formed in the water, and have more regular crest-lines), transverse and longitudinal dunes, from 0.1 to 100 m high, and complex pyramidal types, with heights of 20 to 450 m. The observed dune patterns are extremely complex, even in regions (such as the Fachi-Bilman Erg) characterized by relatively constant (trade) winds. It seems that large accumulations of sand (draas) modify the local wind patterns both by topographic deflection and by extreme heating of the sand surface, which produces convective circulation of air (more than 70% of the heat is transferred from the sand to the air during the day).

Much of the area of ergs is covered by only a thin veneer of sand, locally piled into barchans or longitudinal dunes. For example, Wilson (1973) estimated that in the Simpson Desert (Australia), an erg composed almost exclusively of longitudinal dunes, the *average* thickness of sand was only one metre, though individual dunes rise to 30 m. Eolian sands in such areas are much less likely to be preserved than those in areas dominated by draas, where average sand thickness may be over 100 m.

Our knowledge of the internal structures of modern desert dunes is still somewhat limited, and is due almost entirely to the work of McKee (1966; McKee and Douglass, 1971; McKee and Tibbitts, 1964; McKee and Moiola, 1975; see also Sharp, 1966). For example, the structures of longitudinal dunes, the commonest dune type in modern deserts, are known only from pit sections (each about 1 m^2) of a single small seif dune (14 m high) in the Zallaf sand sea of Libya (McKee and Tibbitts, 1964). The most detailed of McKee's studies have been of various dune types in the White Sands dune field of New Mexico. This dune field is not typical, both because it is small (700 km^2) and because the

sands are composed of gypsum which is relatively easily stabilized by occasional wetting during rains.

For longitudinal dunes, the predominant wind directions are roughly parallel to the crest, though there remain two theories on the origin of this type of dune. Bagnold (1941) among others thought that seifs formed by the amalgamation and modification of barchans, under the influence of winds blowing predominantly from *two* directions (both in the same or adjoining quarters). Hanna (1969) has argued that longitudinal dunes in Australia and elsewhere are parallel to a single predominant wind direction, with sand blown up to the crest from the stony interdune areas by secondary currents. The secondary currents are supposedly spiral vortices whose scale is related to the thickness of the atmospheric boundary layers (about 1 km). On either theory, sand is blown across the crest, first from one side then from another, and McKee and Tibbitts (1964) found that, near the surface, a Libyan seif dune was composed of large-scale cross-strata dipping at high angles in two nearly opposite directions, each roughly normal to the dune crest. It appears, however, that longitudinal dunes are relatively stable types (many show a considerable degree of vegetation, except near the crest) which form in areas of only sparse sand supply: they grow at the downwind end and do not migrate laterally. It is therefore possible that the structures recorded by trenching near the crest are not typical of those, more likely to be preserved, in the deeper parts of the dune.

For whatever reason, there have been very few examples of longitudinal dunes identified in the geological record. Most ancient eolian sandstones are characterized by large scale, high angle (20 to 30°) cross-beds with a single prominent modal direction of dip and a range of about 90 degrees on either side of the mode. These sands, if in fact they are eolian, are generally thought to be deposited from transverse dunes.

Modern transverse dunes are of several different types. Isolated transverse dunes (barchans) are by far the simplest and clearest type; they are crescent-shaped dunes, with the "horns" pointing downwind. McKee's (1966) trenches of a barchan dune showed high angle cross-strata inclined downwind with a rather low spread of

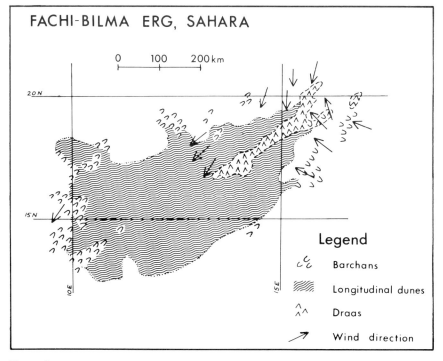

Figure 1

Distribution of dune types and wind directions in the Fachi-Bilma Erg, southeastern Sahara (modified from Mainguet and Callot, 1974).

directions, and cut by numerous, low angle, erosion surfaces, inclined at low angles downwind. This is perhaps not what was previously expected, but in any event, it is unlikely that thick ancient deposits of eolian sand can have been deposited from barchans, because barchans form mainly in peripheral regions of ergs, where sand supply is very limited. Transverse dunes with more or less continuous crests oriented almost normal to the wind direction are also found in modern deserts. The internal structures of one example, about 12 m high, was investigated by McKee at White Sands. Thick sets of tabular, planar cross-beds were revealed, with closely spaced erosion surfaces near the top of the dunes, but with erosional surfaces widely spaced and dipping at moderate angles downwind in the lower part of the dune. This dune did not display the wedge shaped sets typical of some supposed ancient dunes. Wedge shaped sets were shown by parabolic (partly vegetated) dunes at White Sands. Parabolic dunes are more typical of vegetated coastal dunes than of modern desert dunes, but it seems probable that the wedge shape of the dune results more from the irregular, sinuous shape of the crest (which is certainly typical of many modern transverse dunes in deserts) than from any inherent property of the parabolic dune shape.

Up to the present time, the closest matching of a modern dune type with an ancient dune sandstone has been achieved by Thompson (1969). He has described well-exposed, large scale cross-bedding at four localities in the Triassic of Cheshire (England) which closely matches the cross-bedding described by McKee from dome-shaped dunes at White Sands. The dome shaped dunes described by McKee are relatively low types, only a few metres high, without active slip faces: they are thought to develop from transverse dunes in areas where dunes are exposed to strong winds which the dune can grow.

The structures formed in large, complex pyramidal dunes (which are surely the most likely to be preserved in the stratigraphic column) are almost completely unknown from direct investigation. It is generally assumed that, because of their complex morphology, the direction of dip of cross-bedding within these complexes

must be equally disperse. A study of the Kelso Dunes in the Mojave Desert, California (Sharp, 1966) suggests that perhaps this is not necessarily the case. These dunes occupy an area of about 115 km^2 and rise to a height of 200 m above the surrounding land surface. The dune morphology is complex with transverse dunes superimposed on larger, oblique sand ridges: the wind regime is also complex, but predominantly from the west. Net changes in the positions of dunes measured over a 10 year period were small (50-100 m), despite highly active crestal movement related to short term changes in wind direction. Measurement of lee slope orientation at any one time showed a highly dispersed pattern, but trenching and more careful observation suggested that this was only a superficial phenomenon: cross-bedding observed in areas of reversed (westward dipping) lee slope was still predominantly eastward, and when an attempt was made to separate predominant from subordinate lee slopes the orientations corresponded fairly well with the predominant wind towards the east. Thus possibly the apparently chaotic form of draas is not representative of their internal structures. Many draas are found in regions where there is one predominant wind direction. Most sand is moved during relative short periods of strong winds (for a good description of a typical sand storm at White Sands, see McKee and Douglass, 1971), and it is possible that the cross-bedding reflects the orientations of these winds much more faithfully than does the large scale external morphology.

A feature of many ancient sandstones interpreted as eolian is the presence of near-horizontal surfaces that truncate sets of cross-beds and are generally spaced about 0.5 to 15 m apart. Stokes (1968) interpreted these surfaces as due to deflation of sand above a horizontal water table. McKee and Moiola (1975) recorded similar surfaces from interdune areas at White Sands, but pointed out that the water table in dunes is not generally horizontal (this objection does not seem entirely convincing, because progressive removal of sand from dunes would surely lead to the water table gradually approaching horizontality). They suggested that a water-table mechanism was not necessary, and that truncation surfaces

are simply produced by dune migration. Brookfield (1977) has recently reviewed the occurrence and origin of bounding surfaces in eolian sands. He agrees that such surfaces are formed by the migration of large bed forms: the mechanism is basically the same as that which produces bounding surfaces between cross-bed sets in "normal" medium scale, water-laid cross-beds. Generally the spacing between such surfaces is only a fraction of the total height of the bed form. In eolian sandstones, this fact together with the known small rates of supply of wind-blown sand, implies that bounding surfaces spaced a few metres apart can only be formed by very large bed forms that migrate only very slowly, that is by the migration of draas.

Characteristics of Ancient Eolian Deposits
In constructing facies models for submarine fans, fluvial systems and deltas in earlier papers in this series, there was little doubt that, for instance, the ancient examples used to build the fluvial model were in fact fluvial in origin. Because of the current controversy about some of the "classic" ancient eolian units, and their possible shallow marine origin, it is not so certain that the eolian facies model is being constructed exclusively with eolian examples. We do not agree with the shallow marine reinterpretations, but will discuss this problem below. Our description is based principally on the "classic" eolian units of the southwestern U.S.A. listed in Table I.

Table I

Characteristics of classic ancient eolian sandstones of the Western United States

Formation	Age	Approx. Max Thickness (m)	Thickness of Cross-bed Sets	Dip of Cross-beds (degrees)	Type of Cross-bedding	Reference
CASPER	Penn.-Perm.	240	min. 15 m	15-25	mostly troughs	Steidtmann, 1974
COCONINO	Perm.	330	"huge"	25-30	randomly oriented wedge-shaped sets	Baars, 1962
CEDAR MESA	Perm.	450	up to 30 m	up to 25 sweeping toe-sets	mostly planar tabular	Baars, 1962
DE CHELLY	Perm.	330	up to 35 m	15-35	simple planar tabular sets	Baars, 1962
WHITE RIM	Perm.	200	up to 20 m	19-27 long toe-sets	mostly planar tabular	Baars & Seager, 1970
LYONS	Perm.	40	up to 13 m	commonly 25 max. 28	mostly planar tabular	Walker & Harms, 1972
WINGATE	U. Triassic	130	up to 15 m	up to 30		Dane, 1935
NAVAJO	L. Jurassic	700	up to 30 m	20-30 long toe-sets	lower part – mostly planar tabular sets normally 3-10 m upper part – most troughs. Sets normally up to 6 m	Sanderson, 1974; Freeman & Vischer, 1975
ENTRADA (Slick Rock Member)	U. Jurassic	280	up to 8 m		"wedging sets" "sweeping eolian cross-beds"	Craig & Shawe, 1975

Ubiquitous large scale cross-bedding
The major feature of all these units is the presence of extremely large-scale cross-bedding (Fig. 2 and Fig. 3). In fact, the scale and type of the cross-bedding ("sweeping", "wedging" sets) has become so identified with eolian deposition that authors feel they can refer to "obvious eolian cross-stratification" without any other description of the unit in question (e.g., Baars, 1975, p. 125).

Individual cross-bed sets in the units of Table I are up to 35 m thick. Commonly, many sets will be of this thickness, not just the exceptionally thick ones. Reading between the lines of the descriptions, it appears that there are two main types – planar-tabular (Cedar Mesa, De Chelly, White Rim, Lyons, Lower Navajo) and trough or wedge shaped (Casper, Coconino, Upper Navajo, Entrada). The exact

Figure 2
Single set of planar tabular cross bedding, at least 20 m thick from canyon floor to upper truncation surface. White Rim Sandstone (Permian, Utah).

Figure 3
Set of low angle cross bedding showing asymptotic passage of toesets into bottomsets. Note length and thickness of bottomsets: exposed thickness of set about 2m. White Rim Sandstone (Permian, Utah).

three-dimensional geometry of such immense sets is very hard to document, but there is probably a broad correlation between straight crested dunes and planar-tabular sets, and between sinuous crested dunes and trough or wedge-shaped sets.

One of the best descriptions of the detailed structure of the large cross-beds is that of Walker and Harms (1972; Fig. 4). In vertical cross section parallel to flow, the set appears to be planar-tabular. On the exhumed lee face, there are very low amplitude ripples that suggest sand was blown across the dune face. The plan view shows that the entire dune advanced down-dip *not* by sand avalanching down the lee face, but by the addition to the lee face of wedges of sand blown into place across the dune. These features have been illustrated in photographs by Walker and Harms (1972, their Figs, 1, 2, 6, 8), who also suggest that the foresets dip at an angle (25-28 degrees) less than that of the angle of repose of dry sand (about 34 degrees) because sand is blown along the face rather than avalanching over the dune crest.

Eolian trough cross-bedding has been described and illustrated by Knight (1929) from superb, three dimensional outcrops near Sand Creek, Wyoming. Knight's diagrams are now classic examples of symmetrical troughs, and still appear in text books to illustrate this feature. In the Casper Sandstone, these troughs are up to 305 m wide, several times as long, and at least 15 m deep. Excellent photographs are given by Steidtmann (1974).

Several authors make the observation that the large cross-beds commonly have very long, sweeping, asymptotic topsets and bottomsets (Fig. 3). These bottomsets appear to be longer, to

aggrade into thicker units and to occur much more commonly in the units in Table I than in known subaqueous situations.

In summary, the outstanding feature of the cross-bedding is not only the immense scale (a few metres to a maximum of about 35 m), but also the fact that it occurs through great stratigraphic thicknesses (Table I) almost unrelieved by other lithologies or other sedimentary structures. This point will be discussed again later when controversial interpretations are considered.

Minor sedimentary structures. Many minor structures suggestive of subaerial deposition have been recorded from the eolian units listed in Table I. Some of the more important ones occur in the Lyons Sandstone (Walker and Harms, 1972), and include wind ripples, raindrop impressions, animal tracks, avalanche structures, and grain lag layers.

The ripples found in eolian units tend to be low, long wavelength features that may only be visible when sunlight glances across an outcrop. They are commonly oriented with crests parallel to the dip of foreset slopes (Fig. 4). A detailed view of the cross stratification in Figure 2 is shown in Figure 5. At about three levels, low angle cross lamination can be seen in sets each about one cm thick. This cross lamination is climbing *up* the main foreset slope, and is probably the stratigraphic record of small eolian ripples being blown diagonally up the foreset. By contrast, water-formed ripples of both wave and current type have lower ripple indices (wavelength/height), and are rarely oriented in the manner shown in Figure 4.

Animal tracks have been observed in many of the formations listed in Table I. Excellent examples are figured by Walker and Harms (1972, Fig. 12), who

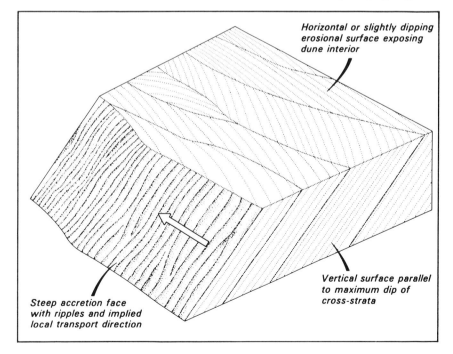

Figure 4
Geometry of eolian cross-bedding as seen in the Lyons Sandstone, Lyons Quarry (Permian, Colorado). From Walker and Harms (1972).

Figure 5
Detail of stratification on large foreset shown in Figure 2. Note the three thin sets of very low angle foresets, interpreted as low amplitude- *long wavelength wind ripples, with sand being blown up the large foresets in Figure 2. White Rim Sandstone (Permian, Utah).*

discuss experimental work indicating an exclusively subaerial, dry sand environment for the preservation of animal tracks.

Soft sediment deformation, in some cases on a very large scale (e.g., Sanderson, 1974) is known from several ancient eolian formations. Minor deformation structures have been described from modern dunes (see Bigarella, 1972). Many deformation structures seem to have been formed after sands become saturated with water, and therefore are not diagnostic of eolian deposition. There is one minor type, however, which does seem to be formed only in *partially* wet sands: small scale soft sediment faulting, (e.g., Steidtman, 1974), and this is found, for example, in the Navajo Sandstone.

Coarse lag deposits are known in modern interdune blowout areas (Bagnold, 1941, his Figs. 10, 53), and have also been described from the Lyons Sandstone (Walker and Harms, 1972, their Figs. 15, 16). The lags consist of coarser sand or gravel, and are characterized by very equal spacing of the grains on the lag surface. The equal spacing is a response to the dispersion of the larger grains in the lag bombardment from saltating sand blowing across the surface.

Within many eolian sandstones, there are thin mudstone, limestone or dolomite horizons representing the deposits of

ephemeral lakes. The upper surfaces are commonly covered in desiccation cracks, ripple marks, and wavy laminae indicative of drying out. In the Navajo Sandstone, these thin fine grained layers are less than a few metres thick, and rarely extend for more than a square kilometre (Freeman and Visher, 1975, p. 661).

Paleontological evidence. The eolian units in Table I are characterized by a variety of dinosaurs, dinosaur tracks, fresh water ostracods, and plant remains. A summary of the Navajo and Nugget fossil occurrences has been given by Picard (1977). No marine fossils have been found within the main, large scale cross-bedded portions of these formations, although some of them intertongue with other marine formations (e.g., the Navajo intertongues in its upper part with marine limestones of the Carmel Formation).

Interpretation of Eolian Sandstones
Until about 1960, the units listed in Table I were widely accepted as being eolian in origin. The interpretations were based essentially upon four criteria: 1) the absence of a marine fauna, and presence of dinosaur bones, dinosaur tracks, and non-marine ostracods, 2) the similarity of the giant cross-bed sets to those found in modern dunes, 3) the abundance of giant cross-bed sets, and

their monotonous occurrence through tens or hundreds of metres of section, 4) associated minor features, e.g., "wind" ripples, raindrop impressions, mud-cracked limestones and mudstone horizons. Criticisms of the eolian interpretation were published by Baars (1962, p. 178, Cedar Mesa) and Baars and Seager (1970, White Rim), who based their arguments on the style of the cross-bedding. Thus the Cedar Mesa was interpreted as "littoral to beach", citing "moderate to low angle cross-stratification, very low angle to horizontal thin sand beds, long sweeping curves from moderately dipping cross strata into horizontal fore-set beds, thin simple sets of low angle cross strata . . ." (Baars, 1962, p. 178). These features can equally well be interpreted as the long, low fore-sets and toe-sets of eolian dunes, the low angles being due partly to non-preservation of the steeper, higher parts of the dunes, and partly to the fact that the fore-set and toe-set was constructed by sand being blown around the dune, and over the dune crest in suspension, rather than avalanching. Baars did not mention, nor attempt to account for, the sets up to 30 m thick, which are unknown in modern littoral and beach environments. Criticism of the interpretation of the White Rim Sandstone (Baars and Seager, 1970) was based upon an irregular bar-like topography on the top of the White Rim. This bar like topography is mantled by a veneer of sand that is wave rippled, almost certainly in a marine environment. Baars and Seager (1970, their Fig. 6) show one veneered bar 1 to 3 km wide, about 15 km long, and with a vertical relief of 15 to 60 m. However, they base their interpretation of the White Rim Sandstone less on the immense cross-bedding within the White Rim (up to 20 m; Fig. 1), than on the eroded topography, the veneer, and the wave ripples *above* the White Rim.

Further criticism of the eolian origin of some of the units in Table I has been based upon comparisons with modern sand waves and tidal sand ridges that are common on many shelf areas (Houbolt, 1968; McCave, 1971; Swift, 1975). Thus Stanley, Jordan and Dott (1971, p. 13) wrote that "it can no longer be assumed *a priori* that large festoon cross strata prove an eolian dune origin for the Navajo or any similar sandstone *because of the essential identity of form and scale of modern submarine dunes or sand waves,* as documented during the last decade (e.g., Jordan, 1962)" (our italics). It is important, therefore that we describe briefly what the scale and form of these features is. We believe that some authors have misinterpreted the marine sand waves and tidal ridges, and hence have mistakenly reinterpreted some of the "classic eolian" units as shallow marine.

Marine Sand Waves and Tidal Ridges

These features have been described in detail from the North Sea (Stride, 1963; Houbolt, 1968; McCave, 1971) and the Atlantic Shelf off New England (Jordan, 1962,; Swift,1975). Houbolt's map of the tidal sand ridges shows that they may be 30 to 40 m high, about one to two km wide and 20 to 60 km long. More importantly, Houbolt made sparker sections across two of the ridges (Fig. 6) showing that they are asymmetrical, with a steep face toward the northeast, and apparent steep stratification surfaces within the sand ridges. However, it is clear in Figure 6 that the vertical scale on these diagrams is exaggerated, and calculation shows that the dip of the "steep" face averages only *5 degrees.* Thus Houbolt's sand ridges are *not* modern examples of cross-bed sets tens of metres thick with dips of 5 to 30 degrees (Table I), and Houbolt does *not* "show cross stratification in the submarine sand ridges of the North Sea to exceed 40 m in thickness" (as claimed by Pryor, 1971, in a reinterpretation of the "classical eolian" Permian Weissliegendes Sandstones of Germany).

Similar problems exist if tidal sand waves are used as a basis for interpretation. Sand waves have been described in the North Sea by McCave (1971) and Terwindt (1971). They are up to seven m high, with wavelengths of 200 to 500 m. McCave's echo soundings (Fig. 7) show that they have asymmetrical profiles, but calculations show average dips of 5 degrees on the "steep" face. Terwindt (1971, his Table 1) gives value for slopes on the sand waves in 21 locations, and the maximum steep face angle is only about 11 degrees. The mean steep face angle, calculated from Terwindt's table, is 4.5 degrees. Thus despite their height and

asymmetry, tidal sand waves of this type are clearly unsuitable models for the reinterpretation of "eolian cross-bedding", despite the brief comment of Stanley, Jordan and Dott (1971) cited earlier, and the more detailed reinterpretation of the Navajo by Freeman and Visher (1975).

The sand waves described by Jordan (1962) from Georges Bank (Gulf of Maine) are probably the largest and steepest reported. On Georges Shoal, the sand waves are up to 13 m high but are mostly symmetrical with calculated dips of "steep" faces of about 2 to 3 degrees. On Cultivator Shoal, the sand waves are asymmetrical, up to about 10 m high with steep faces of about 18 to 20 degrees. In most studies (Houbolt, 1968; McCave, 1971), the sand waves and tidal ridges appear to be covered with small bedforms (megaripples) - their migration would produce cross-bedding of a scale up to about a metre and the probable internal structure of the sand waves and tidal ridges would be complex, medium scale cross-bedding (with sets less than 1 m thick). Due to regular reversal of tidal currents, and the fact that different sides of tidal ridges are generally dominated by different phases (flood or ebb) of the tide, at least some of the cross-beds should be dipping almost opposite directions. One would expect to see abundant internal truncations

("reactivation surfaces") and at least some "herringbone" or bipolar cross-bedding.

The low angle of the "steep" faces of tidal sand ridges and sand waves (commonly about 5 degrees) makes a shallow marine reinterpretation of the units in Table I unlikely. Even if

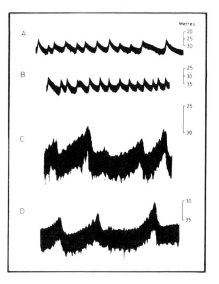

Figure 7

Profiles of sand waves in the southern bight of the North Sea. Vertical scale positioned with respect to sea level = O m. Lengths of sections about 3800 m (A); 2800 m (B); 900 m (C); and 1220 m (D); Calculated dips of "steep" faces average 5 degrees. (From McCave, 1971, p. 206).

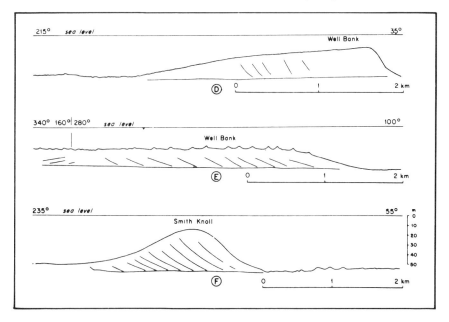

Figure 6

Sparker profiles of tidal sand ridges off southeast England. Although they look like asymmetrical cross-bedded features, compare the vertical and horizontal scales given. Calculated "steep" slopes average only 5 to 6 degrees. (From Houbolt, 1968).

comparison were made with Cultivator Shoal (heights up to 10 m, dips 18 to 20 degrees) the scale is still too small. However, we consider the other evidence to be overwhelmingly in favour of the classic eolian interpretations, namely the scale, abundance and monotony of the immense cross-bedding through tens or hundreds of metres of section, the absence of marine faunas, and the strong indicators of subaerial deposition - dinosaur tracks, raindrop impressions, gravel lags of equally-spaced grains, and interbedded thin mud-cracked layers. The latest marine reinterpretation of the Navajo (Freeman and Visher, 1975) has been positively and convincingly shot down by Picard (1977), Folk (1977), Steidtmann (1977) and Ruzyla (1977). Their arguments along with the excellent eolian documentation of the Lyons Sandstone by Walker and Harms (1972), constitute the most important pro-eolian literature.

Eolian Facies Model

In previous papers in this series, the facies models have been presented not only in terms of types of sedimentary structures, but in terms of their sequence and association. Examples include the fluvial fining-upward sequences and the deltaic coarsening-upward sequences. Based upon the sequences, it has been shown that the facies models can act as a norm (for purposes of comparison), a predictor (in new situations) and a guide for future observations.

By contrast, in eolian systems, there seems to be no preferred vertical sequence of sedimentary structures, nor any consistent lateral changes. This may be due to insufficient study of sequence in ancient eolian sandstones, or a tendency for irregular and unpredictable distribution of dune types in a desert. The result is that the "eolian facies model" can only be stated in terms of an assemblage of characteristic features, which may act as a norm, and as a guide to future observations, but which cannot act successfully as a predictor in new situations.

The single most characteristic feature must still be the large sets of cross-bedding on a scale of 5 to 35 metres, not simply occurring singly, but in monotonous cosets tens or hundreds of metres thick. Other features that strengthen the interpretation are the stratigraphic context and absence of marine fossils in the sandstones, and the minor features that indicate subaerial deposition.

As a guide for future observations, we would stress the importance of hunting for the minor features such as raindrop impressions, soft-sand faults, and animal tracks, that (in conjunction with abundant large scale cross-bedding) indicate subaerial depositon. Eolian sandstones might be expected to be interbedded with, or intertongue with, fluvial and alluvial fan sediments (Laming, 1966), or to overlie coastal sediments in regressive sequences.

References

Short List—Vital Eolian Reading

Bigarella, J. J., 1972, Eolian environments: their characteristics, recognition and importance: in J. K. Rigby and W. K. Hamblin, (eds.), Recognition of ancient sedimentary environments: Soc. Econ. Paleontol. Mineral., Spec. Publ. 16, p. 12-62.

Probably the best general review of modern and ancient eolian environments.

Freeman, W. E. and G. S. Visher, 1975, Stratigraphic analysis of the Navajo Sandstone: Jour. Sed. Petrology, v. 45, p. 651-668.

A very controversial interpretation of the Navajo Sandstone. However by reading this paper, and the discussions by Folk, Picard, Ruzyla and Steidtmann, you will get some idea of the descriptive parameters and their use. None of these discussions debates the scale and dip angle of modern subaqueous sand waves and ridges.

Hunter, R.E. 1977. Basic types of stratification in small eolian dunes. Sedimentology, v. 24, p. 361-387.

This is a very important paper, describing characteristic smaller-scale features of eolian stratification. It is well-illustrated, and doubtless many of the features will be used in the future as evidence in favour of eolian deposition.

Sanderson, I. D., 1974, Sedimentary structures and their environmental significance in the Navajo Sandstone, San Rafael Swell, Utah: Brigham Young Univ. Geol. Studies, v. 21, p. 215-246.

The best modern description of the Navajo Sandstone - probably the classic eolian unit.

Walker, T. R. and J. C. Harms, 1972, Eolian origin of flagstone beds, Lyons Sandstone (Permian), type area, Boulder County, Colorado: Mountain Geol., v. 9, p. 279-288.

Start your eolian reading with this paper - an excellent, concise description of major and minor structures in the Lyons Sandstone, with very good illustrations.

Modern Deserts and Dunes

Bagnold, R. A., 1941, The physics of blown sand and desert dunes: Methuen, London, 265 p.

Cooke, R. V. and A. Warren, 1973, Geomorphology in Deserts: London, Batsford, 394 p.

Hanna, S. R., 1969, The formation of longitudinal sand dunes by large helical eddies in the atmosphere: Jour. Applied Meteorology, v. 8, p. 874-883.

Hunter, R. E., 1977, Basic types of stratification in small eolian dunes: Sedimentology, v. 24, p. 361-387.

Mainguet, Monique, 1976, Propositions pour une nouvelle classifications des édifices sableux éoliens d'après les images des satellites Landsat I, Gemini, Noaa3: Zeitschr. f. Geomorph., v. 20, no. 3, p. 275-296.

Mainguet, Monique et Y. Callot, 1974, Air photo study of typology and interrelations between the texture and structure of dune patterns in the Fachi-Bilma Erg, Sahara: Zeitschr. f. Geomorph., Suppl. Bd. 20, p. 62-69.

McKee, E. D., 1966, Structure of dunes at White Sands National Monument, New Mexico (and a comparison with structures of dunes from other selected areas): Sedimentology, v. 7, no. 1 (special issue), p. 1-70.

McKee, E. D. and J. R. Douglas, 1971, Growth and movement of dunes at White Sands National Monument, New Mexico: U.S. Geol. Survey Prof. Paper 750-D, p. 108-114.

McKee, E. D. and R. J. Moiola, 1975, Geometry and growth of the White Sands dune field, New Mexico: Jour. Research U.S. Geol. Survey, v. 3, p. 59-66.

McKee, E. D. and G. C. Tibbits, Jr., 1964, Primary structures of a seif dune and associated deposits in Libya: Jour. Sed. Petrology, v. 34, p. 5-17.

Sharp, R. P., 1966, Kelso dunes, Mojave Desert, California: Geol. Soc. Amer. Bull., v. 77, p. 1045-1074.

Wilson, I.G., 1971, Desert sandflow basins and a model for the origin of ergs: Geographical Jour., v. 137, p. 180-199.
Wilson, I. G., 1972, Aeolian bedforms - their development and origins: Sedimentology, v. 19, p. 173-210.

Wilson, I. G., 1973, Ergs: Sedimentary Geol., v. 10, p. 77-106.

Descriptions of Ancient Eolian Deposits

Baars, D. L., 1962, Permian System of Colorado Plateau: Amer. Assoc. Petrol. Geol. Bull., v. 46, p. 149-218.

Baars, D. L., 1975, The Permian System of Canyonlands country: in J. E. Fassett, (ed.), Canyonlands Country; Four Corners Geol. Soc., 8th Field conf., p. 123-128.

Brookfield M. E., 1977, The origin of bounding surfaces in ancient aeolian sandstones: Sedimentology, v. 24, p. 303-332

Craig, L. C. and D. R. Shawe, 1975, Jurassic rocks of east-central Utah, in J. E. Fasset (ed.), Canyonlands Country: Four Corners Geol. Soc., 8th Field Conf., p. 157-165.

Dane, C. H., 1935, Geology of the Salt Valley Anticline and adjacent areas, Grand County, Utah: U.S. Geol. Survey Bull. 863, 184 p.

Knight, S. H., 1929, The Fountain and the Casper Formations of the Laramie Basin: a study of the genesis of sediments: Wyoming Univ. Pub. Sci., v. 1, p. 1-82.

Laming, D. J. C., 1966, Imbrication, paleocurrents and other sedimentary features in the lower New Red Sandstone, Devonshire, England: Jour. Sed. Petrology, v. 36, p. 940-959.

Sanderson, I. D., 1974, Sedimentary structures and their environmental significance in the Navajo Sandstone, San Rafael Swell, Utah: Brigham Young Univ. Geol. Studies, v. 21, p. 215-246.

Steidtmann, J. R., 1974, Evidence for eolian origin of cross-stratification in sandstone of the Casper Formation, Southernmost Laramie Basin, Wyoming: Geol. Soc. Amer. Bull., v. 85, p. 1835-1842.

Stokes, S.L., 1968, Multiple parallel-truncation bedding planes - a feature of wind-deposited sandstone formations: Jour. Sed. Petrology, v. 38, p. 510-515.

Thompson, D. B., 1969, Dome-shaped aeolian dunes in the Frodsham member of the so-called "Keuper" Sandstone Formation (Scythian-? Anisian: Triassic) at Frodsham, Cheshire (England): Sidementary Geol., v. 3, p. 263-289.

Walker, T. R. and J. C. Harms, 1972, Eolian origin of flagstone beds, Lyons Sandstone (Permian), type area, Boulder County, Colorado: Mountain Geol., v. 9, p. 279-288.

Controversial Interpretations

Baars, D. L. and W. R. Seager, 1970, Stratigraphic control of petroleum in White Rim Sandstone (Permian) in and near Canyonlands National Park, Utah: Amer. Assoc. Petrol. Geol. Bull., v. 54, p. 709-718.

Folk, R. L., 1977, Stratigraphic analysis of the Navajo Sandstone: a discussion: Jour. Sed. Petrology, v. 47, p. 483-484.

Freeman, W. E. and G. S. Visher, 1975, Stratigraphic analysis of the Navajo Sandstone: Jour. Sed. Petrology, v. 45, p. 651-668.

Picard, M. D., 1977, Statigraphic analysis of the Navajo Sandstone: a discussion: Jour. Sed. Petrology, v. 47, p. 475-483.

Pryor, W. A., 1971, Petrology of the Weissliegendes sandstones in the Harz and Werra-Fulda areas, Germany: Geol. Rundschau, v. 60, p. 524-552.

Rusyla, K., 1977, Stratigraphic analysis of the Navajo Sandstone: a discussion: Jour. Sed. Petrology, v. 47, p. 489-491.

Stanley, K. O., W. M. Jordan and R. H. Dott, Jr., 1971, early Jurassic Paleogeography, Western United States: Amer. Assoc. Petrol. Geol. Bull., v. 55, p. 10-19.

Steidtmann, J. R., 1977, Stratigraphic analysis of the Navajo Sandstone: a discussion: Jour. Sed. Petrology, v. 47, p. 484-489.

Descriptions of Modern Subaqueous Sand Waves and Tidal Ridges.

Houbolt, J. J. H. C., 1968, Recent sediments in the southern bight of the North Sea: Geol. en Mijnbouw, v. 47, p. 245-273.

Jordan, G. F. 1962, Large submarine sand waves: Science, v. 136, p. 839-848.

McCave, I. N., 1971, Sand waves in the North Sea off the coast of Holland: Marine Geol., v. 10, p. 199-225.

Stride, A. H., 1963, Current-swept sea floors near the southern half of Great Britain: Geol. Soc. London, Quart. Jour., v. 119, p. 175-199.

Swift, D. J. P., 1975, Tidal sand ridges and shoal-retreat massifs: Marine Geol., v. 18, p. 105-134.

Terwindt, J. H. J., 1971, Sand waves in the southern bight of the North Sea: Marine Geol., v. 10, p. 51-67.

MS received Sept. 2, 1977.
Reprinted from Geoscience Canada, Vol. 4, No. 4, p. 182-190.

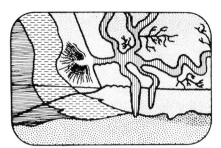

Facies Models 5. Deltas

A. D. Miall
Institute of Sedimentary and Petroleum Geology
Geological Survey of Canada
Calgary, Alberta T2L 2A7

Introduction

Deltaic depositional models differ from those described in earlier papers in the facies models series in that their recognition has not depended on a distillation of observations on ancient rocks but has arisen largely from a study of depositional processes on modern delta systems. A second important difference is that there are at least three distinct delta models, or "norms", to choose from in interpreting ancient rocks; many deltas are combinations of all three.

Definition

The concept of the delta is one of the oldest in geology, dating back, in fact, to about 400 B.C. At that time Herodotus made the observation that the alluvial plain at the mouth of the Nile was similar in shape to the Greek letter Δ, and the term was born.

One of the earliest modern definitions of a delta was provided by Barrell (1912) who stated that is "a deposit, partly subaerial, built by a river into or against a permanent body of water". There is little reason even now to revise this definition. Common usage amongst present-day geologists studying ancient rocks is that the term deltaic deposit is restricted to those bodies of clastic sediment formed in subaerial and shallow water environments (marine or lacustrine) in which the influence of a river or rivers as the main sediment source can be recognized, and in which a gradation into an offshore, generally finer-grained facies, can be traced. As discussed below,

there are many modern deltas where the depositional influence of the river is strongly masked by waves, ocean currents, tidal currents or winds, and the deposits of such deltas may be very hard to recognize in the ancient record.

A Short History of Delta Studies

Modern work in the English-speaking world commenced with the classic studies of Gilbert on the deltas in Lake Bonneville. Gilbert was the first to attempt a hydrodynamic explanation of delta formation, and his ideas dominated thinking on the subject for many years. A classic paper by Barrell (1912) on the ancient Catskill delta also had a far-ranging influence.

Since the nineteen twenties interest in deltas has been stimulated by the fact that the sediments of many ancient deltas contain extremely large deposits of coal, oil and gas. Nowhere is this more true than in the hydrocarbon-rich Gulf Coast of Texas and Louisiana, and research into deltaic sedimentation during the last forty years has been overwhelmingly dominated by studies of Gulf Coast deltas and their Quaternary and Tertiary antecedents. Most attention became focused on the Mississippi, which rapidly replaced the Lake Bonneville deltas of Gilbert as the standard model delta in geology textbooks.

Sedimentological research into the Mississippi commenced with the monumental work of Fisk, who established the depositional framework of the modern delta with the aid of many thousands of shallow boreholes. Subsequently the American Petroleum Institute funded a major research effort named, succinctly, Project 51, the objective of which was the study of modern sediments along the northwest margin of the Gulf of Mexico. The publication which summarizes this work (Shepard et al., 1960) contains landmark papers on depositional processes in the Mississippi, by Shepard and by Scruton. Further publications on the depositional history, depositional environments and cyclic sedimentation in the Mississippi, were provided by Kolb and Van Lopik (in Shirley, 1966) and by Coleman and Gagliano (1964, 1965).

The other delta that was studied extensively at this time was that of the Niger (Allen, in Morgan, 1970a; Oomkens, 1974).

Recently, some very useful compilations of papers on ancient and modern deltas have appeared (Shirley, 1966; Morgan, 1970a; Broussard, 1975), and several series of short-course lecture notes have been published, all of which contain much of value both to the specialist and non-specialist (Fisher et al., 1969; Curtis et al., 1975).

Wright et al. (1974) studied some 400 parameters of 34 modern alluvial-deltaic systems using multivariate statistical techniques in order to determine what controls their geometry, orientation and composition. The unifying concepts which emerged from this study (summaries of which are provided by Coleman in Curtis et al., 1975, and by Coleman and Wright in Broussard, 1975) are of fundamental importance to the geologist dealing with ancient rocks. They indicated that deltas can be divided into at least six types. However, for the purpose of the present paper it is sufficient to use the three main categories defined by Scott and Fisher (in Fisher et al., 1969) and by Galloway (in Broussard, 1975) as shown in Figure 1. These are the three "norms" referred to in the introduction.

Most of the publications referred to above are dominated by Gulf Coast geologists. Several important papers by "outsiders" are included in the compilations, for example, descriptions of the Rhine delta in Lake Constance by Müller (in Shirley, 1966), of the Niger, by Allen (in Morgan, 1970a) and of the Rhône by Oomkens (op. cit.) but, nevertheless, the pre-eminence of Houston- and New Orleans-based oil companies and such organizations as the Coastal Studies Institute of Louisiana State University in delta research, is astonishing. Conversely, contributions by Canadians and about Canadian deltas, ancient and modern, are few and far between. None of the major advances in understanding of deltaic sedimentation were made in this country. Some of the earliest work on modern deltas was carried out by Johnson (1921, 1922) on the Fraser delta, although these publications appear to have had little influence on subsequent research in deltaic sedimentation.

44

Delta formation and classification

The distribution, orientation and internal geometry of deltaic deposits is controlled by a variety of factors, including climate, water discharge, sediment load, river-mouth processes, waves, tides, currents, winds, shelf slope and the tectonics and geometry of the receiving basin (Wright et al., 1974). In a brief paper such as this it is impossible to describe fully the inter-relationships between all these variables, but several generalizations are possible, and these enable a meaningful classification of delta types to be made, as shown in Figure 1:

Variations in sediment input. Climate, water discharge (rate and variability) and sediment load (quantity and grain-size) are to some extent inter-related. In humid, tropical regions precipitation normally is high relative to evapotranspiration; runoff tends to be high and steady. The predominance of chemical over mechanical weathering leads to high dissolved-load sediment yields. These factors give rise to relatively stable, meandering channel patterns.

In Arctic or arid conditions precipitation is erratic, vegetation is sparse, and braided channel patterns with large bedloads tend to occur (Coleman *in* Curtis, 1975, and Coleman and Wright *in* Broussard, 1975 provide a more complete discussion of this topic).

Variations in river-mouth flow behaviour. When a sediment-laden river enters a body of standing water one of three types of flow dispersal may occur, depending on the density differences between the river water and that of the lake or sea into which it flows. Variations in temperature, salinity and sediment load can cause such differences in density.
A) *Inflow more dense:* flow forms a planar jet along the bottom. The result commonly is a turbidity current. The deposits which form from such bottom currents are classified as submarine fans.
B) *Inflow equally dense:* this occurs where rivers enter freshwater lakes. Sediment is dispersed radially and competency is lost rapidly. The result is a narrow, arcuate zone of active deposition and the delta which forms contains distinct topsets, steeply-dipping foresets, and bottomsets. This is the classical Gilbertian delta.
C) *Inflow less dense:* most marine deltas are formed under these conditions because freshwater is less dense than seawater, unless it is unusually cold or sediment-laden. The type of sediment dispersal which takes place depends on the strength of waves, tides and longshore currents, as discussed below.

Variations in transportation patterns on the delta. The type of energy conditions that exist in the sea at the river mouth are of fundamental importance in controlling depositional environments and the geometry of the resulting sediments. In fact the most useful classification of delta types is one based on the relative strengths of fluvial and marine processes (Fig. 1), as shown by Scott and Fisher (*in* Fisher *et al.*, 1969), Coleman (*in* Curtis, 1975), Galloway (*in* Broussard, 1975) and Coleman and Wright (*op. cit.*).

A) *River-dominated deltas:* if waves, tidal currents and longshore currents are weak, rapid seaward progradation takes place, and a variety of characteristic, fluvially dominated depositional environments develops. At the mouth of each distributary subaqueous levees may form as the jet of river water enters the sea (Fig. 2). The main sediment load is deposited in a distributary mouth bar, which becomes finer grained seaward. As progradation proceeds the river slope is flattened and flow becomes less competent. At this stage a breach in the subaerial levee may occur upstream during a period of high discharge. Such a breach is termed a crevasse. The shorter route it offers to the sea via an interdistributary bay generally is the cause of a major flow diversion, and a subdelta (crevasse-splay) deposit may develop rapidly. Eventually the crevasse may become a major distributary and the process is repeated.

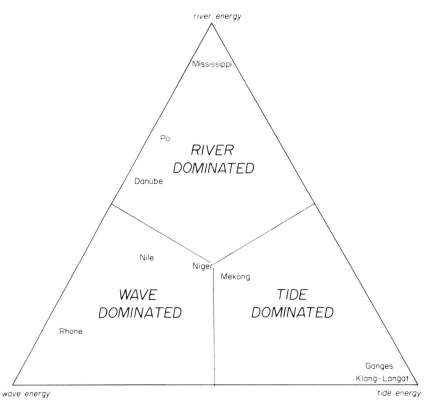

Figure 1
A classification of delta types, based on variations in transportation patterns on the delta (afer Galloway, Fig. 3, in Broussard, 1975).

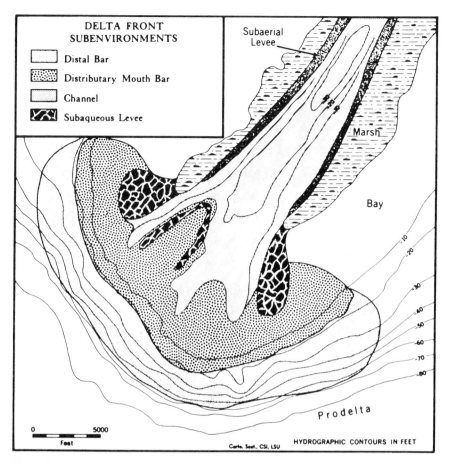

Figure 2
*Subenvironments at a distributary mouth in a
river-dominated delta, South Pass,
Mississippi delta (from Coleman and
Gagliano, 1965, Fig. 9).*

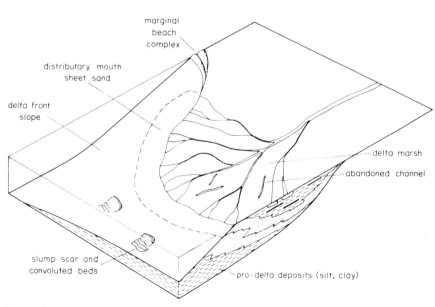

Figure 3
*Block diagram of a lobate, river-dominated
delta, showing the principal environments
and sedimentary facies.*

These are, in very brief outline, the
principal mechanisms that occur in
river-dominated deltas. The sediments,
sedimentary structures and organic
remains they contain will be described later.

There are two main subtypes in this
delta category. It was stated earlier that
the river discharge could be either
steady, generally with a high suspension
load, or fluctuating, with a typically
higher proportion of bedload in the
sediments. The first type tends to form
birdsfoot deltas with few distributaries,
shoestring sands and discrete mouth
bar deposits (Figs. 2, 4). The second
type normally is lobate in outline; there
are a greater number of distributaries,
each of which tends to be more
ephemeral, and the sediments are
coarser grained and the mouth bar
deposits merge laterally into sheet
sands (Fig. 3).
B) *Wave-dominated deltas:* in
environments of strong wave activity
mouth bar deposits are continually
reworked into a series of superimposed
coastal barriers. These may completely
dominate the final sedimentary
succession, and the internal geometry of
the deposits will be quite distinctive.
Sand bodies will tend to parallel the
coastline, in contrast to those of river-
dominated deltas, which are more nearly
perpendicular to the coast (Fig. 5).
C) *Tide-dominated deltas:* where the
tidal range is high the reversing flow that
occurs in the distributary channels
during flood and ebb may become the
principal source of sediment dispersal
energy. Within and seaward of the
distributary mouths the sediment may be
reworked into a series of parallel, linear
or digitate ridges parallel to the direction
of tidal currents (Fig. 6).

In cases where powerful longshore
currents exist the sediments will be
reworked into a series of barrier deposits
and offshore sand ridges parallel to the
coastline. The area of principal sediment
accumulation will be displaced
downcurrent from the main distributary
mouth(s) and, in extreme cases, the
sediment may be completely dispersed
along the shoreline with the development of no recognizable delta. Such
deposits will be described in a
subsequent paper on shoreline sand
models.

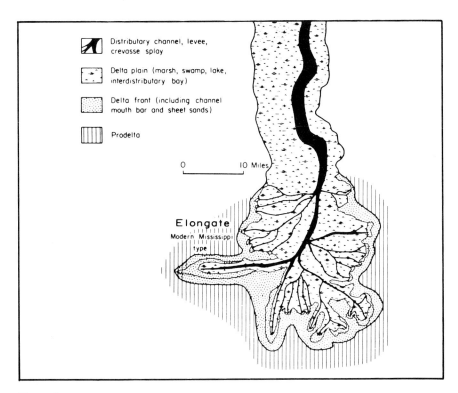

Figure 4
A birdsfoot-type river-dominated delta; the modern Mississippi delta (from Fisher et al., 1969, Fig. 39).

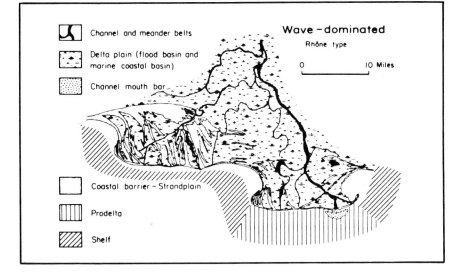

Figure 5
A wave-dominated delta; the modern Rhône delta (from Fisher et al., 1969, Fig. 37).

Deltaic cycles

Scruton (*in* Shepard *et al.*, 1960) was one of the first to point out that the growth of a delta is cyclic. He divided the cycle into two phases:

A) *Constructional phase:* active seaward progradation causes prodelta muds to be overlain by delta-front silts and sands, these in turn by distributary-mouth deposits, mainly sands, and finally top-set delta marsh sediments, possibly including peat beds (Fig. 7).

B) *Destructional phase:* a delta lobe eventually is abandoned if crevassing generates a shorter route to the sea. The topmost beds are then attacked by wave and current activity and may be completely reworked. Compaction may allow a local marine transgression to occur.

This description of the delta cycle is, of course, idealized. Firstly, it is most appropriate only for Mississippi-type deltas. Secondly, different parts of the same delta may be in different stages of development. The terminology is unfortunate; a major suite of superimposed barrier deposits caused by wave-reworking is as much a "constructive" deposit as is a distributary mouth sheet sand. Nevertheless, river- and wave-dominated deltas commonly are referred to as "high-constructive" and "high-destructive" deltas, respectively, in the literature (for example, Fisher *et al.*, 1969).

The complete delta cycle (sometimes termed a megacycle) may generate a stratigraphic succession between 50 and 150 m, or more, in thickness, but it may contain or pass laterally into numerous smaller cycles representing the progradation of individual distributaries or crevasse splays. As shown by Coleman and Gagliano (1964) and Elliott (1974) these can range from approximately two to 14 m in thickness. As in the case of the larger scale cycles they tend to coarsen upward (more complete descriptions later).

The manner in which cyclic deltaic sequences are superimposed upon each other depends on the relative rates of sedimentation and subsidence (including compaction). If the two rates are in approximate balance a delta will tend to build vertically; if subsidence is faster the delta will prograde seaward, and as each part of the depositional basin becomes filled successive progradational events will move laterally. The mechanisms are described by Curtis (*in* Morgan, 1970a, p. 293-297). Figure 8 shows how relatively slow subsidence rates have resulted in a suite of seven separate but partially overlapping lobes at the mouth of the Mississippi during the last 5000 years. The most recent lobe is itself in the process of forming several subdeltas, as shown in Figure 9.

Cyclic processes in other types of deltas are rather different. For example, in wave-dominated delta the sediments consist mainly of superimposed barrier sand deposits. However, far less subsurface information is available for modern wave- and tide-dominated deltas than for the Mississippi, and their internal geometry is, therefore, less well known.

Recognizing ancient deltas in the surface and subsurface

As shown in previous sections, numerous variables affect the nature of a deltaic deposit, and so it is impossible to describe a single delta model in a few brief sentences. In general terms: 1) deltaic deposits tend to be thick (several hundreds or even thousands of metres); 2) they contain considerable volumes of sand and/or silt; 3) coal beds commonly are present; 4) the faunal content of interbedded units may indicate marine, brackish and fresh water depositional environments; 5) sedimentary struc-tures indicate shallow water deposition by traction- rather than turbidity-currents; 6) a gradation into finer-grained clastic deposits of offshore origin should be traceable (criteria 3 and 4 are, or course, of no use in the Precambrian).

Some more specific criteria for the recognition of the principal delta types are described in the following paragraphs.

River-dominated deltas. The rapid seaward progradation of these deltas gives rise to the most characteristic feature of deltaic sediments, the coarsening-upward cycle. The com-plete cycle of a delta lobe (typically 50 to 100 m thick) and the distributary and crevasse cycles which are its component parts are summarized in Figure 10, and Figure 11 is an illustration of lateral changes that have been recognized in the coarsening-upward cycles of a Tertiary deltaic deposit in Banks Island, Arctic Canada.

Each cycle commences with a clay, generally laminated and sparsely fossiliferous. Prodelta clays tend to be organic-rich because of the abundant plankton growth which takes place in response to the influx of nutrient-rich river waters. They therefore make good petroleum source beds. The clays grade up into interbedded clay and silt or very fine sand, in which small-scale ripple marks and bioturbation are common. Distributary mouth sand bars or sheet sands may form the coarsest member of the cycle. The influence of strong unimodal currents near the distributary mouths generally is apparent in the form of abundant planar and trough crossbeds and ripple-marks. Organic remains, other than fragmented and transported debris (including plant material) are rare. The top of the cycle may be formed of delta marsh sediments, including paleosols and coal, or by distributary channel sands. These may be of finger- or shoestring-shape, as described in the classic work of Fisk. In some instances still more regressive facies are preserved, in the form of

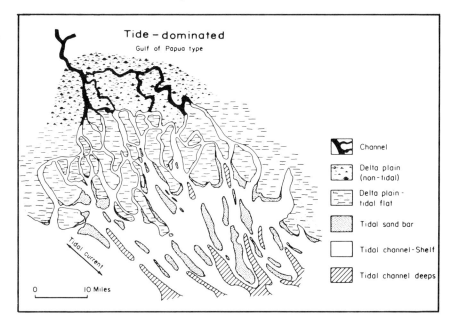

Figure 6
A tide-dominated delta; the modern Gulf of Papua (from Fisher et al., 1969, Fig. 47).

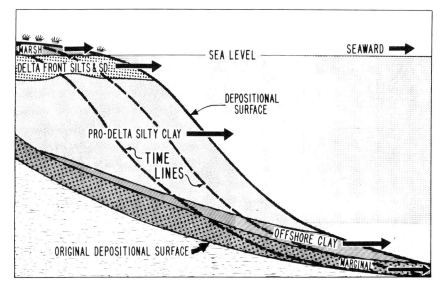

Figure 7
The "constructional" phase of the delta cycle (from Scruton, Fig. 9, in Shepard et al., 1960).

48

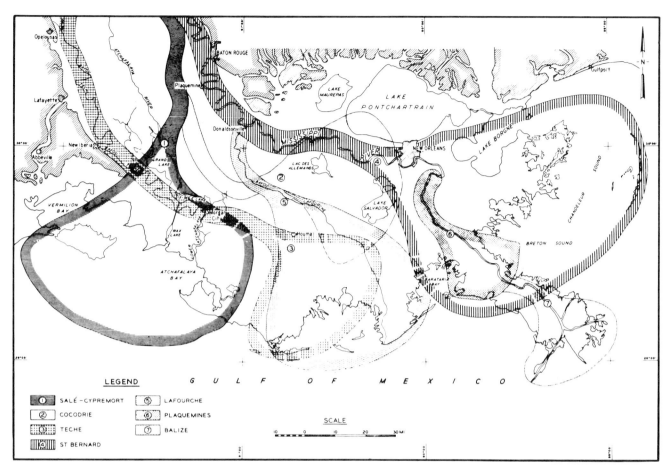

Figure 8
The seven partially overlapping lobes of the Mississippi delta which have developed

during the last 5000 years (from Kolb and Van Lopik, Fig. 2, in Shirley, 1966).

alluvial channel and flood plain sequences (described in the previous paper in this series).

Other facies occurring in river-dominated deltas include inter-distributary bay deposits and marginal, reworked deposits derived from abandoned delta lobes. The former generally are fine grained and in some instances contain shell beds; the latter commonly contain abundant shell debris, are glauconitic and bioturbated, and may be mineralogically and texturally more mature than other delta deposits as a result of wave and current winnowing (Shepard *in* Shepard *et al.*, 1960).

It is important to distinguish deltaic coarsening-upward cycles from those of offshore bar, barrier-bar or shoreface origin. The greatest differences are apparent in the coarse, upper members of the cycles. Deltaic cycles tend to be characterized by high-angle crossbeds with unimodal paleocurrent distribu-

tions. Barrier and shoreface sands generally contain low-angle crossbedding representing wave accretion surfaces, and paleocurrent distributions are bimodal or random. Deltaic sedimentation tends to be more rapid and less bioturbation or sediment sorting takes place. Rapid deltaic loading of sand on to unconsolidated prodelta muds commonly results in convolute bedding or the development of mud lumps or diapirs. Lastly, deltaic cycles of all types commonly are capped by coal beds whereas these are unusual in barrier island and shoreface sequences.

River-dominated deltas can be mapped most readily, particularly in the subsurface, by measuring the total sand content, or the sand/shale ratio in a given stratigraphic unit. Areas of high sand content may outline lobate areas perpendicular to the basin margin, corresponding to the principal paths of deltaic progradation. Figure 12 is an

illustration of a study of this type by Miall (1976b). Many other illustrations are given by Fisher *et al.* (1969).

Wave-dominated deltas. As noted earlier, wave-dominated deltas are characterized by stacked beach-ridge sequences (Fig. 10). Some of the criteria by which to distinguish these from progradation cycles were given in the previous section. Beach-ridge sequences can develop in nondeltaic settings as a result of longshore drift, and additional criteria are necessary in order to identify a specific sequence as deltaic in origin. Bars forming on nondeltaic coastlines commonly are backed by lagoons, the sediments of which may cap the bar sequence, whereas in deltaic settings the bars develop in front of swamps and fluvial channel complexes, the deposits of which should be quite distinctive. Coal may be an important constituent. Figure 13 is a schematic illustration of the sediments and facies relationships occurring in the modern Niger delta, which contains prominent beach-ridge deposits and is cut by tidal channels (Allen *in* Morgan, 1970a).

The geometry of wave-dominated delta deposits is quite different from those where wave influence is low. Beach-ridge sands form linear sand maxima sub-parallel to the basin margin, ideally forming a convex-seaward, arc-, cusp-, or chevron-shaped body. Associated fluvial sands will trend sub-perpendicular to the basin margin. The "classic" delta – that of the Nile – is a good example of a wave-dominated type; the Rhône delta (Fig. 5) is another (Oomkens *in* Morgan, 1970a).

Tide-dominated deltas. Deltas of this type may be difficult to recognize in ancient rocks. The coarser sediments are dispersed by tidal currents into offshore sand ridges parallel to the coastline, such as have been described by Off (1963), and the subaerial part of the delta consists largely of tidal flats comprising mainly fine-grained deposits. Distributaries may contain well-sorted sands, and large quantities of clay and silt will tend to be flushed into the delta marsh environment by overbank flooding during high tides. A typical modern tide-dominated delta is that of the Klang River in Malaysia (Coleman *et al., in* Morgan, 1970a; Coleman and Wright, *in* Broussard, 1975).

None of the characteristics of tidal delta deposits are distinctively "deltaic". Tide-generated sand ridges and tidal flats are widespread at the present day in areas without significant fluvial sediment input. The thickness of the deposit, reflecting the nearby presence of a major river mouth, may be the only clue in the ancient record to the presence of a tide-dominated delta. Few published descriptions of such a deposit are known to the writer. A generalized and partly hypothetical stratigraphic section through a tidal delta is shown in Figure 10.

Concluding remarks

The delta of the Mississippi is pre-eminent in the minds of most geologists as the all-purpose typical delta. There are obvious historical reasons for this, such as the abundance of oil and gas in deltaic deposits in the Gulf Coast, which has stimulated great research efforts into Mississippi sedimentation. The result has been that many river-dominated deltas now are recognized in ancient rocks, whereas the literature on other delta types is sparse. It may be that many beach-ridge and tidal flat sequences are actually deltaic in origin, and more research into wave- and tide-dominated deltas clearly is needed.

Acknowledgements

D. C. Pugh, F. G. Young and R. G. Walker read an earlier version of this paper. Their critical comments are gratefully acknowledged.

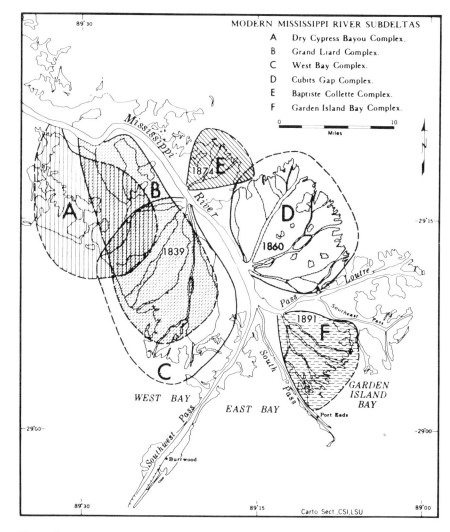

Figure 9
The sub-deltas of the modern Mississippi delta, showing year of initiation, where known (from Coleman and Gagliano, 1964, Fig. 5).

50

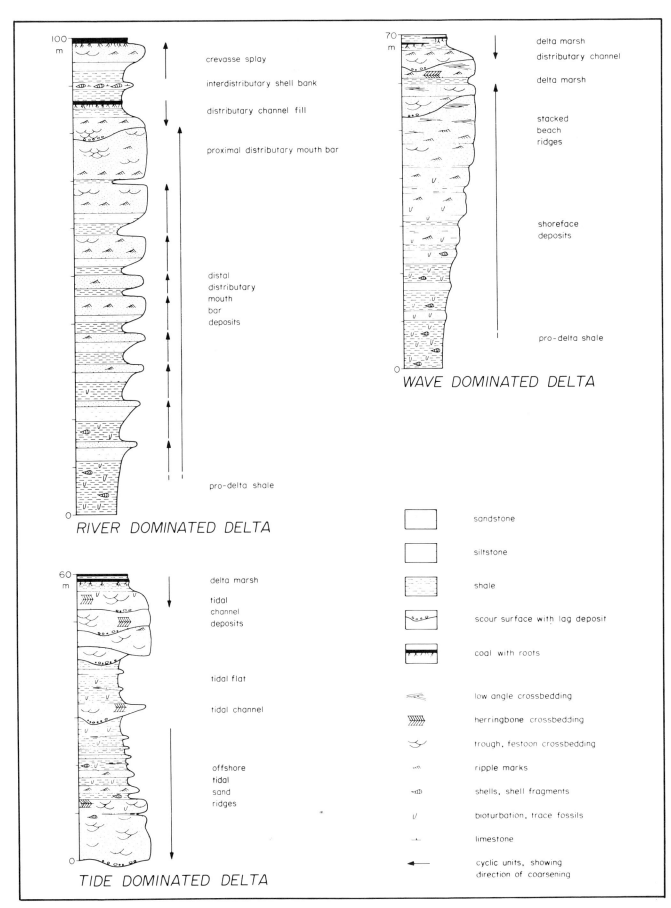

Figure 10
Generalized stratigraphic sections through the three principal types of delta. The stratigraphic order of the various facies types is more or less constant, but their thickness and relative abundance varies markedly from one example to another.

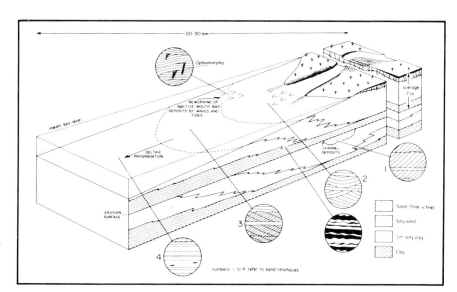

Figure 11
Lateral variability in distributary-mouth sheet sands, Eureka Sound Formation (Tertiary), Banks Island, Arctic Canada (from a subsurface study by Miall, 1976b).

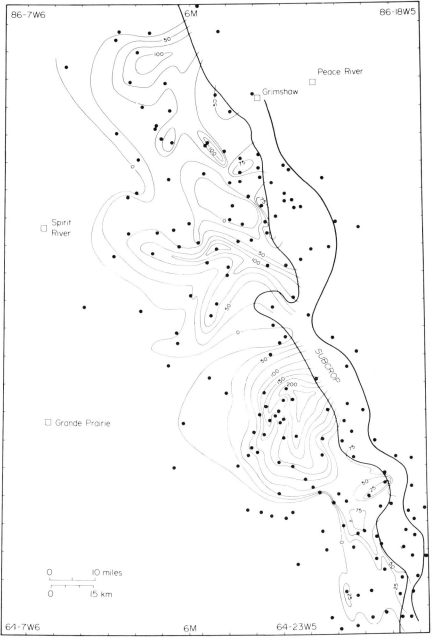

Figure 12
Delta lobes in a member of the Triassic Toad-Grayling Formation, northwestern Alberta. Contours show the distribution of net porous section, in feet. Map location is given by township and range. From a subsurface Study by Miall, (1976a).

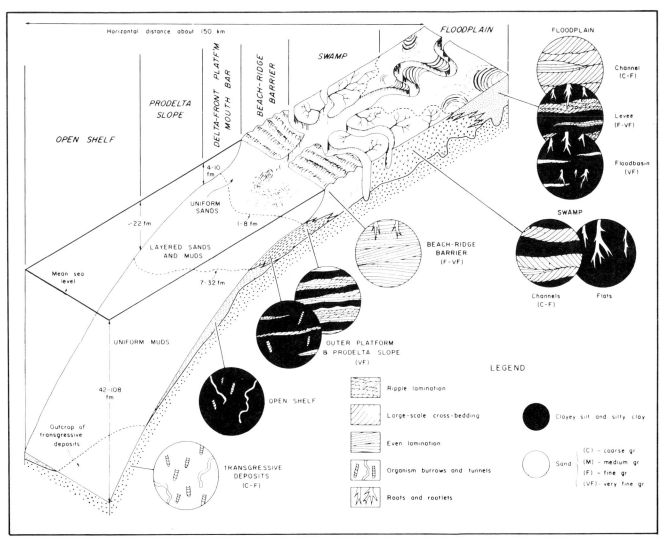

Figure 13
*Depositional environments in the modern
Niger delta (from Allen, Figure 4, in Morgan
1970a).*

Bibliography

Primary Reference List
Treatment of deltas in sedimentological textbooks tends to be sketchy, and the Mississippi is over-emphasized. The papers listed below are "all you really need", especially those in Broussard (1975) and Fisher *et al.* (1969), in which sedimentation models and ancient and modern examples are fully covered.

Broussard, M. L., *ed.*, 1975. Deltas, models for exploration: Houston Geol. Soc., 555 p.

    Papers on process variability and delta classification by Galloway and by Coleman and Wright and a historical survey by Leblanc are the most useful contributions.

Curtis, D. M., J. R. Duncan, and D. S. Gorsline, (organizing committee) 1975. Finding and exploring ancient deltas in the subsurface: Amer. Assoc. Petrol. Geol., Marine Geology Committee workshop, Introductory papers and notes.

    Useful papers on interpretations from geophysical logs and seismic data.

Fisher, W. L., L. F. Brown, Jr., S. J. Scott, and J. H. McGowen, 1969. Delta Systems in the exploration for oil and gas: Bureau Econ. Geol., Texas.

    Modern and ancient deltas are described and classified. The abundant illustrations are particulary useful.

Morgan, J. P., *ed.*, 1970a. Deltaic sedimentation, modern and ancient: Soc. Econ. Paleont. Mineral. Spec. Publ. 15, 312 p.

    A somewhat mixed bag of papers, but including much data unavailable elsewhere, including reviews of the Niger (Allen) and Rhône (Oomkens) deltas and a description of a tidal delta (Coleman *et al.*).

Smith, A. E. and M. L. Broussard, 1971. Deltas of the world: modern and ancient - bibliography: Houston Geol. Soc. Leblanc, R.J., *ed.*, 1976, Modern deltas; ancient deltas: Amer. Assoc. Petrol. Geol., Reprint Series 18 and 19.

    Reprints of 18 excellent papers on deltas, several of them classics.

Secondary Reference List
Much important information is contained in the following references, but most are specialized in scope and can be passed over by the beginner.

Barrell, J., 1912, Criteria for the recognition of ancient delta deposits: Geol. Soc. Amer. Bull., v. 23, p. 377-446.
    Classic delta paper dealing with the Devonian Catskill complex.

Bates, C. C., 1953, Rational theory of delta formation: Amer. Assoc. Petrol. Geol. Bull., v. 37, p. 2119-2162.
    The first description of flow dispersal patterns at river-mouths.

Coleman, J. M., and S. M. Gagliano, 1964, Cyclic sedimentation in the Mississippi river deltaic plain: Gulf Coast Assoc. Geol. Soc. Trans., v. 14, p. 67-80.

Coleman, J. M., and S. M. Gagliano, 1965, Sedimentary structures: Mississippi River deltaic plain: in G. V. Middleton, ed., Primary sedimentary structures and their hydrodynamic interpretation: Soc. Econ. Paleont. Mineral. Spec. Publ. 12, p. 133-148.
    Two important papers on sediments and sedimentary processes in a river-dominated delta.

Elliott, T., 1974, Interdistributary bay sequences and their genesis: Sedimentology, v. 21, p. 611-622.
    A series of vertical profiles for recognizing some of the minor subenvironments in river-dominated deltas.

Fisher, W. L., and J. H. McGowen, 1967, Depositional systems in the Wilcox Group of Texas and their relationship to occurrence of oil and gas: Gulf Coast Assoc. Geol. Soc. Trans., v. 17, p. 287-315.
    Detailed regional subsurface study.

Morgan, J. P., 1970b, Deltas - a résumé: Jour. Geol. Education, v. 18, p. 107-117.
    Brief historical summary of the development of the delta concept and a résumé of sedimentary processes in the Mississippi delta.

Off, T., 1963, Rhythmic linear sand bodies caused by tidal currents: Amer. Assoc. Petrol. Geol. Bull., v. 47, p. 324-340.
    Characteristic structures found in front of tide-dominated deltas.

Oomkens, E., 1974, Lithofacies relations in the Late Quaternary Niger Delta complex: Sedimentology, v. 21, p. 195-222.
    Supplements Allen's work.

Shepard, F. P., F. B. Phleger, and T. H. Van Andel, eds., 1960, Recent sediments, Northwest Gulf of Mexico, 1951-1958: Amer. Assoc. Petrol. Geol., 394 p.
    The results of API Project 51. See especially papers by F. P. Shepard and P. C. Scruton. The latter was the first to describe the delta cycle.

Shirley, M. L., ed., 1966, Deltas and their geologic framework: Houston Geol. Soc.
    The first Houston compilation, now a little dated.

Wright, L. D., J. M. Coleman, and M. W. Erickson, 1974, Analysis of major river systems and their deltas: morphologic and process comparisons: Coastal Studies Institute, Lousiana State University, Technical Report No. 156.
    A study of 34 modern alluvial-deltaic systems using multivariate statistical techniques.

Modern Canadian Deltas
Investigations in this area are rather patchy. The Fraser delta seems to have received the most attention.

Johnson, W. A., 1921, Sedimentation of the Fraser River delta: Geol. Surv. Can., Mem. 125, 46 p.

Johnson, W. A., 1922, The character of the stratification of the sediments in the recent delta of the Fraser River, British Columbia, Canada: Jour. Geol. v. 30, p. 115-129.
    Important early papers on deltaic sedimentation.

Kellerhals, P., and J. W. Murray, 1969, Tidal flats at Boundary Bay, Fraser River delta, British Columbia: Can. Petrol. Geol. Bull., v. 17, p. 67-91.
    Surface geology, including description of sediments, bedforms, flora and fauna.

Luternauer, J. T., and J. W. Murray, 1973, Sedimentation of the western delta-front of the Fraser River, British Columbia: Can. Jour. Earth Sci., v. 10, p. 1642-1663.
    Sedimentology of the intertidal and shallow subtidal portion of the delta.

Mackay, J. R., 1963, The Mackenzie delta area, Northwest Territories: Geol. Surv. Can., Geographical Branch, Mem. 8 (Reprinted 1974 as Geol. Surv. Can., Misc. Rept. 23).
    Concerned primarily with physical geography

Mathews, W. H., and F. P. Shepard, 1962, Sedimentation of Fraser River delta, British Columbia: Amer. Assoc. Petrol. Geol. Bull., v. 46, p. 1416-1438.
    Physiography, submarine surface geology, growth rate of modern delta.

Pezzetta, J. M., 1973, The St. Clair River Delta: sedimentary characteristics and depositional environments: Jour. Sediment Petrol., v. 43, p. 168-187.
    Investigations mainly on U.S. side of delta. Factor analysis and trend surface analysis help discriminate subenvironments in a small, lacustrine birdsfoot delta.

Smith, N. D., 1975, Sedimentary environments and late Quaternary history of a "Low energy" mountain delta: Can. Jour. Earth Sci., v. 12, p. 2004-2013.
    A small modern delta in a freshwater lake in Banff National Park, investigated with the use of auger sampling. Silts and clays predominate but bar-finger channel gravels are also present.

Ancient Deltaic Deposits in Canada
Deltaic deposits are particularly abundant in the Jurassic-Paleogene of the Western Interior, but detailed regional sedimentological studies are sparse. Atlantic Canada appears to be the only major region of the country which lacks any important deltaic deposits.

A. Cordilleran Region
Eisbacher, G. H., 1974a, Deltaic sedimentation on the northeastern Bowser Basin, British Columbia: Geol. Surv. Can., Paper 73-33.
    Brief facies description of river-dominated delta of Jurassic-Cretaceous age in a successor basin.

Eisbacher, G. H., 1976, The successor basins of the western Cordillera: GSC Paper 76-1, Part A., p. 113-116.
    More field data from Bowser Basin (see Eisbacher, 1974a).

Jeletzky, J. A., 1975, Hesquiat Formation (new), a neritic channel and inter-channel deposit of Oligocene age, western Vancouver Island, British Columbia: Geol. Surv. Can. Paper 75-32.
    A shallow-water marine fan deposit.

Muller, J. E. and M. E. Atchison, 1971, Geology, history and potential of Vancouver Island coal deposits: Geol. Surv. Can. Paper 70-53.

54

Muller, J. E. and J. A. Jeletzky, 1970, Geology of the Upper Cretaceous Nanaimo Group, Vancouver Island and Gulf Islands, British Columbia: Geol. Surv. Can. Paper 69-25.

Primarily deals with stratigraphy and biochronology, with two paleogeographic maps. Data in this and the preceeding paper suggest deposition in a wave-dominated delta.

B. Western Interior (Alberta and British Columbia)

Many detailed stratigraphic studies of deltaic rocks are available, but most of these are omitted in the following list. Only those publications which include sedimentological and paleogeographic data are included.

Caldwell, W. G. E., ed., 1975, The Cretaceous system in the western Interior of North America: Geol. Assoc. Canada Spec. Paper 13.

Proceedings of a symposium held at Saskatoon, May 1973. Contains a useful historical paper by K. M. Waage and several excellent stratigraphic-paleogeographic studies.

Eisbacher, G. H., M. A. Carrigy, and R. B. Campbell, 1974, Paleodrainage pattern and late-orogenic basins of the Canadian Cordillera: in W. R. Dickinson, ed., Tectonics and sedimentation: Soc. Econ. Paleont. Mineral. Spec. Publ. 22, p. 143-166.

A regional summary, including a discussion of the two major foreland basin molasse assemblages (Kootenay-Blairmore; Belly River-Paskapoo).

Gibson, D. W., 1974, Triassic rocks of the southern Canadian Rocky Mountains: Geol. Surv. Can. Bulletin 230.

Some distal deltaic rocks outcrop in the Rocky Mountains but the main belt of deltaic rocks is in the subsurface of central Alberta and has yet to be described (work in preparation by Miall).

Jansa, L. F., 1972, Depositional history of the coal-bearing Upper Jurassic-Lower Cretaceous Kootenay Formation, Southern Rocky Mountains, Canada: Geol. Soc. Amer. Bull., v. 83, p. 3199-3222.

Surface and subsurface facies analysis of coal-bearing deltaic rocks and tidal flat deposits.

Jansa, L. F., and N. R. Fischbuch, 1974, Evolution of a Middle and Upper Devonian sequence from a clastic coastal plain-deltaic complex into overlying carbonate reef complexes and banks, Sturgeon-Mitsue area, Alberta: Geol. Surv. Can., Bull. 234.

Facies analysis based on subsurface geophysical logs and cores. Relationship of cementation to depositional environments.

McLean, J. R., 1971, Stratigraphy of the Upper Cretaceous Judith River Formation on the Canadian Great Plains: Sask. Research Council, Geology Division, Rept. No. 11.

Primarily a stratigraphic and petrographic study, but with illustrations of sedimentary stuctures and an environmental interpretation of one fully cored borehole.

Mellon, G. B., 1967, Stratigraphy and petrology of the Lower Cretaceous Blairmore and Mannville Groups, Alberta Foothills and Plains: Research Council Alberta, Bull. 21.

Mainly a stratigraphic and petrologic study, with brief description of sedimentary cycles.

Mellon, G. B., J. W. Kramers, and E. G. Seagel, eds., 1972, Proceedings first geological conference on western Canadian coal: Research Council Alberta, Inf. Series No. 60.

Concerned mainly with stratigraphy and coal petrography, but contains a useful paper on the Early Cretaceous Gething Delta of B.C. by D. F. Stott.

Miall, A.D., 1976a, The Triassic sediments of Sturgeon Lake South and adjacent areas: in M. Lerand, ed., The Sedimentology of selected clastic oil and gas reservoirs in Alberta: Can. Soc. Petrol. Geol., p. 25-43.

Definition of lobate and birdsfoot deltas using subsurface log data.

Shawa, M. S., 1969, Sedimentary history of the Gilwood sandstone (Devonian) Utikuma Lake area, Alberta, Canada: Can. Petrol. Geol. Bull., v. 17, p. 392-409.

A detailed local core study with a discussion of sedimentary structures, grainsize distributions and limited paleocurrent data derived from oriented core.

Shawa, M. S., ed., 1975, Guidebook to selected sedimentary environments in southwestern Alberta, Canada: Can. Soc. Petrol. Geol. Field Conference 1975.

An illustrated guide to several Cretaceous outcrop sections, including several well-exposed deltaic sequences.

Shepheard, W. W., and L. V. Hills, 1970, Depositional environments, Bearpaw-Horseshoe Canyon (Upper Cretaceous) transition zone, Drumheller "Badlands", Alberta: Can. Petrol. Geol. Bull., v. 18, p. 166-215.

Detailed local sedimentological study based on surface mapping.

C. Western Interior (Yukon and Northwest Territories)

Bowerman, J. N., and R. C. Coffman, 1975, The geology of the Taglu gas field in the Beaufort Basin, N.W.T.: in C. J. Yorath, E. R. Parker and D. J. Glass, eds., Canada's continental margins and offshore petroleum exploration: Can. Soc. Petrol. Geol., Mem. 4, p. 649-662.

Brief description of subsurface stratigraphy of Tertiary, gas-bearing deltaic rocks.

Holmes, D. W., and T. A. Oliver, 1973, Source and depositional environments of the Moose Channel Formation, Northwest Territories: Can. Petrol. Geol. Bull., v. 21, p. 435-478.

Deltaic and fluvial facies are described. Emphasis on grainsize distributions using factor analysis and probability plots.

Myhr, D. W. and F. G. Young, 1975, Lower Cretaceous (Neocomian) sandstone sequence of Mackenzie Delta and Richardson Mountains area: Geol. Surv. Can. Paper 75-1, Part C. p. 247-266.

Regional subsurface facies reconstruction, with some core illustrations.

Young, F. G., 1973, Mesozoic epicontinental, flyschoid and molassoid depositional phases of Yukon's north slope: in J. D. Aitken and D. J. Glass, eds., Proc. Symp. Geology of the Canadian Arctic: Geol. Assoc. Can. and Can. Soc. Petrol. Geol., p. 181-202.

Young, F. G., 1975, Upper Cretaceous stratigraphy, Yukon coastal plain and northwestern Mackenzie Delta, Geol. Surv. Can., Bull. 249.
    Alluvial, deltaic and littoral facies are described but little detailed information is available regarding the interrelationships of these facies.

D. Innuitian Region
Agterberg, F. P., L. V. Hills, and H. P. Trettin, 1967, Paleocurrent trend analysis of a delta in the Bjorne Formation (Lower Triassic) of northwestern Melville Island, Arctic Archipelago: Jour. Sed. Petrol., v. 37, p. 852-862.
    Application of a trend-analysis smooths out irregularities and reveals a fan-shaped deltaic dispersal system. (See also Trettin and Hills, 1966).

Dineley, D. L. and B. R. Rust, 1968, Sedimentary and paleontological features of the Tertiary-Cretaceous rocks of Somerset Island, Arctic Canada: Can. Jour. Earth Sci., v. 5, p. 791-799.
    Facies analysis and paleocurrents of a small remnant of a deltaic succession.

Embry, A. F., III, 1976, Middle-Upper Devonian clastic wedge of the Franklinian Geosyncline: Univ. Calgary, Unpublished Ph. D. Thesis.
    A detailed regional stratigraphic and sedimentological study.

Miall, A. D., 1970, Continental-marine transition in the Devonian of Prince of Wales Island, Northwest Territories: Can. Jour. Earth Sci., v. 7, p. 125-144.
    Part of the facies spectrum includes thin deltaic redbeds interbedded with marine shales and carbonates. Brief description of sedimentary structures and fossils.

Miall, A.D.,1976b, Sedimentary structures and paleocurrents in a Tertiary deltaic succession, Northern Banks Basin, Arctic Canada: Can. Jour. Earth Sci., v. 13, p. 1422-1432.
    Facies analysis of a river-dominated delta system. Gross geometry of delta lobes can be outlined from scattered outcrop data.

Roy, K. J., 1973, Isachsen Formation, Amund Ringnes Island, District of Franklin: Geol. Surv. Can., Paper 73-1, Part A, p. 269-273.

Roy, K. J., 1974, Transport directions in the Isachsen Formation (Lower Cretaceous), Sverdrup Islands, District of Franklin: Geol. Surv. Can., Paper 74-1, Part A, p. 351-353.
    Brief facies descriptions. Paleocurrent patterns suggest a fan-shaped deltaic dispersal system.

Trettin, H. P. and L. V. Hills, 1966, Lower Triassic tar sands of north-western Melville Island, Arctic Archipelago, Geol. Surv. Can. Paper 66-34.
    Stratigraphy, petrography, sedimentary structures and paleocurrents, plus descriptions of tar deposits.

Young, G. M., 1974, Stratigraphy, paleocurrents and stromatolites of Hadrynian (Upper Precambrian) rocks of Victoria Island, Arctic Archipelago, Canada: Precamb. Research, v. 1, p. 13-41.

Young, G. M. and C. W. Jefferson, 1975, Late Precambrian shallow water deposits, Banks and Victoria Islands, Arctic Archipelago: Can. Jour. Earth Sci., v. 12, p. 1734-1748.
    Brief facies descriptions and paleocurrent analysis of deltaic rocks interbedded with tidal sequences.

E. Appalachian-St. Lawrence Lowlands Region
Lumsden, D. N. and B. R. Pelletier, 1969, Petrology of the Grimsby sandstone (Lower Silurian) of Ontario and New York: Jour. Sediment. Petrol., v. 39, p. 521-530.
    Grainsize and petrographic summary of a deltaic sandstone.

Martini, I. P., 1971, Regional analysis of sedimentology of Medina Formation (Silurian), Ontario and New York: Amer. Assoc. Petrol. Geol., v. 55, p. 1249-1261.
    Sedimentary petrography, paleocurrent analysis (including grain orientation) and sedimentary structures in a deltaic-tidal flat-longshore bar complex. Interpretations are strictly two dimensional because data were derived solely from outcrops along the nearly straight Niagara escarpment.

Martini, I. P., 1974, Deltaic and shallow marine sediments of the Niagara Escarpment between Hamilton, Ont. and Rochester, N.Y., a field guide: Maritime Sediments, v. 10, p. 52-66.

F. Canadian Shield
Very few of the sedimentary rocks in the Shield have been studied sedimentologically. Many clastic units are described in the literature as being of "shallow-marine" origin, and many of these probably are deltaic rocks.

Donaldson, J. A., 1965, The Dubawnt Group, Districts of Keewatin and Mackenzie: Geol. Surv. Can., Paper 64-20.
    Deltaic and fluvial rocks – brief description and paleocurrent analysis.

Hoffman, P. F., 1969, Proterozoic paleocurrents and depositional history of the East Arm Fold Belt, Great Slave Lake, Northwest Territories: Can. Jour. Earth Sci., v. 6, p. 441-462.

G. Ancient Deltas in the United States
Many excellent papers are included in the primary and secondary reference lists. Below are a few additional references.

Horne, J.C. and J.C. Ferm, 1978, Carboniferous depositional environments: eastern Kentucky and southern West Virginia: Univ. South Carolina, Field Guidebook.
    A well illustrated guide to superb exposures of some deltaic and barrier sequences.

Hubert, J.F., J.G. Butera, and R.F. Rice, 1972, Sedimentology of Upper Cretaceous Cody-Parkman delta, southwestern Powder River Basin, Wyoming: Geol. Soc. Amer., Bull., v. 83, p. 1649-1670.
    A wave dominated, high-destructive delta. Detailed paleocurrent studies, including the measurement and interpretation of oriented pillow structures.

Ricoy, J.V. and L.F. Brown, Jr., 1977, Depositional systems in the Sparta Formation (Eocene) Gulf Coast Basin of Texas: Bur. Econ. Geol., Austin, Texas, Geol. Circ. 77-7.
    High constructive and destructive deltas; strandplain and barrier bar systems. A surface and subsurface study.

Siemers, C.T., 1976, Sedimentology of the Rocktown Channel Sandstone, Upper part of the Dakota Formation (Cretaceous), central Kansas: Jour. Sed. Petrol., v. 46, p. 97-123.
    Deltaic distributory channel sandstone: a detailed study using facies, paleocurrent and grain size data.

H. Ancient Deltas in Other Areas

Collinson, J.D., 1969, The sedimentology of the Grindslow Shales and the Kinderscout Grit: a deltaic complex in the Namurian of northern England: Jour. Sed. Petrol., v. 39, p. 194-221.
    Interbedded fluvial, deltaic and slope deposits.

Collinson, J.D. and N.L. Banks, 1975, The Haslingden Flags (Namurian, G.) of southeast Lancashire; bar-finger sands in the Pennine Basin: Yorkshire Geol. Soc. Proc., v. 40, p. 431-458.
    Bar-finger and distributary mouth deposits.

Elliott, T., 1975, The sedimentary history of a delta lobe from a Yoredale (Carboniferous) cyclothem: Yorkshire Geol. Soc. Proc., v. 40, p. 505-536.
    A river-dominated delta, with progradational and abandonment phases.

Klein, G. de V., V. de. Melo, and J.C.D. Favera, 1972, Subaqueous gravity processes on the front of Cretaceous deltas, Reconcavo Basin, Brazil: Geol. Soc. Amer., Bull., v. 83, p. 1469-1492.
    Mass flow, slump and water escape structures and grouping of sediments into three associations: delta plain, high delta-front, and low delta-front trough.

McCabe, P.J., 1977, Deep distributory channels and giant bedforms in the Upper Carboniferous of the Central Pennines, northern England: Sedimentology, v. 24, p. 271-290.

Vos, R.G., 1977, Sedimentology of an Upper Paleozoic river, wave and tide influenced delta system in southern Morocco: Jour. Sed. Petrol., v. 47, p. 1242-1260.

MS received May 18, 1976.
Revised April, 1979.
Reprinted from Geoscience Canada,
Vol. 3, No. 3, p. 215-227.

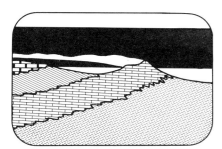

Facies Models 6. Barrier Island Systems

G.E. Reinson
Atlantic Geoscience Centre
Geological Survey of Canada
Bedford Institute of Oceanography
Dartmouth, Nova Scotia B2Y 4A2

Introduction

The AGI *Glossary of Geology* defines a barrier island as a "long, low, narrow, wave-built sandy island representing a broadened barrier beach that is sufficiently above high tide and parallel to the shore, and that commonly has dunes, vegetated zones, and swampy terranes extending lagoonward from the beach." Inherent in this definition is the fact that there have to be three major geomorphic elements in a barrier-island system: 1) the sandy barrier-island chain itself, 2) the enclosed body of water behind it (lagoon or estuary) and, 3) the channels which cut through the barrier and connect the lagoon to the open sea (tidal inlets) (Fig. 1). This tripartite geomorphic framework clearly demonstrates that barrier-island systems are composites of three major clastic depositional environments: 1) the subtidal to subaerial barrier-beach complex, 2) the back-barrier region or subtidal-intertidal

lagoon and, 3) the subtidal-intertidal delta and inlet-channel complex (Fig. 2). The view of the barrier island as a composite depositional system, until recently, has not been fully appreciated by geologists. This lack of appreciation is evident in most of the pre-1970 literature where there was an overwhelming preference for the use of just one barrier-island model (prograding Galveston Island Model) or "norm" for interpreting ancient rocks. If one recognizes the barrier-beach, lagoon, and tidal channel-delta scenario, it should be

obvious that a single model for such a complex system is completely unrealistic.

Fortunately, within the last decade, there has been a renaissance in the interpretation of ancient barrier-island sequences. This has come about largely through the investigations of modern barrier-island systems by numerous workers including M.O. Hayes and J.C. Kraft (Hayes and Kana, 1976; Kraft, 1971, 1978). Because of these modern studies, we are now recognizing that Galveston Island is just one of at least

Figure 1
Oblique aerial view of the barrier-island system at Tracadie, New Brunswick, showing the linear barrier-beach (B), the tidal inlets (I) through the barrier, and the lagoon (L) behind the barrier. Note the flood-tidal delta (F) in left foreground. In the upper part of photo, ice abuts against back-barrier marsh, May 3, 1977. (Photo by R. Belanger, AOL)

three distinct barrier-island stratigraphic models.

In this review I will attempt to synthesize barrier-island stratigraphic sequences into three "end-member" depositional models for use in interpreting ancient rocks. As with deltas (Miall, 1976), our ideas on barrier-island rock deposits stem from the study of modern barrier systems. Consequently, the review draws heavily on examples of modern deposits to develop the "end-member" models.

Origin and Occurrence
Theories regarding the origin of barrier islands have been reviewed at length in the recent geological literature (Schwartz, 1973; Swift, 1975; Wanless, 1976; Field and Duane, 1976). The question of origin is controversial but there are three main hypotheses: 1) the building-up of submarine bars; 2) spit progradation parallel to the coast and segmentation by inlets; and 3) submergence of coastal beach ridges. The controversy remains largely unresolved because most of the evidence pertaining to origin has usually been destroyed by subsequent modification. Extensive modification and evolution of modern barrier islands has been occurring since the early Holocene through a combination of processes including inlet cut and fill, washover deposition, and longshore transport (Field and Duane, 1976).

These processes have been enhanced by the progressive landward retreat of the barrier islands in response to the Holocene transgression (Swift, 1975).

Swift (1975) and Field and Duane (1976) consider that barrier formation by offshore bar emergence is insignificant compared to the other two mechanisms. Swift (1975) favors submergence of mainland beach ridges as the most important mode of formation. Considering the trend of sea level rise throughout the Holocene, it is certainly the most feasible mechanism for explaining the evolution, if not the initial origin of most of the extensive barrier-island regions existing today. However, spit progradation parallel to the coast cannot be completely dismissed as a significant mode of origin, because it is also readily observed to be initiating, as well as modifying, barriers at the present time. Many extensive barrier-island chains of the present day probably have had a composite mode of origin, by both spit progradation and coastal submergence. Variations in sediment supply and wave climate could easily induce periodic spit progradation at specific localities while submergence of coastal ridges was occurring on a more regional scale.

Barrier islands are most prevalent in coastal settings which have the following characteristics: 1) a low-gradient continental shelf adjacent to a low-relief coastal plain, 2) an abundant sediment

supply, and 3) moderate to low tidal ranges (Glaeser, 1978). Both the shelf and the coastal plain are composed of unconsolidated sediments, which are the material source for the building of barrier islands by nearshore processes. Glaeser noted that only 10 per cent of the world's barrier-islands are present along coastlines where tidal ranges exceed three metres. However it was Hayes (1975,1976) who focused attention on the importance of tidal range in controlling the occurrence and morphology of barrier-island systems. Hayes observed not only that barrier islands were rare on macrotidal coastlines (greater than 4 m tidal range), but that there were geomorphological differences between barrier islands of microtidal regions (less than 2 m tidal range) and those of mesotidal regions (2 to 4 m tidal range). In general, microtidal barrier islands are long and linear with extensive storm washover features (Fig. 3), and tidal inlets and deltas are of relatively minor importance. Mesotidal barrier islands are short and stunted, and characterized by large tidal inlets and deltas. Microtidal barriers are overwashed frequently by storm waves because of the lack of large enough tidal inlets to allow storm surges to flow past the barrier, rather than overtopping it (Hayes, 1976). According to Hayes, microtidal barrier islands can be consi-

Figure 3
Oblique aerial photo of the barrier beach at Tabusintac, New Brunswick, illustrating broad washover sand flats extending into the back-barrier marsh. The extent of the washover flats (W) is substantial relative to the dune ridge (D) (Photo by R. Belanger, May 3, 1977).

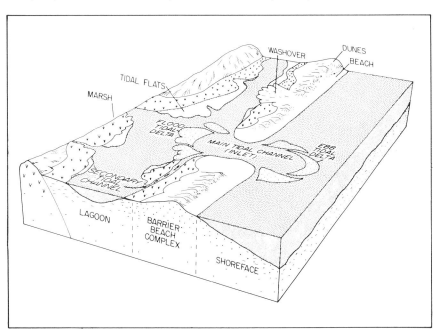

Figure 2
Block diagram illustrating the various subenvironments in a barrier-island system.

dered to be wave dominated as opposed to mesotidal barriers which are affected by both wave and current processes.

Depositional Environments and Lithofacies

The three main environments of a barrier-island system (barrier beach, lagoon, tidal channel-delta complex) are made up of a number of subenvironments (Fig. 2), each of which is characterized by distinct lithofacies. Facies of the barrier-beach and channel-delta environments are mainly sand and gravel, whereas the lagoonal (back-barrier) deposits can consist of both mud and sand. Barrier-beach deposits are elongate bodies which parallel the strandline and enclose finer-grained deposits of the lagoon. Tidal-channel and delta sand deposits, on the other hand, are generally oriented perpendicular or oblique to the barrier complex, and can extend into the lagoon and seaward into the nearshore zone. The transition between lagoon deposits and barrier and channel-delta deposits occurs in the overlapping subenvironments of the back-barrier tidal flats, marsh, washover fans and flood-tidal deltas.

The lateral and vertical extent and the occurrence of specific facies within a barrier-island system is dependent upon tidal range and the relative importance of tidal-current versus wave-generated processes, as discussed previously. For example, tidal-flat deposits will not be an important facies in microtidal environments because of the limited tidal range, whereas they may be extensive in mesotidal environments. Similarly, tidal channel and delta deposits are likely to be more prevalent in mesotidal than in microtidal environments because of the stronger tidal currents generated by the larger tidal range. The following discussion covers all the depositional environments and corresponding deposits of barrier-island systems, but it should be kept in mind that all facies will not necessarily be present in every barrier-island deposit.

Barrier Beach and Related Facies

The depositional subenvironments of a barrier-beach complex include: 1) the subtidal zone or *shoreface,* 2) the intertidal zone or *beach (foreshore),* 3) the subaerial zone or *back shore-dune* landward of the beachface, and 4) the

supratidal to subaerial wave-and wind-formed *washover* flats which extend across the barrier into the lagoon. Shoreface deposits are discussed with the barrier-beach complex because they form the foundation for the barrier, and also are a major source of sediment for barrier-island accretion.

Shoreface Deposits. The shoreface environment is defined as the area seaward of the barrier from low tide mark to a depth of about 10 to 20 m (Fig. 4). The lower limits of the shoreface correspond to the position at which waves begin to affect the sea bed. Hence the shoreface is an environment in which depositional processes are governed by wave energy. The amount of wave energy dissipated on the bottom decreases with increased water depth, and this inverse relationship governs the range of textures and sedimentary structures observed in shoreface deposits.

Lower shoreface deposits occur seaward of the break in the shoreface slope at the toe of the barrier-island sediment prism. The lower shoreface is a relatively low-energy transitional zone, where waves begin to affect the bottom, but where offshore shelf or basinal depositional processes also occur. This is reflected in the sediments which consist generally of very fine to fine-grained sands with intercalated layers of silt and sandy mud. Physical sedimentary structures include mainly planar laminated beds, which are often almost completely obliterated by bioturbation.

Trace-fossil assemblages are abundant in lower shoreface sediments (Howard, 1972).

Middle shoreface deposits extend over most of the shoreface slope (Fig. 4), in the zone of shoaling and breaking waves. This zone is subjected to high wave energy relative to the lower shoreface and is characterized generally by one or more longshore bars. The occurrence of longshore bars is related to a low-gradient shoreface and abundant sediment supply (Davis, 1978); both these conditions favor the landward movement and build-up of linear sand bars by shoaling and breaking waves.

Middle shoreface deposits can be highly variable in terms of sedimentary structures and textures, depending on whether nearshore bars are present or absent. Generally fine- to medium-grained, clean sands predominate, with minor amounts of silt and shell layers. Depositional structures include low-angle wedge-shaped sets of planar laminae, but ripple laminae and trough cross laminae are common (Campbell, 1971; Howard, 1972; Land, 1972). Middle shoreface deposits may be extensively bioturbated, especially in the lower parts, but the biogenic structures are generally less diverse than in deposits of the lower shoreface (Howard, 1972). The facies model proposed by Davidson-Arnott and Greenwood (1976) illustrates the complexity of sedimentary structures that can occur in a barred nearshore zone (Fig. 5). Vertical rock sequences of such deposits could

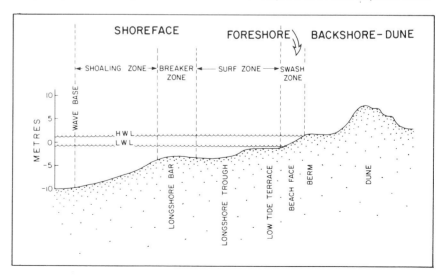

Figure 4
Generalized profile of the barrier beach and shoreface environments.

display interbedded sets of landward dipping ripple cross lamination, seaward dipping low-angle plane bedding, subhorizontal plane laminations and both landward- and seaward-dipping trough cross-bedded sets.

The shoreface environment is subjected to extreme modification by storm processes because effective wave base can be lowered dramatically by larger than normal, storm-generated waves. Truncated laminated bed sets and eroded burrow tops are common in ancient middle shoreface deposits (Howard, 1972), as are thick units (2 m) of subhorizontal laminated sand overlying coarse lag layers (50 cm thick) in modern deposits (Kumar and Sanders, 1976; Reineck and Singh, 1972). These features indicate that very high-amplitude waves periodically scour the bottom, suspending and then redepositing the sediment as the storm wanes. Graded bedding has been documented in some shoreface sediments (Hayes, 1967) and is attributed to storm-generated turbidity currents. Kumar and Sanders (1976) and Davidson-Arnott and Greenwood (1976) suggest that the bulk of shoreface deposits preserved in the rock record may consist of storm deposits rather than fair-weather deposits.

Upper shoreface sediments are closely associated with foreshore deposits, because they are situated in the high-energy surf zone just seaward of the beachface and landward of the breaker zone (Fig. 4). Consequently they have been grouped with foreshore facies in some rock studies (i.e., Davies et al., 1971) and have been considered to represent the shoreface-foreshore transition zone in others (Howard, 1972). The complex hydraulic environment of the surf zone (i.e., shore-normal currents generated by plunging waves superimposed on shore-parallel wave-driven currents) gives rise to the complex sequence of multidirectional sedimentary structures and variable sediment textures characteristic of these deposits. Textures range from fine sand to gravel, and biogenic structures are common but not abundant. The predominant depositional structures are multidirectional trough cross-bed sets (15 to 45 cm thick) (Fig. 6), but low-angle bidirectional planar cross-bedded sets and subhorizontal plane beds may also be present. The trough cross-beds are

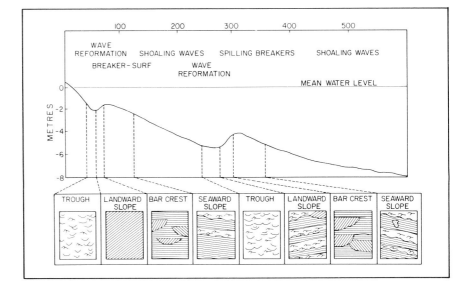

Figure 5
Facies model of nearshore bars in Kouchi-bouguac Bay, New Brunswick, illustrating *characteristic sedimentary structures and wave transformation zones (from Davidson-Arnott and Greenwood, 1976).*

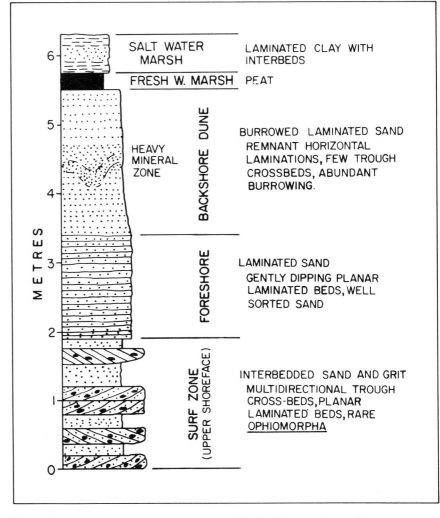

Figure 6
Generalized barrier sequence in the Upper *Tertiary Cohansey Sand of New Jersey (modified from Carter, 1978).*

thought to indicate the multidirectional current flow in the surf zone (Clifton *et al.*, 1971; Carter, 1978;). Predominantly bidirectional trough cross-beds oriented parallel to depositional strike are common in upper shoreface deposits, and may be indicative of deposition under strong longshore current conditions.

The effects of storm activity can be recorded in upper shoreface-foreshore deposits as well as in middle shoreface sediments. This is illustrated by the ridge and runnel sequence depicted in Figure 7. Such a sequence results when storm waves erode the beachface, removing sediment to the shoreface. The sediment is returned to the beach during the post-storm recovery, in the form of a ridge and runnel (bar and trough), which develops on the low tide terrace just seaward of the foreshore (Davis *et al.*, 1972; Owens and Frobel, 1977). The ridge migrates shoreward eventually welding onto the beachface and creating a distinctive sequence of upper shoreface-foreshore deposits.

Foreshore Deposits. The foreshore environment is confined to the intertidal zone, which is usually marked by a sharp change in slope, both at the base and at the top of the beachface (Fig. 4). The foreshore is the zone of wave swash, the surge of water caused by incoming plunging breakers in the surf zone. Swash runup occurs with each wave surge and backwash runoff between each surge. The swash-backwash mechanism is mainly responsible for the distinct subparallel to low-angle seaward-dipping, planar laminations (Fig. 8) which occur as wedge-shaped sets in most beach deposits. The boundaries between sets are generally not truncated, but rather mark the changing slope of the prograding beachface during the accretionary phase. Examples of foreshore deposits in the rock record include those illustrated in Figures 6 and 8 and those proposed by Campbell (1971), Howard (1972), Davies *et al.* (1971), and Land (1972).

Backshore-Dune Deposits. The backshore-dune environment is characterized by subaerial, predominantly wind-generated depositional processes. The backshore seaward of the dunes is a flat-lying to landward-sloping area called the berm; the seaward limit, called

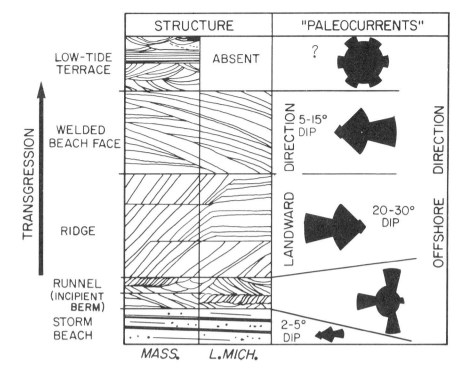

Figure 7
Transgressive sequence formed by the landward migration of ridge-and-runnel during beach constructional phase. Vertical sequence would be about 1 m thick (from Davis et al., 1972).

Figure 8
Horizontal even parallel laminae and low-angle inclined laminae in beach deposits of the Foremost Formation, southern Alberta (from Ogunyomi and Hills, 1977, plate 1C).

the berm crest, is well defined by the marked change in slope at the top of the beachface (Fig. 4). Sediment is transported to the berm crest by high spring tides or storms and is distributed over the backshore area by winds and washover (discussed below). Subhorizontal to landward-dipping plane beds characterize the backshore (Fig. 9) and may be interbedded or overlain by small- to medium-scale trough cross beds of incipient dune origin. Trough cross-stratified sets, up to 2 m in thickness, are characteristic of dune deposits, but planar cross-stratified sets are also common. The trough cross strata may be multidirectional in orientation and bounded by curved bedding surfaces (Campbell, 1971). Dune beds are commonly extensively disturbed by root growth (Figs. 9, 10) and may contain small paleosol horizons and isolated organic debris. Other biogenic structures such as decapod burrows may also occur in backshore-foredune deposits (Fig. 6).

Washover Deposits. Washover deposits result when wind-generated storm surges overtop and cut through barriers creating lobate or sheet deposits of sand which extend into the lagoon (Figs. 2, 3). These washover flats then provide corridors for transferring wind-transported sand across the foredune belt to form back-barrier sand flats (Fig. 9). This mechanism increases the

width of the barrier, providing environments favorable for stabilization by marsh growth.

Modern studies of washover deposits indicate that there are two dominant sedimentary structures, subhorizontal (planar) stratification, and small- to medium-scale delta foreset strata where the washover detritus protrudes into the

lagoon (Fig. 11). Textural and heavy mineral laminations and graded bedding can also occur (Andrews, 1970; Schwartz, 1975) depending on the nature of the source material. Textures may range from fine sand to gravel, but generally fine- to medium-grained sand forms the bulk of washover deposits.

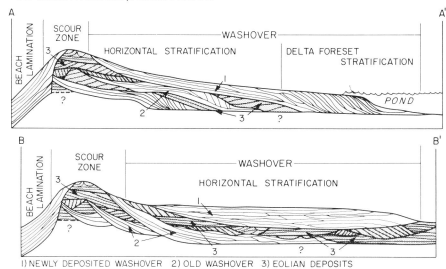

1) NEWLY DEPOSITED WASHOVER   2) OLD WASHOVER   3) EOLIAN DEPOSITS

Figure 11

Schematic cross-sections through two washover fans showing sequences of sedimentary structures. A "transgressive" situation is depicted. The sedimentary structures represent the upper few meters or less of the sand-body complex. Section A-A' shows a horizontal stratification to delta-foreset structural sequence resulting from flow across a subaerial surface into a body of standing

water. Section B-B' shows occurrence of horizontal stratification resulting from flow across a subaerial surface. Eolian processes may modify or bury washover deposits to various degrees (from Schwartz, 1975).

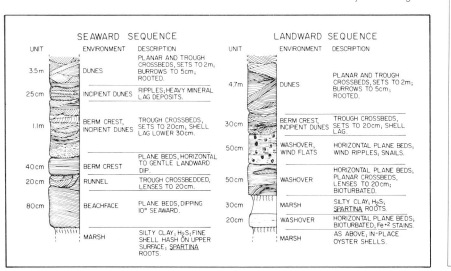

Figure 9

Lithologic sequences observed in the landward edge, and in the seaward edge of a Kiawah Island beach ridge (from Barwis, 1976, Figs. 2 and 3).

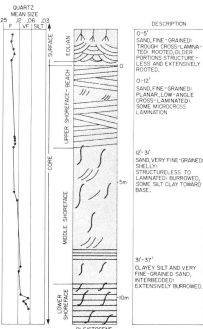

Figure 10

Sequence of sedimentary structures, textures and lithology in a core through Galveston Island (From Davies et al., 1971, Fig. 6).

Washover deposits are generally thin, ranging from a few centimeters to two metres for each overwash event. In plan form they form elongate, semi-circular, sheet-like or tabular bodies a few hundred metres in width and oriented normal to the shoreline. Coalescing washover fans can be in the order of kilometres in width (Fig 3), creating extensive washover flats which cover large tracts of the barrier.

Recent studies on modern barrier-island systems have illustrated that washover deposits form a significant portion of barrier sand bodies, especially in microtidal regions. Under transgressing conditions washover is one of the main processes by which the barrier island migrates landward, and probably is one of the main mechanisms responsible for the initiation of new tidal inlets. It is likely that washover deposits are more prevalent in ancient barrier sequences than has been recognized to date.

Tidal Channel (Inlet) and Tidal-Delta Facies

Tidal channel and tidal-delta sand bodies are intricately associated facies both with respect to their close proximity to one another, and with regard to their internal sedimentary structures and textures. This is because their formation is governed primarily by tidal-current processes directed normal or oblique to the sand barrier. The *ebb-tidal delta,* the sand accumulation formed seaward of the barrier by ebb-tidal currents, is affected by longshore and wave-generated currents, whereas the *flood-tidal delta,* the sand body deposited landward of the barrier by flood-tidal currents, is little influenced by wave and wind-generated processes (Figs. 2, 12).

There are two types of tidal-channel environments, the main channels, or tidal inlets connecting the lagoon to the ocean, and the secondary channels located adjacent to the tidal deltas and back barrier lagoon margins. Secondary tidal channels are sometimes so closely related to the formation of tidal delta complexes that the resultant facies are difficult to differentiate. Tidal channel and tidal delta deposits are separated here mainly for ease in discussion. However, this separation serves also to emphasize the fact that channel facies can occur independent of tidal deltas, whereas the occurrence of tidal delta facies is dependent on the presence of tidal channels.

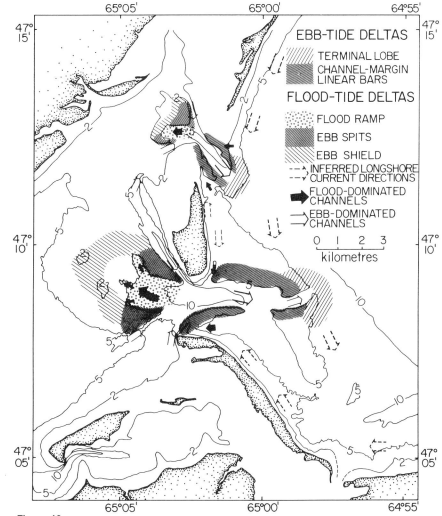

Figure 12

Morphology of tidal deltas and inferred tidal- and longshore-current patterns at the mouth of the Miramichi estuary, New Brunswick *(from Reinson, 1977).*

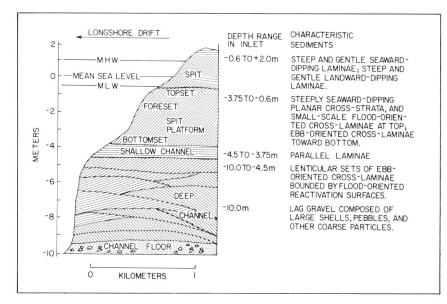

Figure 13

Vertical sequence of sedimentary structures formed by the migration of Fire Island Inlet, New York (modified from Kumar and Sanders, 1974).

Tidal Channel Deposits. Tidal channel deposits form mainly by lateral migration, as in a meander bend in a river. The best known and most important channel deposits are tidal-inlet fill sequences, which result from the shore-parallel migration of tidal inlets (Fig. 13). The direction and rate of inlet-channel migration is controlled by the magnitude of net longshore sediment supply. Barrier extension occurs by spit accretion on the updrift side of an inlet, with a corresponding erosion of the downdrift channel margin (Fig. 14). The shifting of the main inlet through a barrier causes the tidal channels both landward and seaward of the barrier, and the tidal deltas, to shift position also. The sand body that is deposited by inlet migration will be elongated parallel to the barrier island, having a length equal to the distance the inlet has migrated (Fig. 14). The thickness of the inlet lithosome will be equal to the depth of the inlet, if no subsequent erosion of the upper boundary occurs during deposition of dune, beach and washover deposits of the overlying accreting barrier.

The studies of Land (1972), Kumar and Sanders (1974), Hubbard and Barwis (1976), Hayes (1976), Barwis and Makurath (1978), and Carter (1978) indicate that channel-fill sequences resulting from barrier-inlet (or tidal channel) migration have the following general characteristics: 1) an erosional base often marked by a coarse lag deposit; 2) a deep channel facies consisting of bidirectional large-scale planar and/or medium-scale trough cross-beds; 3) a shallow channel facies consisting of bidirectional small- to medium-scale trough cross-beds and/or plane-beds and "washed-out" ripple laminae; 4) a fining-upward textural trend and a thinning-upward of cross-bed set thickness. The difference in size, orientation and type of sedimentary structures in the deep channel and shallow channel deposits generally reflects an increase in current-flow conditions in the shallow channel relative to the deep-channel environment.

The modern inlet-fill sequence (Fig. 13) described by Kumar and Sanders (1974) has a deep channel facies characterized by ebb-oriented planar cross-laminae; this reflects the predominance of sand-wave bedforms deposited under lower flow regime conditions in an ebb-current dominated environ-

ment. The overlying shallow channel facies is characterized by plane-parallel laminae and "washed-out" ripple laminae, reflecting plane bed deposition under "transitional" or upper flow regime conditions.

The studies of Barwis and Makurath (1978) and Land (1972) serve as comparative rock analogs to the Fire Island Inlet deposits (Fig. 13). The Silurian inlet sequence of Barwis and Makurath consists of a channel lag deposit overlain by 4.1 m of bidirectional trough and planar cross-bedded, medium-grained sandstone, with set thickness averaging 15 cm. The cross-bed orientations reflect deposition under tidal-current transport reversals along an axis oblique to the paleostrand. This deep channel facies is overlain gradationally by a fine-grained sandstone unit (1.8 m thick) dominated by bidirectional trough cross-bed sets averaging 2.5 cm in thickness, and "washed-out" ripples. The tidal channel sequence described by Land (1972) averages 8 m in thickness and consists of bimodal to polymodal trough cross-bed sets (ranging from 10 cm to 90 cm in thickness) in the lower 5 to 6 m, and subparallel beds in the upper 2 to 3 m.

The hypothetical tidal-inlet sequences proposed by Hayes (1976) and Hubbard and Barwis (1976), based on their study of mesotidal inlets of South Carolina, differ slightly from the Kumar and Sanders model (Fig. 13) with regard to the vertical sequence of sedimentary structures. However, the inference of sequential deposition under increasing flow conditions is still evident. Their inlet sequence is as follows: 1) a basal lag or disconformable bottom; 2) a deep channel deposit consisting of bidirectional large-scale planar cross-beds that have a slight seaward dominance, inter-

layered with bidirectional medium-scale trough cross-beds; 3) a shallow channel deposit consisting predominantly of small- to medium-scale bidirectional trough cross-beds. The planar cross-beds are suggestive of sand-wave deposition under ebb-dominant channel flow, whereas the trough cross-beds record deposition as megaripples under stronger currents and alternating reversals of flow directions.

Rock sequences similar to the hypothetical inlet sequences of Hayes (1976) and Hubbard and Barwis (1976) are illustrated in Figures 15 and 16. Carter (1978) interprets the sequence in Figure 15 as a back-barrier tidal channel deposit, and the presence of the interbedded sand and clay facies seems to preclude an inlet-fill origin. This example illustrates the similarity between back-barrier tidal channel deposits and tidal inlet deposits, and also points to a similar mode of origin, that of lateral channel migration concomitant with barrier-inlet migration. The tidal channel sequence in Figure 15 could also be interpreted as part of a flood-tidal delta complex, and the reasons for this alternate hypothesis will become evident in the following discussion on tidal delta deposits.

Tidal Delta Deposits. Hayes (1975), based mainly on his work in New England and Alaska, recognized that tidal deltas display a common morphological pattern governed by segregated zones of ebb and flood flow. This recognition prompted him to propose generalized models for both ebb- and flood-tidal delta deposition. Subsequent studies by Hayes and co-workers (Hayes and Kana, 1976) and others (Reinson, 1977; Armon, 1979), indicate that the models are generally applicable elsewhere, in both microtidal and meso-

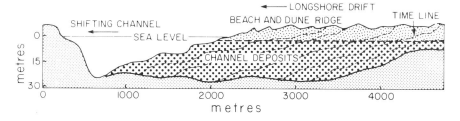

Figure 14
Generalized cross-section parallel to shoreline illustrating the development of a barrier-inlet sand body by lateral inlet migration (modified from Hoyt and Henry, 1965).

tidal regions. Tidal deltas can occur in a variety of forms (from linear shoals to complex channel-shoal systems) depending on tidal range, wave climate, and sediment supply, but the basic morphological pattern as illustrated in Figure 12 is generally clearly evident.

The typical morphology of a flood-tidal delta, that of a seaward-opening parabola bounded by marginal channels, is related to the segregation of tidal-current flow paths during ebb and flood phases. This flow segregation results from the time-velocity asymmetry of the tidal currents; that is, maximum flood and ebb flows occur near high water and low water respectively. Maximum flood flow traverses through the flood ramp and over the shoal, whereas maximum ebb flow is diverted around the shoal because of the drop in water level. This flow segregation gives rise to a distinct pattern of bedform distribution (Booth-royd and Hubbard, 1975; Hubbard and Barwis, 1976; Reinson, 1979), with predominantly flood-oriented sand waves covering the flood ramp and centre of the shoal, bidirectional mega-ripples on the ebb-shield and ebb spits, and ebb-oriented sand waves in the adjacent channels.

The deposits resulting from flood-tidal delta formation will be characterized by a varied sequence of planar cross-beds and trough cross-beds. The preponderance of one bedform over the other, and their orientation and position in vertical sequence, will depend on the locality at which the sequence is located within the tidal-delta complex. Hubbard and Barwis (1976) proposed a lithologic sequence for a flood-tidal delta as follows: 1) basal bidirectional cross-strata (megaripples) - represents early phases of deposition; 2) interbedded seaward-oriented trough cross-strata (megaripples) and landward-oriented planar cross-beds (sand waves) - represents deposition prior to ebb-shield development; and 3) landward-oriented planar cross-strata with upward-decreasing set thickness (sand waves) - represents deposition on flood ramp. Deposits adjacent to this sequence would be characterized by bidirectional trough cross-strata (megaripples) - representing ebb-shield and ebb-spit deposition. The total thickness of such a sequence would be in the order of 10 m. Hayes (1976) proposed a stratigraphic sequence for a regressive flood-tidal delta situation (Fig. 17). This sequence is

dominated by planar bidirectional cross-strata.

The morphology of ebb-tidal deltas is controlled largely by tidal-current segregation during different phases of the tidal cycle, but the interaction of waves with tidal currents is also important in the formation of ebb deltas. This interaction is reflected in the complex bedform distribution, which consists of ebb-oriented sand waves or megaripples in the main ebb-channel, with flood-oriented sand waves or megaripples in the marginal flood channels (Fig. 12). Channel-margin, linear bars and swash bars (areas of intense wave and current interaction) are characterized by multi-directional megaripples and plane beds. As in flood deltas, the vertical sequences resulting from ebb-delta formation would exhibit extreme variations in sedimentary structures from one locality to another, within a specific ebb-delta deposit. Ebb-delta deposits are so dependent on inlet conditions and wave climate that it is imposssible to characterize them in a specific sequence. Perhaps the major difference between ebb-tidal delta deposits and flood-delta deposits is the occurrence of multidirectional cross-beds in ebb delta sequen-

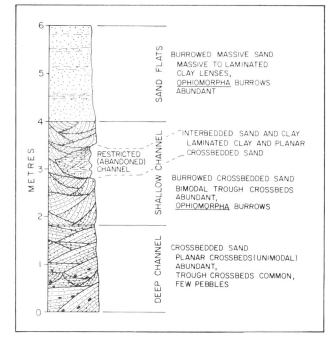

Figure 15
Generalized barrier-protected sequence in the Upper Tertiary Cohansey Sand of New Jersey (modified from Carter, 1978).

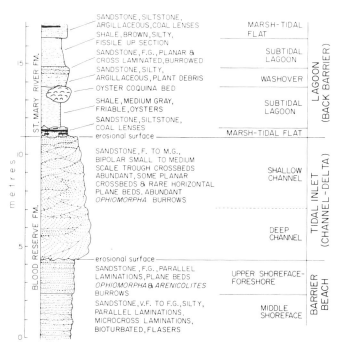

Figure 16
Composite stratigraphic section of the Blood Reserve - St. Mary River Formations, southern Alberta (modified from Young and Reinson, 1975).

ces, as opposed to the predominantly flood-oriented or bidirectional cross-beds of flood-tidal delta sequences.

As mentioned earlier, flood- and ebb-delta deposits have textures and sedimentary structures similar to inlet fill sequences, and therefore the identification of delta sand bodies in the rock record may depend more on their geometry and stratigraphic position relative to surrounding facies. In modern barrier systems, tidal inlet-delta associated deposits are integral parts of barrier-island sand bodies. By analogy such deposits shoud be expected to occur in ancient barrier sequences, yet they have been little recognized up until very recently. The studies of Barwis and Makurath (1978), Horne and Ferm (1978), Hobday and Horne (1977) and Land (1972) amply illustrate the importance of channel-delta deposits in ancient barrier sequences, and also lead one to suspect that such deposits

have been misinterpreted in many rock sequences in the past.

Lagoonal (Back Barrier) Facies

Lagoonal sequences generally consist of interbedded and interfingering sandstone, shale, siltstone and coal facies characteristic of a number of overlapping subenvironments (Figs. 16, 18). Sand facies include *washover* sheet deposits and sheet and channel-fill deposits of *flood-tidal delta* origin. Fine-grained facies include those of the *subaqueous lagoon* and the *tidal flats*, which are situated adjacent to the barrier or on the landward side of the lagoon abutting the hinterland marsh and swamp flatland (Fig. 2). Organic deposits of coal, peat, etc., record *marsh* and *swamp* environments, and usually are very thin, having formed on sand and mud flats of the lagoonal margin, and on emergent washover flats. Abandoned or mature flood-tidal deltas can become

stabilized by marsh vegetation also; this situation and that of the vegetated washover flat can lead to the presence of very thin coal lenses overlying organic-rich sheet sandstones in the rock record (Fig. 18). Subaqueous shale and siltstone facies are often characterized by brackish water macroinvertebrate shells, and in Cretaceous lagoonal deposits, coquinid oyster beds up to 1 m thick, are common (Fig. 16 and Land, 1972). Disseminated carbonaceous material, imprints of plant remains, and root and reed fragments are common in some shale beds, indicating the interfingering of proximal marsh and subaqueous lagoonal environments.

The topic of tidal-flats cannot be given justice here, but some mention is made of these deposits because they do occur in the barrier-island setting. The extent of tidal flat environments in a barrier-island system is a function of tidal range, the greater the tidal range, the more extensive are the flats. So in mesotidal barrier systems we may expect to find sequences similar to the classical tidal flat deposits described by van Straaten (1961), Evans (1965), Reineck and Singh (1975) and Klein (1977). The low tidal flats would be characterized by fine- to medium-grained ripple-laminated sand, the mid flats by interbedded sand and mud containing flasers and lenticular layers, and the high tidal flats by layered mud. The high tidal flats would be succeeded landward (and upwards in a prograding situation) by salt marsh. In most microtidal and mesotidal barrier-island systems the above tidal-flat sequence is attenuated because of the limiting conditions of tidal range.

Lagoonal or back-barrier sequences present a marked contrast to the predominantly clean sandstone sequences of the barrier-beach and inlet-delta environments (Fig. 19). Although the sandstone deposits interfinger with the fine-grained lagoonal deposits in the back-barrier marsh, tidal flat, washover, and flood-tidal environments, this lateral facies change from sandstone to siltstone and shale is still relatively abrupt.

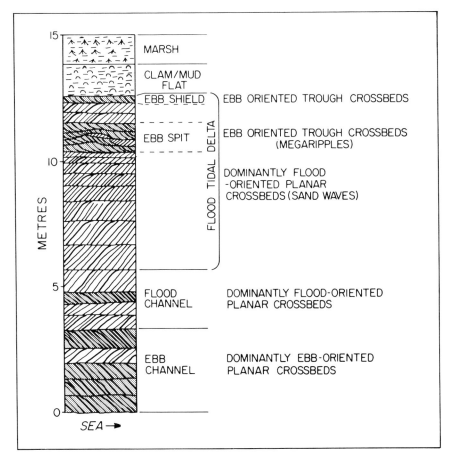

Figure 17
Hypothetical regressive sequence for a mesotidal flood-tidal delta complex (modified from Hayes, 1976).

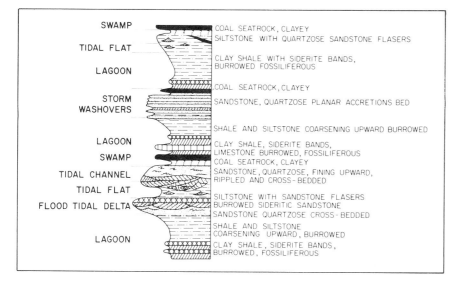

Figure 18
Generalized lagoonal sequence through back-barrier deposits in the Carboniferous of eastern Kentucky and southern West Virginia. Such sequences range from 7.5 to 24 m thick (from Horne and Ferm, 1978, Fig. 11).

Figure 19
Photo showing the sharp contact between the clean sandstones of the Blood Reserve Formation and the overlying finer-grained lagoonal deposits of the St. Mary River Formation (vertical sequence illustrated in Fig. 16).

Stratigraphic Sequences and Depositional Models

Transgression, Regression, and Preservation of Facies

The preservation of specific barrier facies is dependent upon a number of factors including sea level fluctuations, sediment supply, inlet conditions (migrating or stable) and wave climate. The most important condition, dependent largely on sea-level fluctuations, is the nature of the shoreline in terms of transgression or regression. The concepts of transgression and regression as used by geologists usually refer to the overlapping of deeper water deposits over more landward or shallower-water deposits (transgressive sequence), or shallow water deposits over more marine or deep-water facies (regressive). The terms "transgression" and "regression" are also used to imply the process of migration of the shoreline of a water body, in a landward direction (transgression), or in a seaward direction (regression) (Curray, 1964). Generally, transgressive and regressive barrier-shoreline migrations produce corresponding simple transgressive and regressive overlapping sequences (Figs. 20, 21), but this is not always the case. This is because shoreline migrational trends can be "regional" or they can be "local". Regional transgressive shoreline trends can be caused by relative sea-level rise such as is now occurring on the Atlantic coast of the United States, or they can occur by shoreline erosion under relatively stable sea level conditions, in areas where sediment supply is cut off and wave attack is intensified (Kraft, 1978).

Klein (1974) suggests that transgressive sequences have a low preservation potential relative to regressive sequences. Klein bases his suggestion on the fact that along transgressive coasts a thin basal transgressive interval (ravinement deposit) is often preserved and buried by regressive sediments. However, Kraft (1971) contends that the possibility exists for a complete transgressive sequence to be preserved, and that the relative rate of sea level rise will govern the amount of preservation. If transgression occurs largely by shoreface erosion with little or no relative sea level rise, almost total loss will result (Fig. 22). Conversely, if sea level rise is rapid, almost total retention of the sedimentary sequence can be expect-

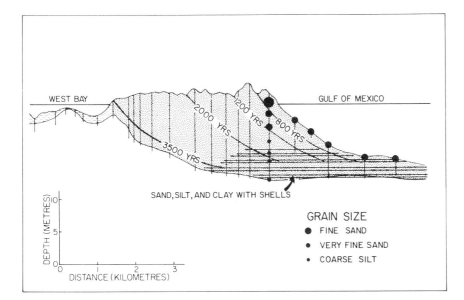

Figure 20
Cross section of the prograding Galveston barrier island (from Bernard et al., 1962).

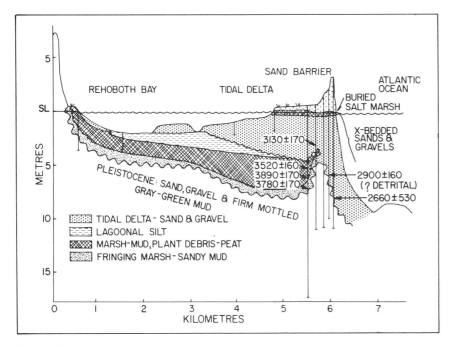

Figure 21
Cross-section of the Delaware barrier coast in the vicinity of a tidal delta, showing the transgressive nature of the Holocene sequence (modified from Kraft, 1971, Fig. 16).

ed. Therefore, under rapid relative sea level rise, it is conceivable that most of the facies, save for the upper portions of the back shore dunes, could be preserved. Under conditions of erosive shoreline retreat (transgression by erosion) most of the barrier facies will be destroyed, save for the reworked sediments of the ravinement lag deposit.

Dominantly transgressive shorelines can have "local" regressive segments within them. Such situations are caused by short-term temporal variations in depositional conditions along the barrier-island strandline. Longshore sediment supply, local wave climate, and number and location of tidal inlets are some of the conditions which can change significantly and can effect both progradational and erosional trends in near juxtaposition. This is illustrated by the beach sequences in Figure 9, the landward sequence being transgressive and the seaward sequence progradational or regressive. The Holocene studies of Kraft *et al.* (1978) also indicate that both transgressive and regressive shoreline trends could be inferred by two different vertical sequences in proximity. Given the presence of "local" regressive sequences in Holocene deposits, the possibility exists for their preservation in the rock record under conditions of rapid sea level rise. If such isolated stratigraphic sequences were encountered in the rock record they could be wrongly interpreted as being representative of the regional paleogeographic submergent or emergent conditions.

Certain facies have a higher potential for preservation than others because of their vertical position with respect to the intertidal zone (i.e., subtidal, intertidal), and their lateral position relative to the wave-dominated open coast or to a migratory inlet. Tidal inlet channel facies will probably have the highest preservation potential of all the sand facies because, depending on the depth of the inlet, they may extend well below low tide level, their basal portion thus being protected from shoreface erosion during transgressive conditions. In addition they would be protected above by the overlying beach-dune facies (Fig. 14). Flood-tidal delta deposits would have a high preservation potential as well because they are situated in the back barrier protected region (for the most part in the subtidal zone), and under

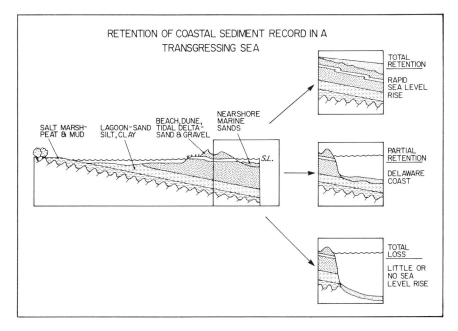

Figure 22
*Schematic diagram illustrating the variation in
retention of the barrier-island sequence in a
transgressing sea. Situations shown are*
*those for rapid sea level rise, little or no sea
level rise, and relative sea level rise-marine
transgression as is occurring on the Delaware coast (from Kraft, 1971, Fig. 24).*

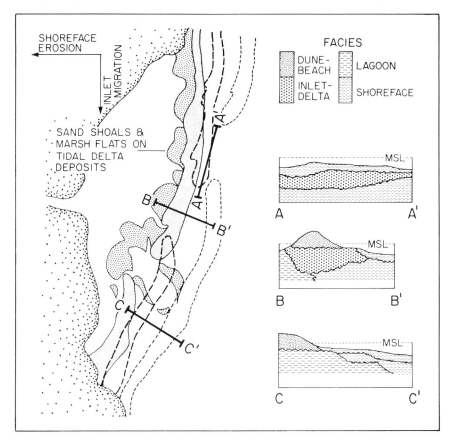

Figure 23
*Schematic diagram illustrating how an inlet-
delta sand body bounded by disconformities
could be formed under conditions of inlet
migration and concomitant shoreface ero-*
*sion. The occurrence and shape of the sandy
body, and the nature of the enclosing
sediments, vary with position in the barrier-
island complex.*

migrating inlet conditions could form relict sand shoals disconnected permanently from the tidal-current conduit (Fig. 23). The distal portions of washover deposits, where they interfinger with lagoonal fine-grained facies, would have a high preservation potential, as would most of the lagoonal facies which are relatively protected by the seaward barrier. These deposits would be the last "to go" under intense shoreline retreat. Ebb-tidal delta deposits would have a low preservation potential under both migrating inlet conditions and transgression, because of exposure to reworking by longshore currents and onshore wave processes.

Regardless of potential for preservation, given the right combination of sediment supply, inlet stability, wave climate and sea-level fluctuations, any one of the barrier-island facies could be preserved in barrier-island stratigraphic sequences; and we should be prepared to encounter all of them in the rock record.

Stratigraphic Models
From the examples of modern and ancient sequences, it is obvious that there cannot be just one generalized facies model for barrier-island deposits. If we apply the facies model criteria of Walker (1976), three "end-member" models for barrier-island stratigraphic sequences can be recognized; the regressive (prograding) model, the transgressive model, and the migrating barrier-inlet model (Fig. 24).

Regressive Model. The distillation of the generalized regressive facies model has come from modern examples, particularly the Galveston Island model (Fig. 20). As mentioned previously, prior to 1970 this example was the "one and only" model accepted for use in interpreting ancient sequences. Such a situation arose because the study of Bernard *et al.* (1962) was one of the first to present a detailed stratigraphic model for a barrier-island system. The situation became analogous to the thinking on deltas; when the word "delta" was mentioned, geologists thought of the "Mississippi" (Miall, 1976). Similarly the "Galveston Island" model came to mind immediately at the mention of "barrier island". Some of the earlier literature depicted other stratigraphic models (i.e., Hoyt, 1967 – transgressive barrier, and

70

Hoyt and Henry, 1965 – migrating inlet barrier), but these were largely ignored in the wave of enthusiasm for the Galveston model. Galveston Island should be recognized for what it is, a good example of the regressive facies model. It is not adequate as a "norm", because it does not include the essential characteristics displayed by most modern middle and upper shoreface deposits (Figs. 5, 7).

The regressive facies model in Figure 24 serves as a norm for interpreting ancient regressive barrier sequences only. It is a gradational-based, coarsening upwards sequence, dominated by shoreface, foreshore, and backshore-dune facies of the barrier-beach complex. Ancient examples of regressive

barriers include those of Davies *et al.* (1971), and Carter (1978) (Fig. 6).

Transgressive Model. The distillation of the generalized "end-member" sequence for the transgressive facies model comes also from modern examples, such as that depicted in Figure 21. This facies model is more complicated than the regressive model in terms of interbedding of facies and alternating lithologies. It is characterized by subtidal and intertidal back-barrier facies and does not show a fining-upwards or coarsening-upwards trend. The contact between some facies may be sharp or erosional. Many ancient sequences will deviate substantially from the normative model, because the facies stacking in

transgressive sequences is quite variable, due to the rapid response of depositional environments to change in sediment supply and inlet conditions in transgressive situations. A good example of an ancient transgressive sequence is the study of Bridges (1976). However, well-documented ancient examples of the transgressive facies model are few. This may reflect an overemphasis on the use of the regressive model in past literature.

One of the main differences between the regressive model and the transgressive model lies in the relationship with lagoonal facies. In the regressive sequence lagoonal deposits overlie the sand facies, whereas in the transgressive model lagoonal facies underlie, or are incorporated within the lower to middle portions, of the sequence (Fig. 24).

Barrier-Inlet Model. The distillation of the barrier-inlet model derives from very recent, well-documented modern and ancient examples (Figs. 13, 14, 16) including the studies of Land (1972), Hayes (1976), and Barwis and Makurath (1978). The barrier-inlet facies model is a fining-upwards sequence with a thinning-upwards trend in cross-bed set thickness (Fig. 24). It is characterized by an erosional base and dominated by sand facies of tidal-channel and marginal spit-beach environments.

Examples of Hybrid Models. Transgressive, regressive, and barrier-inlet depositional conditions can occur in combination to produce mixed sequences which have affinities with more than one "end-member" norm.

The so-called vertical build-up barrier of Padre Island is really a combination of the regressive and transgressive models, with the landward side of the barrier migrating into the lagoon by washover deposition, and the seaward side prograding outwards by beach-ridge accretion (Hayes, 1976; Dickinson *et al.,* 1972). Matagorda Island, situated near Galveston Island, has been shown by Wilkinson (1975) to have formed during both a transgressive phase and a subsequent regressive phase. Vertical sequences from the landward side of the Padre and Matagorda barriers are comparable to the transgressive model, whereas sequences from the seaward side are comparable to the regressive model.

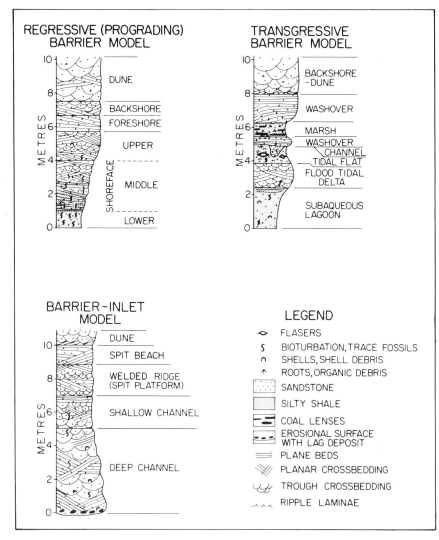

Figure 24
The three "end-member" facies models of barrier-island stratigraphic sequences.

An example of a combined regressive and barrier-inlet model may be the sequence illustrated in Figure 16. The top of the channel sequence is truncated by the overlying lagoonal deposits, and only the lower part of the barrier-inlet "end-member" model is preserved. The barrier-inlet "end-member" model (Fig. 24) may in itself be considered in part regressive because of the progradation of beach and dune facies over deeper-water channel deposits.

Perhaps the most complicated hybrid sequence that could occur is the situation of transgression concomitant with, or just after, barrier-inlet migration. Barwis and Makurath (1978) discuss the stratigraphic implications of such a setting in some detail. Basically, if transgression is occurring largely by shoreface erosion, and the migrating inlet is deep enough to produce an inlet deposit whose base is substantially below the foreshore-shoreface boundary, an inlet-delta sand body, bounded above and below by disconformities, could occur (Fig. 23). In vertical sequence such a deposit would be comparable to the lower part of the barrier-inlet "end-member" model, with an erosional surface similar to the basal lag, situated at the top. Kumar and Sanders (1970) consider that this dual migration setting could be the origin of many linear sand bodies on the inner shelf. They further suggest that submergence of migrating barrier-inlet shorelines may account for some of the basal transgressive sands in the geological record, the sands being of inlet-fill as opposed to barrier-beach or offshore bar origin.

Summary
Prior to the 1970s, the prograding Galveston Island depositional model was in the forefront in the minds of most geologists, as the "one and only" facies model for use in interpreting ancient barrier-island sequences. Studies conducted within the last decade on modern and ancient barrier-island deposits, indicate that the "regressive facies model" cannot be applied to a number of barrier-island sequences, and therefore the use of one normative model is unrealistic. Three generalized facies models or "end-member" norms can be recognized: the regressive barrier model, the transgressive barrier model,

and the barrier-inlet model (Fig. 24). Most sequences can be explained through comparative analysis with individual "end-member" models, or combinations of them, and this is emphasized by the recent studies of Horne and Ferm (1978) (Fig. 25), Barwis and Makurath (1978), and others.

Acknowledgements
I thank R.G. Walker and S.B. McCann for their valuable comments on the initial draft. O. Ogunyomi supplied the photograph for Figure 8. A. Cosgrove, H. Slade and K. Hale provided drafting services and C. Middleton typed the manuscript.

Selected Bibliography
The following reference list purposely emphasizes the most recent and summary-type papers on barrier-island systems. The reference lists contained in some of these articles are exhaustive, so the reader can become familiar with nearly all of the barrier-island literature. Unfortunately, some of the excellent summary studies on both modern and ancient barrier-island deposits are contained in field and course guidebooks; these are not easily accessible to the interested geologist.

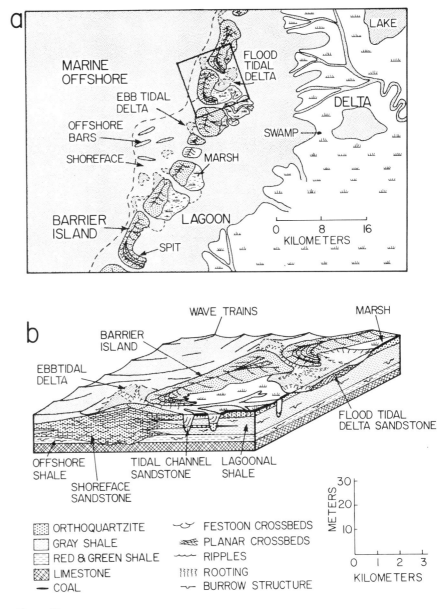

Figure 25

Reconstruction of the depositional environments (from stratigraphic sequences and lateral relationships) in the Carboniferous

Carter Caves Sandstone of Kentucky (from Horne and Ferm, 1978, composite of Figs. 21 and 22).

72

Essential Reference List

A. Holocene Barrier-Island Deposits

Bernard, H.A., R.J. Leblanc, and C.F. Major, 1962, Recent and Pleistocene geology of southeast Texas: *in* Geology of the Gulf Coast and Central Texas and Guidebook of Excursions: Houston Geol. Soc. - Geol. Soc. Amer. Ann. Mtg., 1962, Houston, Texas, p. 175-205.

Documentation of the oft-cited (and illustrated) prograding Galveston barrier-island model. The material in this original paper is more accessible in the following summary article.

Bernard, H.A., and R.J. Leblanc, 1965, Resume of the Quaternary geology of the northwestern Gulf of Mexico Province: *in* H.E. Wright Jr., and D.G. Frey, eds., The Quaternary of the United States: Princeton University Press, Princeton, N.J., p. 137-185.

Davis, R.A. Jr., ed., 1978, Coastal Sedimentary Environments: Springer-Verlag, New York, 420 p.

Contains excellent review chapters on modern beach deposits by R.A. Davis and Holocene coastal stratigraphic sequences by J.C. Kraft.

Hayes, M.O., and T.W. Kana, ed., 1976, Terrigenous Clastic Depositional Environments, Some Modern Examples: Amer. Assoc. Petrol. Geol. Field Course, Univ. South Carolina, Tech. Rept. No. 11-CRD, 315 p.

This field course guidebook is a must for geologists who wish to become aware of the latest thinking on the evolution of barrier-island deposits. Contains an excellent summary on transitional-coastal depositional environments by Hayes, an extensive bibliography on barrier-island and related environments, and some other useful papers (cited in text) on South Carolina barrier-island environments.

Kraft, J.C., 1971, Sedimentary facies patterns and geologic history of a Holocene marine transgression: Geol. Soc. Amer. Bull., v. 82, p. 2131-2158.

Discussion and documentation of the stratigraphy and facies distribution of a modern transgressive barrier-island system.

Kumar, N., and J.E. Sanders, 1974, Inlet sequence: a vertical succession of sedimentary structures and textures created by the lateral migration of tidal inlets: Sedimentology, v. 21, p. 491-532.

An excellent study of an inlet sequence and development of an inlet-fill model.

Schwartz, M.L., ed., 1973, Barrier Islands: Stroudsburg, Penn., Dowden, Hutchinson and Ross, 451 p.

A reprint collection of the classical historical and contemporary papers dealing with the origin and geomorphic development of barrier-island shorelines.

Swift, D.J.P., 1975, Barrier island genesis: Evidence from the Middle Atlantic Shelf of North America: Sediment. Geol., v. 14, p. 1-43.

This paper discusses the origin and evolution of barrier islands within the context of the east-coast setting of the United States.

B. Ancient Barrier-Island Deposits

Barwis, J.H., and J.H. Makurath, 1978, Recognition of ancient tidal inlet sequences: an example from the Upper Silurian Keyser Limestone in Virginia: Sedimentology, v. 25, p. 61-82.

Excellent documentation of an ancient tidal inlet-fill deposit. Also contains succinct reviews of modern tidal inlet studies and of published studies on ancient barrier deposits which appear to have tidal-inlet affinities.

Briggs, G., ed., 1974, Carboniferous of the Southeastern United States: Geol. Soc. Amer. Spec. Paper 148.

Papers by D.K. Hobday, J.C. Ferm, J.C. Horne, Ferm and J.P. Swinchatt, and R.C. Milici deal with the interpretation of Carboniferous orthoquartzites and related deposits as being of barrier-island, tidal channel-delta, and back-barrier origin.

Campbell, C.V., 1971, Depositional model – Upper Cretaceous Gallup beach shoreline, Ship Rock area, Northwestern New Mexico: Jour. Sed. Petrology, v. 41, p. 395-409.

Contains detailed descriptions of the geometry and sedimentary structures of Cretaceous barrier-beach facies.

Carter, C.H., 1978, A regressive barrier and barrier-protected deposit: Depositional environment and geographic setting of the Late Tertiary Cohansey Sand: Jour. Sed. Petrology, v. 48, p. 933-950.

Includes ancient sequences interpreted as containing surf-zone deposits and tidal channel deposits.

Davies, D.K., F.G. Ethridge, and R.R. Berg, 1971, Recognition of barrier environments: Amer. Assoc. Petrol. Geol. Bull., v. 55, p. 550-565.

Comparative study of the Galveston barrier model with Cretaceous and Jurassic sandstone deposits. Illustrates the use of the modern analogue in interpretation of ancient barrier-island deposits, and presents a model for barrier-island sedimentation. An excellent paper, but we must remember that the model presented is only one type of barrier island, the Galveston type. There are others.

Fisher, W.L., and J.H. McGowen, 1969, Depositional systems in Wilcox Group (Eocene) of Texas and their relation to occurrence of oil and gas: Amer. Assoc. Petrol. Geol. Bull., v. 53, p. 30-54.

Stratigraphic study using outcrop and subsurface data which proposes a regional depositional model consisting of deltaic environments flanked by barrier-island, lagoonal, and strandplain sediments.

Hobday, D.K. and J.C. Horne, 1977, Tidally influenced barrier-island and estuarine sedimentation in the Upper Carboniferous of southern West Virginia: Sedimentary Geol., v. 18, p. 97-122.

Interpretation of Upper Carboniferous sandstone units as barrier-island deposits, with the recognition of tidal-delta, tidal-channel and washover facies.

Horne, J.C. and J.C. Ferm, 1978, Carboniferous Depositional Environments: Eastern Kentucky and Southern West Virginia: Dept. of Geol., Univ. South Carolina, 151 p.

This excellent field guidebook summarizes the studies of Horne, Ferm and co-workers on the regional stratigraphy and paleoenvironmental interpretation of Carboniferous deltaic and barrier-island deposits of southeastern United States. Contains numerous outcrop illustrations and environmental reconstructions of various barrier-island facies and synthesizes these into a composite depositional framework.

Some of this material has recently been published in the following Paper:

Horne, J.C., J.C. Ferm, F.T. Caruccio, and B.P. Baganz, 1978, Depositional models in coal exploration and mine planning in Appalachian region: Amer. Assoc. Petrol. Geol. Bull., v. 62, p. 2379-2411.

LeBlanc, R.J., 1972, Geometry of sandstone bodies: *in* T.D. Cook, ed., Underground Waste Management and Environmental Implications: Amer. Assoc. Petrol. Geol., Mem. 18, p. 133-189.

One of the best of many articles dealing with the classification of clastic depositional environments, with a good introductory review of the barrier-island system.

Land, C.B. Jr., 1972, Stratigraphy of Fox Hills Sandstone and associated formations, Rock Springs uplift and Wamsutter arch area, Sweetwater County, Wyoming: a shoreline-estuary sandstone model for the Late Cretaceous: Quart. Colorado School of Mines, v. 67, no. 2, 69 p.

Well-documented sedimentological and stratigraphic study of Upper Cretaceous tidal-channel, barrier-beach, and lagoonal deposits.

Secondary Reference List

A. Holocene Barrier-Island Deposits

Andrews, P.B., 1970, Facies and genesis of a hurricane-washover fan, St. Joseph Island, Central Texas coast: Rept. Invest. No. 67; Bur. Econ. Geol., Univ. Texas, Austin, Texas, 147 p.

Thorough study of a modern washover fan deposit.

Boothroyd, J.C. and D.K. Hubbard, 1975, Genesis of bedforms in mesotidal estuaries: *in* L.E. Cronin, ed., Estuarine Research, v. 2, Geology and Engineering: Academic Press, New York, p. 167-182.

Illustrates the patterns of bedform development and distribution on tidal deltas in some New England estuaries.

Clifton, H.E., R.E. Hunter and R.L. Phillipps, 1971, Depositional structures and processes in the non-barred high-energy nearshore: Jour. Sed. Petrology, v. 41, p. 651-670.

Presents a sequence of bedforms for the upper shoreface-foreshore in a modern high-energy shoreline setting.

Davidson-Arnott, R.G.D. and B. Greenwood, 1976, Facies relationships on a barred coast, Kouchibouguac Bay, New Brunswick, Canada: *in* R.J. Davis, Jr., ed., Beach and Nearshore Sedimentation: Soc. Econ. Paleontol. Mineral. Spec. Publ. No. 24, p. 149-168.

Excellent study of the internal structures and facies in longshore bars of the shoreface environment.

Davis, R.A. Jr., W.T. Fox, M.O. Hayes, and J.C. Boothroyd, 1972, Comparison of ridge-and-runnel systems in tidal and non-tidal environments: Jour. Sed. Petrology, v. 32, p. 413-421.

Documents the vertical sequence that results when bars migrate shoreward and weld onto the beach face. This sequence of sedimentary structures is far more complicated than the usual seaward-dipping beach laminated deposit.

Fischer, A.G., 1961, Stratigraphic record of transgressing seas in light of sedimentation on Atlantic Coast of New Jersey: Amer. Assoc. Petrol. Geol., Bull., v. 45, p. 1656-1666.

Classic early paper which, by way of illustration of a Holocene transgressive situation, depicts implications of transgressive stratigraphy for interpreting ancient sedimentary sequences.

Hayes, M.O., 1975, Morphology of sand accumulations in estuaries: *in* L.E. Cronin, ed., Estuarine Research, v. 2, Geology and Engineering: New York, Academic Press, p. 3-22.

Outlines morphological models for flood and ebb-tidal deltas, and presents observations on the differences between microtidal and mesotidal barrier-island coastlines.

Kraft, J.C., E.A. Allen, and E.M. Maurmeyer, 1978, The geological and paleogeomorphological evolution of a spit system and its associated coastal environments: Cape Henlopen spit, Delaware: Jour. Sed. Petrology, v. 48, p. 211-226.

Documentation of vertical sedimentary sequences in a Holocene spit complex, with examples of both "regressive" and "transgressive" sequences occurring in the same region.

Kumar, N., and J.E. Sanders, 1976, Characteristics of shoreface storm deposits: modern and ancient examples: Jour. Sed. Petrology, v. 46, p. 145-162.

Emphasizes the possible significance of storm deposits in the geological record, by illustration of modern examples.

McCann, S.B., ed., 1979. The Coastline of Canada: Geol. Survey Canada Paper (in press).

This volume is a basic reference for most of the studies undertaken on Canadian barrier-island systems.

Morton, R.A., and Donaldson, A.C., 1973, Sediment distribution and evolution of tidal deltas along a tide-dominated shoreline, Wachapreague, Virginia: Sedimentary Geol., v. 10, p. 285-299.

A documentation of the three-dimensional development and stabilization of tidal delta deposits in a Holocene Barrier-island system.

Owens, E.H., 1977, Temporal variations in beach and nearshore dynamics: Jour. Sed. Petrology, v. 47, p. 168-190.

Detailed study of the sedimentary processes operative on the opposite-facing barrier beaches of the Magdalen Islands, Gulf of St. Lawrence.

Reineck, H.E., and I.B. Singh, 1972, Genesis of laminated sand and graded rhythmites in storm-sand layers of shelf mud: Sedimentology, v. 18, p. 123-128.

Gives an explanation for the origin of planar laminated sand units (devoid of ripple laminae) in lower shoreface deposits.

Schwartz, R.K., 1975, Nature and genesis of some storm washover deposits: U.S. Army, Corps of Engineers, Tech. Memo. No. 61, 69 p.

Detailed study of the stratigraphy, morphology, and internal geometry of some modern barrier-island washover deposits.

B. Ancient Barrier-Island Deposits

Bridges, P.H., 1976, Lower Silurian transgressive barrier islands, southwest Wales: Sedimentology, v. 23, p. 347-362.

Sandstone-mudstone sequences are interpreted as transgressive barrier-island deposits. The case for transgressive sequences is well-documented.

Cotter, E., 1975, Late Cretaceous sedimentation in a low-energy coastal zone: The Ferron sandstone of Utah: Jour. Sed. Petrology, v. 45, p. 669-685.

Interpretation of shoreface and tidal inlet deposits, with a short review of papers in which ancient tidal inlet deposits have been inferred.

Davies, D.K., and F.G. Ethridge, 1971, The Claiborne Group of central Texas: a record of Middle Eocene marine and coastal plain deposition: Trans. Gulf Coast Assoc. Geol. Societies, v. XXI, p. 115-124.

Presents a regressive barrier-island sequence that contains (from the base up) middle shoreface, tidal channel, washover, subaqueous lagoon, and tidal flat deposits.

Dickinson, K.A., H.L. Berryhill Jr., and C.W. Holmes, 1972, Criteria for recognizing ancient barrier coastlines: *in* J.K. Rigby, and Wm. K. Hamblin, eds., Recognition of Ancient Sedimentary Environments: Soc. Econ. Paleontol. mineral. Spec. Publ. No. 16, p. 192-214.

Would be more properly titled "Holocene barrier-island environments of the Gulf of Mexico, with some ancient examples."

Embry, A.F., G.E. Reinson, and P.R. Schluger, 1974, Shallow marine sandstones: a brief review: *in* M.S. Shawa, ed., Use of Sedimentary Structures for Recognition of Clastic Environments: Can. Soc. Petrol. Geol., Core Conference Guidebook, p. 53-66.

Contains some illustrations of subsurface cores from Canada which could be interpreted as being of barrier-island origin. The reference list includes some earlier Canadian studies on subsurface shallow marine and "barrier" like sand bodies.

Howard, J.D., 1972, Trace fossils as criteria for recognizing shorelines in stratigraphic record: *in* J.K. Rigby, and W.K. Hamblin, eds., Recognition of Ancient Sedimentary Environments: Soc. Econ. Paleontol. Mineral. Spec. Publ. No. 16, p. 215-225.

This article attempts to demonstrate the usefulness of trace fossils for interpreting ancient barrier-island environments, and succeeds admirably.

McGregor, A.A., and C.A. Biggs, 1968, Bell Creek field, Montana: a rich stratigraphic trap: Amer. Assoc. Petrol. Geol. Bull., v. 52, p. 1869-1887.

An excellent subsurface stratigraphic study of the Lower Cretaceous Muddy Sandstone. It is interpreted as a shallow-water nearshore sand deposit, but as pointed out by R.G. Walker (pers. commun.), the isopach map of the Muddy Sandstone resembles a barrier-island sand body cut by tidal inlets.

Ogunyomi, O., and L.V. Hills, 1977, Depositional environments, Foremost Formation (Late Cretaceous), Milk River area, southern Alberta: Canadian Petrol. Geol. Bull., v. 25, p. 929-968.

The Foremost Formation is interpreted as a cyclic sequence of regressive barrier-island deposits.

74

Sabins, F.F., 1963, Anatomy of stratigraphic trap, Bisti Field, New Mexico: Amer. Assoc. Petrol. Geol. Bull., v. 47, p. 193-228.
   An excellent subsurface stratigraphic study of the Gallup Beach Sandstone (Late Cretaceous). The producing sandstones are considered to be offshore bar facies. Their morphology and adjacent facies indicate they could also be interpreted as linear (possibly microtidal) barrier-island deposits.

Other References Cited in Text

Armon, J.W., 1979, Changeability in small flood tidal deltas and its effects, the Malpeque barrier system, Prince Edward Island: *in* S.B. McCann, ed., The Coastline of Canada: Geol. Survey Canada Paper, 1979.

Barwis, J.H., 1976, Internal geometry of Kiawah Island beach ridges: *in* M.O. Hayes and T.W. Kana, eds., Terrigenous Clastic Depositional Environments, Some Modern Examples: Amer. Assoc. Petrol. Geol. Field Course, Univ. South Carolina, Tech. Rept. No. 11 - CRD, P. II 115 - II 125.

Curray, J.R., 1964, Transgressions and regressions: *in* R.C. Miller, ed., Papers in Marine Geology - Shepard Commemorative Volume: MacMillan and Co., New York, p. 175-203.

Evans, G., 1965, Intertidal flat sediments and their environments of deposition in the Wash: Quart. Jour. Geol. Soc., v. 121, p. 209-245.

Field, M.E., and D.B. Duane, 1976, Post-Pleistocene history of the United States inner continental shelf: Significance to origin of barrier islands: Geol. Soc. Amer. Bull., v. 87, p. 691-702.

Glaeser, J.D., 1978, Global distribution of barrier islands in terms of tectonic setting: Jour. Geol., v. 86, p. 283-297.

Hayes, M.O., 1967, Hurricanes as geological agents: case studies of Hurricanes Carla, 1961, and Cindy, 1963: Bur. Econ. Geol., Rept. Invest. No. 61, Univ. Texas, Austin, 56 p.

Hayes, M.O., 1976, Transitional-coastal depositional environments: *in* M.O. Hayes, and T.W. Kana, eds., Terrigenous Clastic Depositional Environments: Amer. Assoc. Petrol. Geol. Field Course, Univ. South Carolina, Tech. Rept. No. 11 - CRD, p. I 32 - I 111.

Hoyt, J.H., 1967, Barrier island formation: Geol. Soc. Amer. Bull., v. 78, p. 1125-1136.

Hoyt, J.H., and V.J. Henry Jr., 1965, Significance of inlet sedimentation in the recognition of ancient barrier islands: Geol. Assoc. Wyoming 19th Field Conference Guidebook, p. 190-194.

Hubbard, D.K., and J.H. Barwis, 1976, Discussion of tidal inlet sand deposits: examples from the South Carolina coast: *in* M.O. Hayes, and T.W. Kana, eds., Terrigenous Clastic Depositional Environments, Some Modern Examples: Amer. Assoc. Petrol. Geol., Field Course, Univ. South Carolina, Tech. Rept. No. 11 - CRD, p. II 128 - II 142.

Klein, G. de V. 1974, Estimating water depths from analysis of barrier island and deltaic sedimentary sequences: Geology, v. 2, p. 409-412.

Klein, G. de V., 1977, Clastic Tidal Facies: Continuing Education Publ. Co., Champaign, Illinois, 149 p.

Kraft, J.C., 1978, Coastal stratigraphic sequences: *in* R.A. Davis Jr., ed., Coastal Sedimentary Environments: Springer-Verlag, New York, p. 361-384.

Kumar, N., and J.E. Sanders, 1970, Are basal transgressive sands chiefly inlet-filling sands: Maritime Sediments, v. 6, p. 12-14.

Miall, A.D., 1976, Facies models 4. Deltas: Geosci. Canada, v. 3, p. 215-227.

Owens, E.H., and D. Frobel, 1977, Ridge and runnel systems in the Magdalen Islands, Quebec: Jour. Sed. Petrology, v. 47, p. 191-198.

Reineck, H.-E., and I.B. Singh, 1975, Depositional Sedimentary Environments with Reference to Terrigenous Clastics: Berlin, Springer-Verlag, 439 p.

Reinson, G.E., 1977, Tidal current control of submarine morphology at the mouth of the Miramichi esturay, New Brunswick: Can. Jour. Earth Sci., v. 14, p. 2524-2532.

Reinson, G.E., 1979, Longitudinal and transverse bedforms on a large tidal delta, Gulf of St. Lawrence, Canada: Marine Geol. (in press).

Straaten, L.M.J.U. van, 1961, Sedimentation in tidal flat areas: Jour. Alberta Soc. Petrol. Geol., v. 9, p. 204-226.

Walker, R.G., 1976, Facies models 1. General introduction: Geosci. Canada, v. 3, p. 21-24.

Wanless, H.R., 1976, Intracoastal sedimentation: *in* D.J. Stanley, and D.J.P. Swift, eds., Marine Sediment Transport and Environmental Management: New York, John Wiley and Sons, p. 221-240.

Wilkinson, B.H., 1975, Matagorda Island, Texas: The evolution of a Gulf Coast barrier complex: Geol. Soc. Amer. Bull., v. 86, p. 959-967.

Young, F.G. and G.E. Reinson, 1975, Sedimentology of Blood Reserve and adjacent formations (Upper Cretaceous), St. Mary River, Southern Alberta: *in* Shawa, M.S., ed., Guidebook to Selected Sedimentary Environments in Southwestern Alberta, Canada: Can. Soc. Petrol. Geol., Field Conference, p. 10-20.

MS received February 7, 1979.

Reprinted from Geoscience Canada, v. 6, No. 2, 1979.

Facies Models 7. Shallow Marine Sands

Roger G. Walker
Department of Geology
McMaster University
Hamilton, Ontario L8S 4M1

Introduction

Shallow seas, with depths in the range of 10 to 200 m, can be divided into two types – those which cover the present continental shelves, and the epeiric seas which cover interior parts of cratons, such as Hudson Bay and the major north-south trending Cretaceous seaway of western North America. Despite the differences in tectonic setting, the processes operating and the resulting sand bodies appear to be similar.

Most work on recent sediments has been done on the present continental shelves, whereas much of the work on ancient shallow marine sandstones has been in epeiric sea settings. It has proved difficult to interpret ancient deposits in terms of what is known about modern shelves, and not surprisingly, there is no well developed shallow-marine facies model. Indeed, there are few reviews that attempt to compare ancient and recent - the best is that of Johnson (1978).

One of the problems of developing a shallow marine model (or models), and of structuring this review, is that there is no simple correlation between processes and results. Modern sand bodies in shallow water can be grouped into a heirarchical scheme of: 1) shoal retreat massifs (tens of km long perpendicular to shoreline, and a few tens of km wide); 2) linear ridges a few tens of km long, a few km wide, a few tens of metres thick; and 3) bedforms such as sand waves (sometimes called dunes or megaripples) up to a few metres in height.

Processes operating in shallow marine situations include: 1) intruding ocean currents; 2) tidal currents; 3) meteorological currents (wind, wave and storm currents); and 4) density currents. Of these, storm and tidal currents seem to be the most important, and studies of these processes in recent sediments have given rise to the concept of *storm-dominated* and *tide-dominated* shelves.

The lack of direct correlation between process and result is due to the fact that different processes may lead to similar sand bodies (e.g., linear sand ridges may be tidally or storm controlled) and that the same process may result in different deposits (for example, tidal currents can form linear sand ridges and smaller bedforms).

Shallow marine sands are of considerable economic importance, and form major hydrocarbon reservoirs in Canada, and the U.S.A. For example, the Cardium Formation (Cretaceous, Alberta) is a major oil reservoir. Immense reserves of natural gas are trapped in similar Cretaceous sands in the Milk River Field of Alberta (9 Tcf, trillion cubic feet; Masters, 1979), in the deep basin gas trap of Alberta and B.C. (recoverable reserves perhaps reaching 150 Tcf; Masters, 1979) and in Cretaceous shallow marine and barrier sandstones of the San Juan Basin of New Mexico (25 Tcf).

After highlighting some of the problems of shallow marine sands, I will consider the first, second and third-order modern sand bodies and the processes affecting them. I will then discuss ancient shallow marine sandstones.

Major Problems

The two major problems of modern shallow marine sands can be stated quite simply: 1) how is the sand transported into depths of 10 to 200 m, and 2) to what extent is it remolded by tidal, storm and oceanographic currents?

Until the early 1970s, the sand and gravel on the present continental shelves was regarded as *relict* - left over from the Pleistocene low stand of sea level (Emery, 1968). It was deposited in fluvial and deltaic environments, and was abandoned without much reworking during the Holocene transgression. However, in 1971, Swift, Stanley and Curray suggested that there had been extensive reworking of the sands during the transgression and that in places they were approaching equilibrium with the present hydraulic regime. Further work, particularly in the North Sea, and on the Atlantic and Pacific shelves of North America, has tended to reinforce the idea that the major shelf sand bodies were formed during the Holocene transgression (from left-over Pleistocene sand), and are presently being maintained and/or reworked by tidal, storm and oceanographic currents.

This interpretation highlights the problem of whether new sand is being transported out onto the shelves, or whether the deposit consists solely of remolded, left-over Pleistocene sand. If it is purely left-over, it will remain more or less at its present thickness of up to a few tens of metres. Yet the geological record suggests that some shallow marine sandstones were deposited in actively aggrading areas - the Late Precambrian Jura Quartzite of Scotland,

76

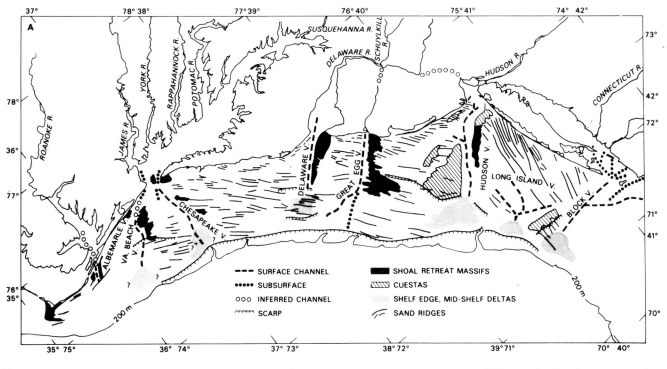

Figure 1
Major morphological features of the Middle Atlantic Bight. Note relationship of shoal

retreat massifs to present estuaries, and to capes (e.g., Cape Hatteras, lower left corner of diagram). Note consistent angle of linear

sand ridges to shoreline (average 22°). From Johnson (1978), after Swift et al., 1973.

---

for example, is over 5000 m thick (Anderton, 1976).

The best studied areas of modern sand accumulation are the continental shelves off Eastern North America, and off the British Coast (including the North Sea). Studies by Swift and associates (Duane *et al.*, 1972; Swift and Sears, 1974; Swift, 1975) on the Atlantic shelf of North America indicate that three scales of morphological features exist: shoal retreat massifs, linear sand ridges, and large bedforms (Fig. 1). These are considered below.

Shoal Retreat Massifs
The *first-order, shoal-retreat massifs* can be traced landward into areas of present-day preferred sand accumulation in estuaries and capes. In transgressive situations, the estuaries act as sand sinks (e.g., Hudson, Delaware, Chesapeake; Fig. 1), and along the Atlantic coast, littoral drift convergence also makes the capes areas of preferred sand accumulation (e.g., Cape Hatteras at southern end of Fig. 1). The massifs are up to 72 km long and 21 km wide. Similar massifs would be predicted in areas where the sea transgressed a coastline with local sand sinks. Despite their size, there has been little work

specifically on cape- and shoal-retreat massifs. Thickness, for example, appears to be in the 10 to 30 m range, but has not been explicitly stated. Preservable sedimentary features would seem to be those associated with the superimposed, second order features. Initiation of shoal-retreat massifs is not a function of shelf processes so much as the transgression of local sand sinks at the shoreline.

Linear Sand Ridges
A *second-order* morphology, consisting of linear ridges, is developed on the shoal-retreat massifs, and upon shelf areas between massifs (Fig. 1). The ridges have an average spacing of about three km and extend for tens of km diagonally across the shelf, making an average angle with the shoreline of about 22° (Fig. 1). The height of the ridge above the adjacent swale is up to 10 m. Flanks of the ridges have dips of a few degrees or less, commonly with a slightly steeper seaward face. Many ridges are attached to the shoreface, and it is now believed that they form in response to modern dynamic processes (Swift *et al.*, 1973).

Storm Control of Linear Ridges. It has been shown that the troughs landward of the ridges are undergoing scour and headward erosion "in response to south-trending wind set-up currents associated with mid-latitude lows [northeaster storms]" (Swift *et al.*, 1973). At the same time, southwest advancing waves drive water over the ridge crests, which are areas of winnowing (Fig. 2). The net result is that the coarsest sediment occurs as lags on the ridge crests and in swales between ridges. Many examples of such ridges have been given by Duane *et al.* (1972), and these authors also suggested the evolutionary sequence shown in Figure 2, in which the longshore currents finally truncate a ridge at its neck, essentially isolating it on the shelf. The ridges and swales are then maintained and slightly modified by storm-generated currents that flow parallel to the swales (Stubblefield and Swift, 1976). Stubblefield *et al.* (1975) have also suggested a model with three stages of activity. During fair weather, most of the activity on the ridges is biogenic. The second stage occurs during storms which are not intense enough to entrain the entire water column. Wave surge on the crest results in winnowing, with fine sand

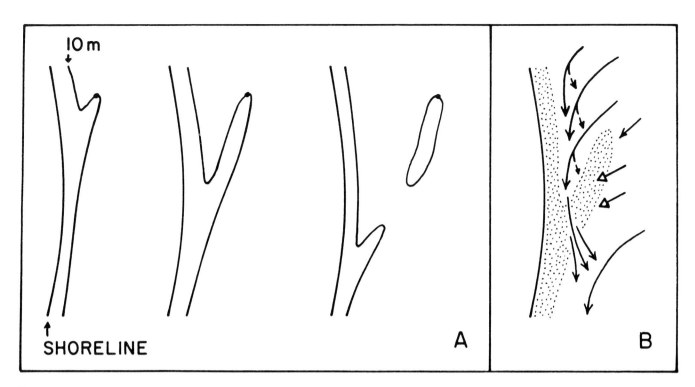

10 m

↑ SHORELINE

A

B

Figure 2
Detachment of linear sand ridges during transgression. A) the black spot is a fixed point - note transgression of shoreline,

growth of swale behind sand ridge by headward erosion, and eventual detachment. B) inferred flow pattern around linear sand ridge; solid arrow - surface wind-driven

current; dotted arrow - bottom wind driven current; open head arrow - surface wave driven current. Redrawn from Swift, 1976.

moving off the crests onto the flanks and into the troughs. The third stage occurs during major storms when the entire water column is in motion. Large scale secondary circulation may develop, with water descending in the troughs, scouring the bottom, and moving sand toward ridge flanks and crests. They suggest that coarse sand is preferentially transported toward the crests, and that the finer sand remains on ridge flanks, developing a coarsening-upward textural gradient through the ridge (Stubblefield *et al.,* 1975). Inasmuch as major currents flow parallel to the ridges, both during initial formation and subsequent modification, I suggest that a ridge preserved in the stratigraphic record might have a paleoflow vector mean parallel to the ridge, and subparallel to the shoreline. The paleoflow indicators would be bedforms, mostly dunes, giving rise to medium scale cross bedding.

Tidal Control of Linear Ridges. I have discussed the evidence for storm controlled ridges, and will now consider those that appear to be controlled by tidal currents, such as those in the North Sea (Fig. 3; Swift, 1975). The ridges, first described in detail by Houbolt (1968),

are 10 to 40 m high, 1 to 2 km wide, and up to 60 km long. Some of them appear to be shoreface-attached, and may develop in a manner analogous to that shown in Figure 2, except that in the North Sea the southward flowing alongshore current is a tidal current rather than a storm current. Once the ridges have been isolated from the shoreline by transgression, their continued development appears to be tidally controlled. The tidal flows tend to be asymmetrical, with one side of a ridge dominated by the flood tide, the other side by the ebb. The result is that "megaripples" (third order features up to a few m high) move from the swales up onto the ridges. In the swales, megaripple crests lie at about 90° to the ridge crest, but this angle decreases to about 50 to 60° on the flank of the ridge (Caston, 1972). Caston also suggested that the Z-shapes of some of the ridges closer to shore were a function of tidal current modification, with unequal rates of cross-ridge sediment transport leading to a kink in the ridge (Fig. 3) (Caston, 1972, Fig. 10). The sonar records of Caston (1972) also demonstrate that sand is moving toward the ridge crest from both sides, contrary to the earlier interpretation by Houbolt (1968).

The internal structure of the sand ridges is poorly known. Houbolt (1968) showed internal stratification planes (see Fig. 6 of Paper 4 in this volume) dipping at about 5° in the direction of the "steep" face of the ridge. These can probably be interpreted as set or coset boundaries of internal cross-bedding - the cross-bedding is probably variable in direction, with some directions subparallel to the ridge but reflecting both ebb and flood flows, and other directions oblique to the ridge reflecting sandwaves moving toward the crest.

Similar large sand ridges with smaller bedforms being driven up onto them exist in other parts of the North Sea (Houbolt, 1968) and in the Gulf of Maine (Jordan, 1962).

Tidal sand ridges have been described from other parts of the world by Off (1963). Many of his examples were from funnel-mouthed estuarine areas where a high tidal range generated powerful tidal currents, which in turn maintained the sand ridges more or less parallel to the elongation of the estuary (e.g., the Thames, Swift, 1975; and the Bay of Fundy). Swift suggested that as such an estuary was drowned during transgression, the elongate sand bodies

78

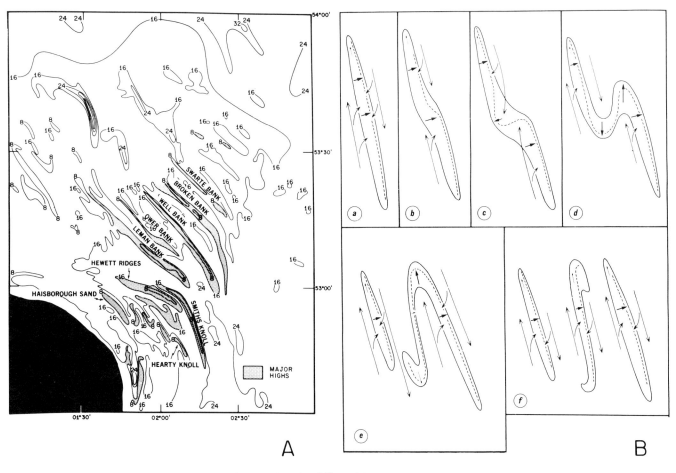

Figure 3
A) Sand ridges in the North Sea, from Swift,
1975, after Houbolt, 1968. The whole group of
linear ridges forms a shoal retreat massif
perpendicular to the coast.
B) Evolution of sand ridges, showing preferred tidal current paths, ebb on one side and

flood on the other. Differential sand transport
leads to the Z shapes. a) ridge straight (e.g.,
Swarte Bank); b) slight kink develops (e.g.,
Leman Bank beginning to enter this stage); c)
larger kink develops (e.g., Ower Bank); d)
incipient pair of ebb and flood channels
forming ; e) new channels lengthen, bank

between straightens out (e.g., Haisborough
Sand to Hearty Knoll); f) cycle complete, three
banks where there was only one. Dotted line
marks ridge crest, open arrows indicate tidal
currents, and solid-head arrows indicate
direction of bank movement. From
Caston, 1972.

would gradually swing-around to a
roughly shore-parallel trend (Swift, 1975,
p. 116-126).

Sand Waves
Third-order features resembling
periodically-spaced bedforms occur on
sandy areas of the shallow sea floor, and
may also be superimposed on linear
ridges and shoal-retreat massifs. They
are commonly termed *sand-waves*
(Caston, 1972; McCave, 1971; Terwindt,
1971), but are not the same features as
the sand waves described from flume
experiments and some rivers. In modern
shallow seas, the sand waves appear to
be controlled by tidal currents or by
intruding ocean currents.

Tidal Control of Sand Waves. In open
shallow seas, tidal ranges tend to be low,
and tidally-generated currents weak.

Coriolis force acting on the tides tends to
result in a gradual change of flow
direction during each tidal cycle, and
water particles trace an elliptical path (in
a horizontal plane).

By contrast, in enclosed seas and
embayments, the tidal range is ampli-
fied, as in the Bay of Fundy and the
southern part of the North Sea. Tidal
currents are much stronger, and the tidal
ellipse tends to be very narrow. In the
southern part of the North Sea, there is a
roughly northward directed tidal flow that
reverses very rapidly to a southward
directed flow during each tidal cycle.
Calculated velocities one m above the
bed are of the order of one m/sec.

These strong tidal currents control the
movement of sand waves in a large
(15,000 km^2) field off the coast of the
Netherlands (McCave, 1971; Terwindt;
1971). Sand wave height varies from

about two to seven m, with wavelengths
of about 200 to 500 m. The "steep" faces
of the sand waves have dips of about 5°
(see Fig. 7 of Paper 4 in this volume).
Sand waves with heights of about five m
or more tend to have megaripples on
their backs. By McCave's (1971) definition, the megaripple heights are up to 1.5
m with wavelengths up to 30 m. Crests
are all perpendicular to principal ebb
and flood directions, and net transport
appears to be mostly toward north-
northeast, but with some south-
southwest transport in the southwestern
part of the sand wave field. Because of
the gentle, 5° dips of the "steep" faces,
the 5 to 7 m high sand waves will not
produce angle-of-repose sets of cross
bedding several metres in thickness
(see Paper 4 in this volume). The
megaripples on the backs of the sand
waves probably form sets of cross

bedding up to a metre or so in thickness, and hence the larger sand waves may have an internal structure consisting of many smaller sets of cross bedding.

Although tidal currents are clearly important in developing and maintaining the sand wave field, there is no evidence that they introduce new material into shallow marine areas from adjacent coastal sources. This generalization applies to the North Sea (where the sand waves consist of reworked Pleistocene sand), and to all other areas of strong tidal current action.

Sand Waves Formed by Intruding Ocean Currents. Recently, another large field of sand waves (termed dunes by Flemming, 1978) has been discovered on the edge of the shelf off Southeast Africa (Fig. 4). In this situation, the controlling current appears to be an intruding oceanic current (Agulhas Current; Flemming, 1978). The sand waves were observed on side-scan sonar, and occur in depths greater than 50 m. They occur in continuous fields up to 20 km long and 10 km wide, and in different parts of the area, maximum heights are given as up to three and eight m. The eight m sand waves have wavelengths in excess of 200 m. Although not discussed in the paper, Flemming (1979, pers. commun.) now recognises sharp-crested sand waves up to 17 m high, with wavelengths of up to 700 m, and lee faces with angles in excess of 25° (Flemming, in press). This latter observation is extremely important, and constitutes the only observation from modern oceans of angle-of-repose slip faces many metres high. It would be surprising if there are not other examples still to be discovered of reworking by intruding ocean currents. Their geological influence has yet to be evaluated, but the 20 m cross bed sets described by Nio (1976) and discussed below could perhaps have formed in a similar manner.

Storm Currents

Some of the effects of storm currents in reworking sand ridges have been discussed above, and further information can be found in Kumar and Sanders (1976) and Lavelle *et al.* (1978). However, the role of storms in transporting new sediment from the coastal zone into shallow seas has probably been underestimated. The most important work has been done on the Texas shelf by Hayes (1967) and R.A. Morton (pers. commun., 1979).

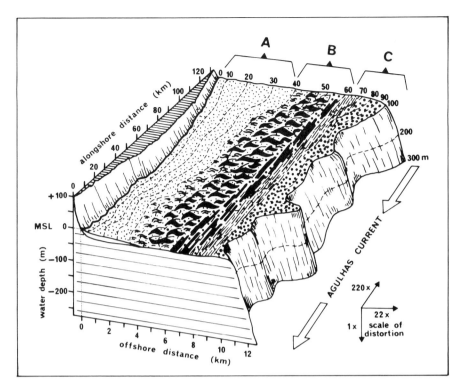

Figure 4
Sand waves (here termed dunes), as observed on side scan sonar, on the shelf edge off the southeast tip of Africa. Sand waves appear to be driven by an intruding ocean current – the Agulhas Current of the Indian Ocean. Individual sand wave fields up to 20 km long, 10 km wide. From Flemming, in press.

Hurricane Carla, in 1961, was one of the worst hurricanes on record on the Texas shelf. It eroded the shelf, and transported "rock fragments and coral blocks" (Hayes, 1967) from depths of 16 to 26 m up onto the barrier beaches. Storm surge tides (due to the wind piling up water in the nearshore area) reached 6.5 m above normal in Matagordo Bay (about 250 km north of Laguna Madre, where Hayes was working). Following passage of the hurricane, cores taken on the shelf demonstrated the presence of a graded bed (Fig. 5) in places more than nine cm thick, in water up to 36 m deep (the limit of the coring program). The bed has now been destroyed by bioturbation. Hayes (1967) noted the correlation between: 1) hurricane channels cut into the Padre Island beach-barrier complex, 2) areas of thicker deposition of the Carla graded bed, and 3) the absence of the graded bed on the seaward side of submerged ridges (shown by 13 to 17 m, and 27 to 29 m isobaths, Fig. 5). Hayes' inference was that returning storm surge entrained sand and developed into a density (turbidity) current that carried sand out onto the shelf for at least 15 km, into depths of at least 36 m. This explanation

may have to be slightly modified, because R.A. Morton (pers. commun. 1979) has quoted U.S. Corps of Engineers data that show "slight" to low" tides in Laguna Madre area. He suggested "large-scale bottom return flows analogous to rip currents" as the "most likely explanation for the graded bed reported by Hayes". Highest velocities recorded on this part of the Texas shelf are "of the order of two m/sec, and are directed alongshore or offshore" (Morton, pers. commun., 1979). Whether or not the Carla current was generated by a hydraulic head of water (Hayes), or "large scale bottom return flows" (Morton), the presence of a graded bed in water at least 36 m deep suggests that sand was entrained and carried in suspension to the depositional site – that is, that Hurricane Carla generated a turbidity current.

Although Carla is the best-documented modern shallow marine turbidite, there is a steadily growing body of geological evidence to suggest that storm-generated density currents can entrain sand and then flow seaward for tens of km, producing graded beds in shallow marine environments. This evidence is reviewed below.

Summary of Processes and Deposits, Recent Sediments

The evidence presented so far suggests the following general conclusions:

1) Shoal-retreat massifs represent local shoreline depocentres that have been transgressed, producing an elongate sand body perpendicular to shore.

2) on the Mid-Atlantic Shelf of North America, the linear sand ridges and smaller superimposed sand waves are storm dominated, both in their initiation and subsequent maintenance during transgression.

3) in the southern North Sea, the linear sand ridges and sand wave field are tidally dominated.

4) intruding ocean currents can also form sand wave fields, as on the shelf off Southeast Africa.

5) the only major process that can transport significant amounts of sand from the shoreline out onto the shelf is a storm-generated density (turbidity) current.

Geological Aspects of Shallow Marine Deposition

One of the difficulties in developing facies models is the problem of correlating large scale morphological features seen in recent sediments (e.g., shoal retreat massifs, or submarine fans (Paper 8 of this volume) with the smaller scale facies that combine to form the large features. It is most unlikely that a shoal retreat massif could be identified in the geological record, although there may be examples of linear sand ridges. Features of the scale of individual bedforms and deposits of individual storms are common.

Studies of recent sediments, with emphasis on the storm-dominated mid-Atlantic Shelf and the tide-dominated North Sea, have suggested to geologists that ancient shallow marine rocks might be interpreted along similar lines. I will discuss situations interpreted as tide-dominated and storm-dominated first.

Tide-dominated systems. The tidal influence on small bedforms, sand waves and sand ridges in the North Sea is now so well established that geologists regard the *abundance* of small and medium scale cross bedding (sets up to one or two metres) as one of the major indicators of tidally-dominated shallow marine deposition. In recent publications (Nio, 1976; Nio and Siegenthaler, 1978), it has been suggested that

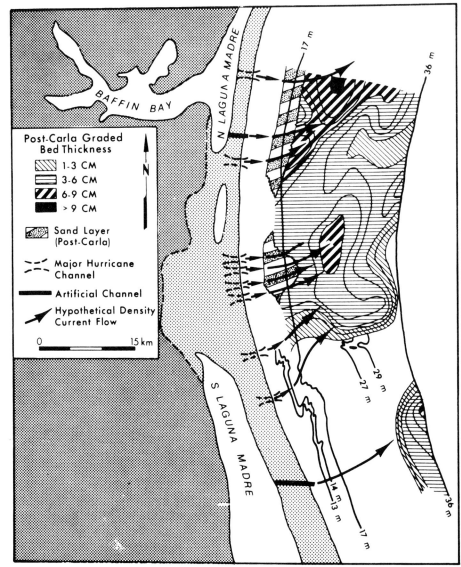

Figure 5
Thickness of graded bed deposited by seaward-directed density current following Hurricane Carla, 1961. Note relationship of thicker depocentres to major hurricane channels in Padre Island. Note also the inferred deflection of density current flow by bottom topography (ridges shown by the 13 to 17 m, and 27 to 29 m isobaths). From Hayes, 1967.

ancient "sandwave complexes" are the result of initial tidal domination, followed by marine transgression to preserve the deposits.

Nio (1976) gives three examples of transgressive sand-wave complexes, with the best data on the Eocene Roda sandwave complex of northern Spain and the Cretaceous Lower Greensand of southern Britain. In the Roda complex (Nio and Siegenthaler, 1978), there are five sandstone bodies 10 to 30 m thick with sandwaves, separated by fossiliferous marls, bioclastic limestones and bioturbated silty sandstones (5 to 30 m). The sandstones contain abundant marine shell debris. Nio and Siegentha-

ler (1978, p. 16) discussed sandwave development in five stages (Fig. 6). The *initial sandwave facies* contains cross-bed sets up to 1.5 m, and develops laterally into larger cross beds. The *sandwave facies* consists of giant cross beds that increase in size downcurrent, reaching a maximum of 20 m (Nio, 1976, Fig. 6). The *slope facies* (formed on the eroded lee slope of the 20 m high sandwaves) consists of cross bed sets up to three m thick, with set boundaries dipping at about 15°. Traced downcurrent, the cross beds gradually become thinner (the proximal to distal change of Fig. 6). The *abandonment facies* consists of an abundantly fossiliferous bed

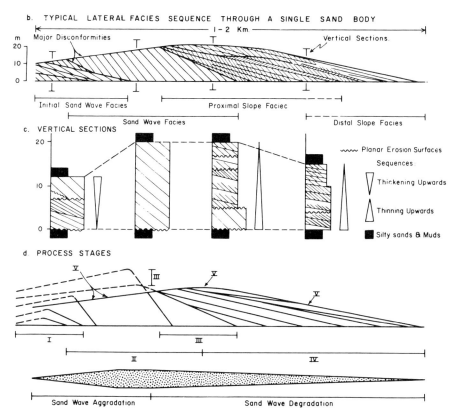

b. TYPICAL LATERAL FACIES SEQUENCE THROUGH A SINGLE SAND BODY

c. VERTICAL SECTIONS

d. PROCESS STAGES

Figure 6

Lateral facies sequence through a typical sandwave complex of the Roda Sandstone, Eocene, northern Spain. Cross section b is

reconstructed from the measured sections shown in c, and d shows the inferred stages of evolution. From Nio and Siegenthaler, 1978, Figure 8.

---

sea that result in maximum amplification of tidal range. Any change of depth, by continued transgression or the beginnings of regression, would result in depth and basin configuration changes, and hence weaker currents. Thus both processes would probably lead to the burial of the sand wave complex by finer sediment, completing the fining-upward sequence.

Storm-Dominated Systems. It has been suggested above that storms can generate unidirectional currents that winnow the big linear sand ridges and drive bedforms along the troughs and up onto the ridges (Stubblefield and Swift, 1976). The resulting cross-bedding might be very difficult to distinguish from tidally-formed cross-bedding. Recently, however, a sedimentary structure has been described that appears to be formed by storm waves *below* normal wave base ("normal" or fairweather wave base depends upon the overall geometry of the shallow sea and the normal wind climate - it can be taken at 10 to 15 m on the Atlantic shelf).

Hummocky Cross-Stratification (H.C.S.) is a term given to gently-undulating sets of cross-stratification that cut each other at low angles – typically three to six degrees. The sets can be concave-up or convex-up (Figs. 11, 12), and it is the convex-up stratification that typically defines the hummocks. In large outcrops, where the H.C.S. can be traced laterally (Fig. 12), it is seen to have wavelengths of one to five m, and individual sets have amplitudes of 10 to 20 cm. Lamination typically conforms to the curved shapes of hummocks and troughs. In small outcrops, H.C.S. is typified by low angle intersecting curved surfaces, with the convex-up stratification serving to distinguish H.C.S. from very low angle trough cross-bedding. An interpretive block diagram of H.C.S. is shown in Figure 13.

The term H.C.S. was created by Harms *et al.* (1975), but identical structures have been termed "truncated wave ripple lamination" (Campbell, 1966) and "sublittoral sheet sandstones" (Goldring and Bridges, 1972). One important extra descriptive feature emphasized by Hamblin and Walker (1979) is the occurence of H.C.S. in sharp-based beds with directional sole marks, interbedded with shales. In this study, from the top of the Jurassic Fernie Formation in southern Alberta, it was

---

10 to 50 cm thick that drapes the sandwave. The conceptual model of Nio and Siegenthaler (1978, p. 18) is shown in Figure 7. Sand is introduced into a marly environment (Fig. 7, 1) and the sand waves gradually grow in size (Fig. 7, 2). Flow directions on the large cross beds are consistently southwestward. During the slope phase (Fig. 7, 3), the sand waves are planed off by dunes moving down the lee faces, with more variable flow directions, some mud drapes and bioturbation. During abandonment (Fig. 7, 4), there is concentration of fossil debris, dying away of the currents, and return to open marine marls and limestones. This example is very significant in that it is the largest demonstrably marine angle-of-repose set of cross bedding yet described, and the sand wave heights exceed anything recorded in modern shallow seas for angle-of-repose bedforms. Similar sand waves, of rather smaller scale, are shown in Figures 8 and 9.

Although there are other descriptions of shallow-marine, tidally controlled cross-bedded sandstones in the litera-

ture (Banks, 1973a, facies 4; Anderton, 1976, coarse facies), there is no good description of an entire, tidally dominated stratigraphic unit. I suspect that many of the Cambrian transgressive sandstones of North America might qualify – certainly the Gog Sandstone of the Kicking Horse Pass section, B.C., is dominated by cross-bedding in sets up to a few metres thick (Fig. 10). More significantly, the Gog in the Kicking Horse Pass lacks evidence of storm influence, whereas other thick quartzose sandstones (e.g., Jura Sandstone, Anderton, 1976) show abundant cross-bedding (?tidal influence) along with evidence for storm activity.

A model of tidally-controlled shelf deposits based upon the present North Sea might predict the development of an overall *fining-upward* sequence (Middleton, pers. commun. 1979). The base of the sequence would consist of a coarser lag, overlain by sand waves winnowed and reworked during the transgression. The sand waves reflect the strongest currents, that is, the depth of water and geometry of the enclosed

82

shown that the interbedded shales and
tops of the H.C.S. beds were bioturbated,
but that the bulk of the H.C.S. was not.
The sharp base and lack of bioturbation
in the H.C.S. indicate a sudden influx of
sediment followed by rapid deposition,
no subsequent reworking of the H.C.S.,
and a return to slow mud deposition. The
lack of reworking, into wave-formed
symmetrical ripples or into current-
formed small scale cross-bedding,
suggests deposition below fair-weather
wave base. The hummocks and troughs
were interpreted as forms produced by
the oscillatory motion of storm waves
feeling the bottom (Harms et al., 1975;
Hamblin and Walker, 1979), and H.C.S.
is now taken as a good indicator of
deposition *below* fairweather wave base
but *above* storm wave base, with the
hummocky topography being controlled
by storm waves.

In the Fernie study (Hamblin, 1978;
Hamblin and Walker, 1979), a broad,
three-part stratigraphic sequence was
demonstrated (Fig. 14); turbidites,
followed stratigraphically upward by
H.C.S. sandstones, overlain in turn by
beach facies. In the absence of stratigraphic record of slope deposits in the
uppermost Fernie beds (i.e., the slope on
which the turbidity currents could have
been generated), it was suggested that
density currents formed near the shoreline, in a manner similar to the Hurricane
Carla density current described above.
Thus a single storm (Fig. 15) could: 1)
develop a density current and 2) form
H.C.S. as sand was dropped from the
density current in shallow water below
fairweather wave base. If the density
current flowed into water below storm
wave base, a normal turbidite (Paper 8 of
this volume) would be deposited.

By reading between the lines in the
literature, it is clear that interbedded
sandstones (with H.C.S.) and shales are
common in shallow marine situations.
Examples include Devonian sandstones
in Britain (Goldring and Bridges, 1972),
the Late Precambrian Innerelv Member
in North Norway (facies 4 of Banks,
1973a), the Cambrian Duolbasgaissa
Formation of North Norway (facies 1, 2
and 3 and Fig. 7 of Banks, 1973b), the
Precambrian Jura Quartzite of Scotland
(fine facies of Anderton, 1976), Upper
Jurassic coquinoid sandstones of
Wyoming (Brenner and Davies, 1973),
Mississippian sandstones and shales in
southern Ireland (de Raaf et al., 1977),
the Upper Ordovician of the Oslo region,

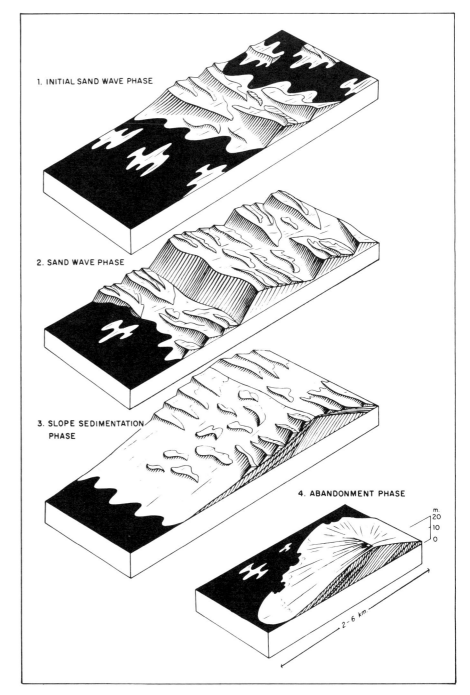

Figure 7
Conceptual model of the evolution of the
Roda sandwave complex. See Figure 6 for
details. From Nio and Siegenthaler, 1978,
Figure 9.

Norway (Brenchley et al., 1979), and
Silurian sandstones and shales of the
Arisaig Group, Nova Scotia (Cant,
in press).

I therefore suggest that the storm
generation of density currents (Fig. 15) is
a major geological process. With a
variety of names for H.C.S., and so few
examples from recent sediments, the
widespread applicability of this process

to the geological record is only just
becoming apparent. I predict that in the
next few years there will be many more
examples of storm-dominated interbedded sands and shales in shallow
marine settings.

Large, Ancient, Shallow Marine Sand Bodies

There are several examples of elongate offshore sand bodies in the Jurassic and Cretaceous shallow marine seas of western North America. Most of the data on their geometry are from subsurface investigations. A list of dimensions is given in Table I. Excluding the Viking in Alberta, these bodies average about 60 km long, 10 km wide, and about 15 m thick, although because of the scarcity and variability of the data, these numbers give only a "feel" for the dimensions.

Most of these sand bodies are surrounded by shales, and were demonstrably deposited away from the shoreline in fully marine situations. The local evidence for offshore deposition is given in the references cited. There is little data on depth of deposition, although forams suggest depths of about 40 to 70 m for the Shannon Sandstone.

The major feature that most of these sand bodies have in common is a coarsening-upward sequence without any signs of emergence, and in several examples, there is a conglomeratic zone at the top (Shannon, Sussex, Cardium, Viking). The lower part of the sequence tends to be shaly and extensively bioturbated, and the central part commonly consists of thinly bedded fine sandstones with symmetrical wave ripples. The upper part contains cross-bedded sandstones (Sussex, Oxfordian sandstones, Cardium), and the conglomerate, if present, caps this sequence. The pebbles are mostly cherty, and reach several cm in diameter. There is evidence that the pebbly layers lie preferentially on one side of these linear sand bodies (Berg, 1975, Fig. 7; Swagor et al., 1976, Figs. 2 and 5A; Evans, 1970, Fig. 5). In the case of the Viking Formation in Saskatchewan (Evans, 1970), the linear sand bodies shift progressively southward, with the conglomerate lying on the southern flank.

The orientation of many of these sand bodies is sub-parallel to the estimated regional shoreline (e.g., Shannon, Sussex, Cardium). Unfortunately, of the list given in Table I, there are only paleoflow data for the Shannon and the Oxfordian sandstones. In the case of the Shannon, paleoflow is dominantly southward, subparallel to the elongation of the sand bodies, but there are also some northward flow directions. In the Oxfordian sandstones, there is a "variety of crest

Figure 8
Large shallow marine sand wave in the lower Greensand, Leighton Buzzard, England. Note large scale (up to 5 m) of cross bedding in lower set, with smaller sets truncating the top, and partly the foreset slope of the sand wave. Compare with sand wave facies and proximal slope facies of Figure 6b.

Figure 9
Detail of part of Figure 8, emphasizing the scale of cross bedding. Note (upper right) that some of the apparent foresets are actually dipping sets of cross bedding. Compare with proximal slope facies of Figure 6b.

orientations for individual bars, as well as a wide variety of sediment transport directions within each bar" (Brenner and Davies, 1974, p. 425).

In the case of the Shannon and Sussex, and the Oxfordian sandstones, the authors agree that sand is being transported tens of kilometres offshore by combinations of storm and tidal currents. There is evidence for lateral shifting of some of the sand bodies (Brenner, 1978, Fig. 19), and hence a very appealing comparison between these sand bodies and the linear sand ridges of the Atlantic shelf, discussed earlier.

The major problem of such a comparison concerns the initial place of formation of the sand bodies. If they initially formed tens of km offshore (as appears the case for the Shannon and Sussex), they cannot be compared with the linear sand ridges that form very close to shore (Fig. 2) and become isolated during transgression. Thus, if they initially formed tens of km offshore, there is no direct modern comparison except for the fields of sand waves (Fig. 4) described by Flemming (1978) as being controlled by an intruding ocean current. The conglomerate remains an outstanding problem, in terms of its derivation, transport across and/or along the shelf, and occurrence at the top of coarsening-upward sequences.

84

One possible mechanism to explain the coarsening upward is related to differential rates of sediment transport in a downcurrent direction. Belderson and Stride (1966) showed that in the shallow marine areas off the British Coast, sediment became finer in the downcurrent direction. In an area where there is not a continuous supply of all grain sizes (for example, on the modern shelves where there is sorting and unmixing of glacial debris), the slower transport of the coarser material might result in a coarsening-upward sequence.

The Cardium poses even greater problems. In outcrop in the S. Ram River (Alberta), there are five coarsening-upward sequences superimposed. The dominant sedimentary structure is hummocky cross stratification, and hence it appears that storm waves have had a major effect on forming sedimentary structures, and perhaps also in generating the flows that transported sediment into the basin (W.L. Duke, pers. commun., 1978). It is not clear whether this process forms elongate sand bodies, nor is it clear why superimposed coarsening-upward sequences should be formed tens of km from the shoreline.

Facies Models for Shallow Marine Sands

This review has shown that many different processes, and sand body morphologies, can interact to form various shallow marine deposits. At present, no one model of deposition can be suggested, nor can "end points" in a spectrum of models be proposed (see Papers 2, 3 and 5). One theme running through recent literature on ancient and modern sands is the distinction between tide- and storm-dominated systems (Johnson, 1978), but even this distinction is not necessarily the best guide or framework for further work. I suggest this because of the very few examples of *purely* tide-dominated (e.g., Nio, 1976) or *purely* storm-dominated (e.g., Brenchley *et al.,* 1979; Hamblin and Walker, 1979) systems. Most deposits seem to indicate both tide and storm influences.

In shallow marine systems, it is instructive to examine whether a facies model can be distilled from modern linear sand ridges, and ancient elongate sand bodies. Gross dimensions and thicknesses are similar, as are the trends subparallel to shore. In ancient sand bodies paleoflow is variable, but

Table I
Dimensions of shallow marine sand bodies, Mesozoic of western North America

Formation	Thickness of sand body (m)	Length of sand body (km)	Width	Author
Shannon, Upper ss.	15	50	30	Spearing, 1976
Lower ss.	21	100	50	
Shannon	20	17.6	2.4	Seeling, 1979
Sussex	12	40	1.6	Berg, 1975
Sussex	c. 30	c. 50	up to 10	Brenner, 1978
Oxfordian Sandstones	c. 15	>5	0.2 to 2	Brenner and Davies, 1974
Cardium	8-13 cgl.	5	1.6 to 3.2	Swagor *et al.,* 1976
Cardium Crossfield	6	96	1.6 to 2.8	Berven, 1966
Garrington	6	128	1.6 to 2.8	
Viking (Lower)	<38	144	128	Tizzard and Lerbekmo, 1975
(Upper) E. Alberta	<35	136	104	
Viking Dodsland-Hoosier	8	up to 80	c. 10	Evans, 1970

Figure 10
Abundant planar-tabular and trough cross bedding in the Gog orthoquartzitic sandstone, Cambrian, Kicking Horse Pass, B.C.

commonly with a component subparallel to the trend of the sand body. This would seem to be compatible with flow patterns and bed form migration on modern linear sand ridges. A coarsening-upward sequence, documented in ancient sand bodies, has been postulated in modern

ridges (Swift *et al.,* 1973; Stubblefield *et al.,* 1975; Belderson and Stride, 1966).

Despite these comparisons, which could be taken as the basis for a facies model, there is one very significant difference. The modern linear sand ridges are formed near the shoreline

(Fig. 2), and are modified on the shelf during transgression. The ancient sand bodies (Shannon, Sussex, etc.) appear to have formed tens of km offshore in water several tens of metres deep by along-shelf transport of sand. This introduces severe problems of forming coarsening-upward sequences, and of transporting gravel to offshore depositional sites. If it were possible to interpret some of the ancient sand bodies in terms of formation at the shoreline, followed by detachment during transgression, and reworking by storm and tidal currents in deeper water (and this would involve the reconsideration of a lot of local geology), then it might be possible to construct a "linear sandstone" shallow marine model.

Apart from the "linear sandstones", two other general facies associations begin to emerge – the abundantly cross-bedded shallow marine orthoquartzites (e.g., the Gog in B.C.), and the thinly interbedded shales and sandstones with hummocky cross stratification (Brenchley *et al.*, 1979; Hamblin and Walker, 1979; and see bibliography). However, there is presently no basis for *prediction*, *no norm*, and a rather vaguely established *framework* for making future observations.

Figure 11

Hummocky cross stratification from the Passage Beds (Fernie Formation), Banff traffic circle section, Alberta. Photo shows stratification correct way up (locally, these beds are overturned). Note convex-up hummock immediately to left of notebook, and more prominently, to right and just above notebook.*

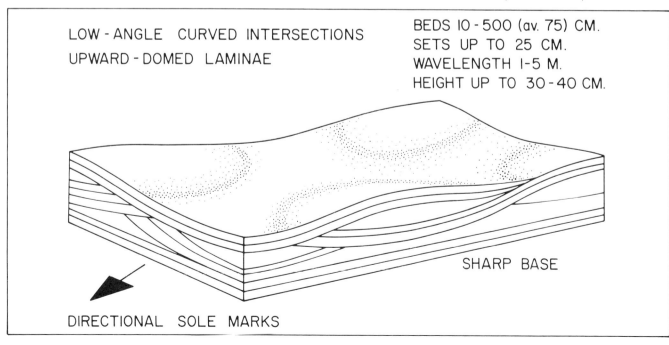

Figure 13
Block diagram showing major features of hummocky cross stratification as seen in the Fernie Formation, Southern Alberta and B.C.

Figure 12
Plan view of hummocky cross stratification, Cardium Formation at Seebe, Alberta. *Note broad gentle curvature of stratification. Arrows point to convex-up hummocks.*

LOW - ANGLE CURVED INTERSECTIONS
UPWARD - DOMED LAMINAE

BEDS 10 - 500 (av. 75) CM.
SETS UP TO 25 CM.
WAVELENGTH 1-5 M.
HEIGHT UP TO 30-40 CM.

SHARP BASE

DIRECTIONAL SOLE MARKS

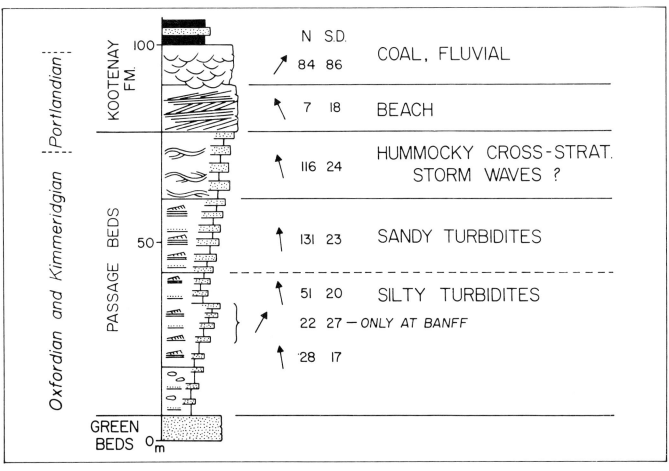

Figure 14

Generalized stratigraphic sequence of the uppermost Fernie (the Passage Beds) and lowermost Kootenay Formations. Arrows show paleoflow vector means, N = sample size, S.D. = standard deviation. Turbidites are characterized by Bouma sequences (note coarsening- and thickening-upward sequence at Banff), and are overlain by hummocky cross-stratified sandstones. These in turn are overlain by beach facies. Modified from Hamblin, 1978.

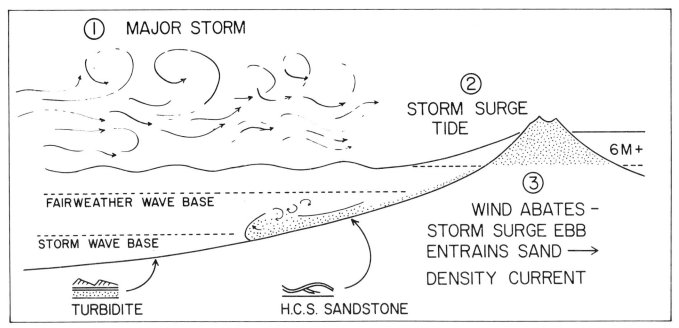

Figure 15

Conceptual diagram relating major storm (1) to storm surge tide (2). When wind abates, a seaward-flowing density current is generated (3). Above storm wave base but below fairweather wave base, the storm waves are still affecting the bottom, and as deposition from the density current takes place, hummocky cross stratification is formed. Below storm wave base, a turbidite with Bouma divisions is formed.

Bibliography

Basic Reading

The basic reading list that follows includes examples of storm- and tide-dominated modern systems, and type examples of ancient systems. There are few good technical reviews that combine modern and ancient observations.

Johnson, H.D., 1978, Shallow siliciclastic seas, in H.G. Reading, ed., Sedimentary Environments and Facies: Blackwell, Oxford, p. 207-258.
This is the best available overall review of ancient and modern shallow siliciclastic seas.

Swift, D.J.P., 1976a, Coastal sedimentation, in D.J. Stanley, and D.J.P. Swift, eds., Marine Sediment Transport and Environmental Management: New York, Wiley, p. 255-310.
A good review of Swift's ideas on coastal sedimentation, summarizing many earlier papers.

Swift, D.J.P., 1976b, Continental shelf sedimentation, in D.J. Stanley, and D.J.P. Swift, eds., Marine Sediment Transport and Environmental Management: New York, Wiley, p. 311-350.
A good summary of Swift's views of the shelf.

McCave, I.N., 1971, Sand waves in the North Sea off the coast of Holland: Marine Geol., v. 10, p. 199-225.
Good survey of the sand wave field, with tidal current data, sand wave sizes and sand flow paths. The model (Fig. 17) is incorrect in suggesting large sets of angle-of-repose cross bedding because maximum "steep" face angle of the sand waves is only about 5°.

Spearing, D.R., 1976, Upper Cretaceous Shannon Sandstone: an offshore, shallow-marine sand body: Wyoming Geol. Assoc., 28th Ann. Field Conf., Guidebook, p. 65-72.
The Shannon presents the same problems as the Sussex (see Berg, 1975; Brenner, 1978) – an offshore shelf sandstone that coarsens upward, and fingers out into shales. Transport process perhaps includes storm enhancement of oceanic or tidal currents.

Nio, S.D., 1976, Marine transgressions as a factor in the formation of sandwave complexes: Geol. Mijnbouw, v. 55, p. 18-40.
Nio uses three examples to support the idea that major ancient sandwave complexes are transgressive in origin, and tidally controlled. Insufficient data is presented for the Burdigalian (Swiss Molasse), but the Eocene Roda Complex (Spain) and Lower Greensand (England) contain giant cross-beds up to 20 and 5 m thick respectively. No examples are given of modern angle-of-repose structures of this size.

Nio, S.D. and J.C. Siegenthaler, 1978, A lower Eocene estuarine-shelf complex in the Isabena Valley: State Univ. Utrecht, Sedimen-tology Group, Report no. 18, p. 1-44.
Good descriptions and diagrams - a more useful reference on the Roda sandwave complex than Nio, 1976, but probably a rather inaccessible publication.

Brenchley, P.J., G. Newall, and I.G. Stanistreet, 1979, A storm surge origin for sandstone beds in an epicontinental platform sequence, Ordovician, Norway: Sediment. Geol., v. 22, p. 185-217.
An excellent description of thin bedded sandstones and shales, with good discussions of the storm generation of offshore flows. The flows appear to have moved at least 40 km offshore, and had a periodicity of about I per 10000-15000 years.

Hamblin, A.P. and R.G. Walker, 1979, Storm-dominated shelf deposits: the Fernie-Kootenay (Jurassic) transition, southern Rocky Mountains: Can. Jour. Earth Sci., v. 16.
Storm-emplaced turbidites, and sandstones with hummocky cross stratification, with the HCS confined stratigraphically between the turbidites and overlying beach.

Symposia Volumes

Swift, D.J.P., D.B. Duane, and O.H. Pilkey, eds., 1972, Shelf Sediment Transport, Process and Pattern: Stroudsberg, Pa., Dowden, Hutchinson and Ross, 656 p.
A good collection of technical papers, grouped into 1) Water motion and process of sediment entrainment; 2) Patterns of fine sediment dispersal; 3) Patterns of coarse sediment dispersal. There are no papers on ancient rocks.

Stanley, D.J. and D.J.P. Swift, eds., 1976, Marine Sediment Transport and Environmental Management: New York, Wiley, 602 p.
This is an updated version of the AGI Short Course notes on "The new concepts of continental margin sedimentation". The papers on water and sediment motion are highly mathematical, and outnumber the contributions on shelf sediments, facies and morphology.

Studies of Recent Shallow Marine Sediments

Belderson, R.H. and A.H. Stride, 1966, Tidal current fashioning of a basal bed: Marine Geol., v. 4, p. 237-257.
First description of the sequence of bed configurations (gravel pavement ⟶ sand ribbons ⟶ sand waves ⟶ sand and muddy sand ⟶ sand patches) as observed on the shelf off the British Coast.

Caston, V.N.D., 1972, Linear sand banks in the southern North Sea: Sedimentology, v. 18, p. 63-78.
Development of Houbolt's (1968) work, showing influence of tides in developing Z-shaped sand ridges.

Duane, D.B., M.E. Field, E.P. Meisburger, D.J.P. Swift, and S.J. Williams, 1972, Linear shoals on the Atlantic inner shelf, Florida to Long Island, in D.J.P. Swift, D.B. Duane, and O.H. Pilkey, eds., Shelf Sediment Transport, Process and Pattern: Dowden, Hutchinson and Ross, Stroudsberg, Pa., p. 447-498.
Interpretation of the shelf ridge and swale topography in terms of modern coastal and nearshore processes. Many good examples of linear ridges.

Emery, K.O., 1968, Relict sediments of continental shelves of the world: Amer. Assoc. Petrol. Geol., Bull., v. 52, p. 445-464.
Shows that about 70 per cent of continental shelves are covered with sand and gravel of Pleistocene origin, considered to be relict and not in equilibrium with modern processes.

Flemming, B.W., 1978, Underwater sand dunes along the southeast African continental margin - observations and implications: Marine Geol., v. 26, p. 177-198.
Important descriptions of dunes up to eight m high seen on side scan sonar, and apparently controlled by the Agulhas oceanic current spilling up onto the edge of the continental shelf.

Flemming, B.W., in press. Sand transport and bedform patterns on the Continental Shelf between Durban and Port Elizabeth (Southeast African Continental Margin). Sediment Geol.
A follow-up of the 1978 paper with new data and illustrations.

Hayes, M.O., 1967, Hurricanes as geological agents: case studies of hurricanes Carla, 1961, and Cindy, 1963: Texas Bur. Econ. Geol., Rept. Invest. 61, 54 p.
Excellent documentation of the geological results of hurricanes, emphasizing development of a graded bed on the shelf.

Houbolt, J.J.H.C., 1968, Recent sediments in the southern bight of the North Sea: Geol. Mijnbouw, v. 47, p. 245-273.
An important paper on tidal sand ridges, with integration of current data. Proposes model of closed-system sand circulation around ridges. Mis-interpretation of the exaggerated scale of the sparker profiles by later workers gave rise to the idea that these sand ridges contain large sets of crossbedding many metres thick.

Jordan, G.F., 1962, Large submarine sand waves: Science, v. 136, p. 839-848.
Documentation of large submarine sand ridges with superimposed bed forms in the Gulf of Maine.

Kenyon, N.H., 1970, Sand ribbons of European tidal seas: Marine Geol., v. 9, p. 25-39.
Side scan sonar pictures of sand ribbons, with a proposed evolutionary sequence of four types of ribbons.

Kulm, L.D., R.C. Roush, J.C. Harlett, R.H. Neudeck, D.M. Chambers, and E.J. Runge, 1975, Oregon Continental Shelf sedimentation: interrelationships of facies distribution and sedimentary processes: Jour. Geol., v. 83, p. 145-175.
Excellent discussion of the transport of silt and clay, and effects of bioturbation. Much of the sand appears to be relict – modern sand is being trapped in estuaries. A different approach to shelf sedimentation from that of the Swift group.

Kumar, N. and J.E. Sanders, 1976, Characteristics of shore-face storm deposits: modern and ancient examples: Jour. Sed. Pet., v. 46, p. 145-162.
Observations of storm sand movement on the shelf, and interpretation of some ancient storm deposits. Emphasizes importance of storms in the geological record.

Lavelle, J.W., D.J.P. Swift, P.E. Gadd, W.L. Stubblefield, F.N. Case, H.R. Brashear, and K.W. Haff, 1978, Fairweather and storm sand transport on the Long Island, New York, inner shelf: Sedimentology, v. 25, p. 823-842.
Sand movement monitored using radioisotope tracers, showing maximum transport and dispersal during winter storms.

Off, T., 1963, Rhythmic linear sand bodies caused by currents: Amer. Assoc. Petrol. Geol., Bull., v. 47, p. 324-341.
Survey of linear sand bodies from maps and charts, demonstrating their common occurrence in funnel-shaped estuaries with high tidal range.

Stubblefield, W.L., J.W. Lavelle, D.J.P. Swift, and T.F. McKinney, 1975, Sediment response to the present hydraulic regime on the central New Jersey shelf: Jour. Sed. Pet., v. 45, p. 337-358.
Presents a model for grain movement during fairweather, intense storms, and less intense storms.

Stubblefield, W.L. and D.J.P. Swift, 1976, Ridge development as revealed by sub-bottom profiles on the central New Jersey shelf: Marine Geol., v. 20, p. 315-334.
Internal ridge structure shown by seismic profiling, with model of ridge development based upon the seismic results.

Swift, D.J.P., 1975, Tidal sand ridges and shoal-retreat massifs: Marine Geol., v. 18, p. 105-134.
A good discussion of tidally controlled ridges and massifs. Makes a good contrasting companion paper to Swift, Duane and McKinney (1973).

Swift, D.J.P., D.B. Duane, and T.F. McKinney, 1973, Ridge and swale topography of the Middle Atlantic Bight, North America: secular response to the Holocene hydraulic regime: Marine Geol., v. 15, p. 227-247.
Development of Duane et al., 1972, giving an interpretation of shelf ridge and swale

topography in terms of modern storm controlled processes along the Atlantic coast.

Swift, D.J.P. and P. Sears, 1974, Estuarine and littoral patterns in the surficial sand sheet, central and southern Atlantic shelf of North America, in G.P. Allen, ed., Shelf and Estuarine Sedimentation: A Symposium: Univ. Bordeaux, Inst. Geol. Bassin d'Aquitaine, Talence.

Swift, D.J.P., D.J. Stanley, and J.R. Curray, 1971, Relict sediments on continental shelves: a reconsideration: Jour. Geol., v. 79, p. 322-346.
Demonstrates that many "relict" Pleistocene sediments are in fact being reworked by modern processes. This is the first major discussion of modern processes acting on shelves.

Terwindt, J.H.J., 1971, Sand waves in the Southern bight of the North Sea: Marine Geol., v. 10, p. 51-67.
Description of the same sand waves as surveyed by McCave (1971) with a little extra data.

Studies of Ancient Shallow Marine Deposits

Anderton, R., 1976, Tidal-shelf sedimentation: an example from the Scottish Dalradian: Sedimentology, v. 23, p. 429-458.
The coarse facies of the Jura Quartzite is interpreted as tidal, and the laminated and rippled finer facies are interpreted as storm dominated. There is no preferred facies sequence, suggesting irregular reworking of tidal deposits by storms, and perhaps also the reworking of storm deposits by tides.

Banks, N.L., 1973a, Innerelv Member: late Precambrian marine shelf deposit, East Finnmark: Norges geol. Unders., v. 288, p. 7-25.
Two marine shelf coarsening upward sequences are interpreted as the deposits mainly of storm surge currents. Their facies 4 and 6 may contain HCS. Transport directions are unimodal and many beds are sharp based. There is no evidence of emergence – the beds appear to represent a storm-dominated prograding (coarsening-upward) shelf sand body.

Banks, N.L., 1973b, Tide dominated offshore sedimentation, Lower Cambrian, North Norway: Sedimentology, v. 20, p. 213-228.
Four facies are described, with numbers 1, 2 and 3 being characterized by sharp-based sole-marked sandstones interbedded with finer sediments. HCS appears to be present in each (e.g., Fig. 7). Facies 4 is trough cross bedded. Banks suggests a tide-dominated environment. This seems unlikely for facies 1-3 which could probably also be interpreted as storm generated density current deposits (see Hamblin and Walker, 1979). One overall sequence 500 m thick suggests increasing tidal dominance (facies 4) upward, but the

scale is much greater than in sand bodies such as the Shannon, Sussex or Cardium.

Campbell, C.V., 1966, Truncated wave ripple laminae: Jour. Sed. Pet., v. 36, p. 825-828.
Describes and figures what would now be termed hummocky cross stratification. Curiously, the stratigraphic units which are described and photographed are never identified.

Cant, D.J., in press, Storm dominated shallow marine sediments of the Arisaig Group (Silurian-Devonian) of Nova Scotia: Can. Jour. Earth Sci.
Storm emplaced sandstones with transported fossils and hummocky cross stratification are described. The Silurian shelf has been zoned paleontologically using brachiopods, and this paper points out the danger of such zonation if the fossils have been transported.

DeRaaf, J.F.M., J.R. Boersma, and A. van Gelder, 1977, Wave-generated structures and sequences from a shallow marine succession, Lower Carboniferous, County Cork, Ireland: Sedimentology, v 24, p. 451-483.
A long description of various types of wave-rippled sandstones, with some HCS described by name. Some coarsening-upward sequences (about 10 m thick) with HCS toward the top – these might be similar to the Cardium, but not the Shannon or Sussex. Similarly, coarsening-and-fining (CuFu) units might resemble some of the Cardium sequences from the S. Ram River. The overall interpretation is one of wind, wave and storm domination and low tidal influence. Discussion of shoals (incipient, submerged and emergent bars) is disappointingly brief.

Goldring, R., 1971, Shallow water sedimentation as illustrated in the Upper Devonian Baggy Beds: Geol. Soc. London, Memoir 5, 80 p.
Detailed descriptions of various shallow marine facies, with excellent and abundant illustrations and photos.

Goldring, R. and P. Bridges, 1973, Sublittoral sheet sandstones: Jour. Sed. Pet., v. 43, p. 736-747.
Excellent descriptions of various stratigraphic units, with sharp-based sandstones interbedded with shales. Many undoubtedly contain HCS, and a general storm dominated process and environment is suggested.

Harms, J.C., J.B. Southard, D.R. Spearing, and R.G. Walker, 1975, Depositional environments as interpreted from primary sedimentary structures and stratification sequences: Soc. Econ. Paleont. Mineral., Tulsa, Short Course 2, 161 p.
Introduces the term hummocky cross stratification (p. 87-89), describes the feature, and interprets in terms of storm waves.

Hobday, D.K. and H.G. Reading, 1972, Fairweather versus storm processes in

shallow marine sand bar sequences in Late Precambrian of Finnmark, North Norway: Jour. Sed. Pet., v. 42, p. 318-324.
An enigmatic deposit. Cross bedding and small sole marks suggest northward flow, yet the sequence is dominated by westward dipping surfaces (6 to 17 degrees) a few metres high. These were interpreted as due to fairweather westward buildout, and erosion during storms. The various paleoflow directions are not satisfactorily explained, but the "bars" are compared with the submarine ridges with superimposed ridges and swales of Swift. Accretion is a fairweather process, and destruction is the result of storms.

Jurassic-Cretaceous, Western North America

Berg, R.R., 1975, Depositional environment of Upper Cretaceous Sussex Sandstone, House Creek Field, Wyoming: Amer. Assoc. Petrol. Geol., Bull., v. 59, p. 2099-2110.
The Sussex occurs encased in marine shales, and was deposited in coarsening-upward sequences in 100-200 ft. of water. In House Creek Field, the sand body is about 1.6 km wide, 40 km long and 5 m thick. Flow appears to have been parallel to the sand body along "flow paths" on the shelf, but transport processes are not specified.

Berven, R.J., 1966, Cardium sandstone bodies, Crossfield-Garrington area, Alberta: Can. Petrol. Geol. Bull., v. 14, p. 208-240.
Elongate Cardium sand bodies interpreted as offshore bars, preserved by widespread transgression.

Brenner, R.L., 1978, Sussex Sandstone of Wyoming - example of Cretaceous offshore sedimentation: Amer. Assoc. Petrol. Geol., Bull., v. 62, p. 181-200.
The coarsening-upward Sussex sst. was deposited tens of km offshore on a muddy shelf (c.f. Shannon: Spearing 1976). There is a series of NW trending bars or ridges in the upper 60 m of Sussex interval. The bars are interpreted to have cross-cutting channels with "tidal deltas" or "washovers" at the ends (although all deposition was fully marine). By comparison with the Shannon, the Sussex is interpreted to consist of progressively eastward-displaced sediment sheets. The data is from cores and seismic - perhaps in outcrop the progressive eastward displacement might show laterally accreting surfaces dipping east with regional flow to south (c.f. Hobday and Reading, 1972). Control of overall southward movement of sand is not explained.

Brenner, R.L. and D.K. Davies, 1973, Storm generated coquinoid sandstone: genesis of high-energy marine sediments from the Upper Jurassic of Wyoming and Montana: Geol. Soc. Amer., Bull., v. 84, p. 1685-1698.
Coquinoid sandstones are interpreted as subaqueous channel lags (thick 1-4 m channelized lags), storm lags (graded layers 3-30 cm thick) and swell lags of shells in mudstones. This paper highlights a detail commonly overlooked or underplayed in studies of shallow marine sandbodies.

Brenner, R.L. and D.K. Davies, 1974, Oxfordian sedimentation in the Western Interior United States: Amer. Assoc. Petrol. Geol., Bull., v. 58, p. 407-428.
Description and interpretation of coarsening-upward marine bars. The interpretation is somewhat vague - "the similarity [not specified] between the sand shoals . . . and areas of sand accumulation on Holocene shelves reflects a similarity in process . . . normal tidal currents, storm driven currents and regional circulatory currents interacted to produce a characteristic motif".

Evans, W.E., 1970, Imbricate linear sandstone bodies of Viking Formation in Dodsland-Hoosier area of southwestern Saskatchewan, Canada: Amer. Assoc. Petrol. Geol., Bull., v. 54, p. 469-486.
Describes elongate sand bodies with conglomeratic flanks. Sand bodies accreted laterally to south. Interpreted to have been deposited far from shore by east-flowing tidal currents.

Hamblin, A.P., 1978, Sedimentology of a prograding shallow marine slope and shelf sequence, Upper Jurassic Fernie-Kootenay transition, southern Front Ranges: M.Sc. Thesis, McMaster University, 196 p.
Storm-emplaced turbidites, and sandstones with hummocky cross stratification, with the HCS confined stratigraphically between the turbidites and overlying beach.

Koldijk, W.S., 1976, Gilby Viking "B": a storm deposit, in M. Lerand, ed., The Sedimentology of Selected Oil and Gas Reservoirs in Alberta: Can. Soc. Petrol. Geol., Calgary, p. 62-77.
The Viking in this area is interpreted as an offshore marine sand with a capping of about 2 m of conglomerate (clasts up to 2.5 cm). Seaward transport of gravel is interpreted as the result of storms.

Masters, J.A., 1979, Deep basin gas trap, Western Canada: Amer. Assoc. Petrol. Geol. Bull., v. 63, p. 152-181.
A good source of information on deep gas, its occurrence and production. The paper is not intended to be a discussion of depositional environments.

Seeling, A., 1979, The Shannon Sandstone, a further look at the environment of deposition at Heldt Draw Field, Wyoming: Mountain Geologist, v. 15, p. 133-144.
Describes more Shannon Sandstone linear ridges, and compares them with the Atlantic Shelf and the North Sea.

Swagor, N.S., T.A. Oliver, and B.A. Johnson, 1976, Carrot Creek Field, Central Alberta, in M. Lerand ed., The Sedimentology of Selected Clastic Oil and Gas Reservoirs in Alberta: Can. Soc. Petrol. Geol., Calgary, p. 78-95
An interpretation of coarsening-upward Cardium sequences in terms of deposition in the lee of terraces on a shelf. Pebbles in the conglomerates reach 45 mm - it is suggested they are transported seaward by storms.

Tizzard, P.G. and J.F. Lerbekmo, 1975, Depositional history of the Viking Formation, Suffield area, Alberta, Canada: Can. Petr. Geol., Bull., v. 23, p. 715-752.
Description of elongate Viking sand bodies from subsurface studies.

MS received May 28, 1979.

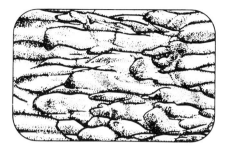

Facies Models 8. Turbidites and Associated Coarse Clastic Deposits

Roger G. Walker
Department of Geology
McMaster University
Hamilton, Ontario L8S 4M1

Introduction

To the sedimentologist, the turbidity current concept is both simple and elegant. Each *turbidite* (defined as the deposit of a turbidity current) is the result of a single, short lived event, and once deposited, it is extremely unlikely to be reworked by other currents. The concept is elegant because it allows the interpretation of thousands of graded sandstone beds, alternating with shales, as the result of a series of similar events, and it can safely be stated that no similar volume of clastic rock can be interpreted so simply.

In this review, I will begin by studying the "classical" turbidite, and will then gradually broaden the scale to encompass turbidites and related coarse clastic rocks in their typical depositional environments - deep sea fans and abyssal plains.

The concept of turbidites was introduced to the geological profession in 1950. At that time, nobody had observed a modern turbidity current in the ocean, yet the evidence for density currents had become overwhelming. The concept accounted for graded sandstone beds that lacked evidence of shallow water reworking, and it accounted for transported shallow water forams in the sandstones, yet bathyal or abyssal benthonic forams in interbedded shales.

Low density currents were known in lakes and reservoirs, and they appeared to be competent to transport sediment for fairly long distances. Many of these different lines of evidence were pulled together by Kuenen and Migliorini in 1950 when they published their experimental results in a now classic paper on "Turbidity currents as a cause of graded bedding". A full review of why and how the concept was established in geology was published by Walker (1973).

After its introduction in 1950, the turbidity current interpretation was applied to rocks of many different ages, in many different places. Emphasis was laid upon describing a vast and new assemblage of sedimentary structures, and using those structures to interpret paleocurrent directions. In the absence of a turbidite facies model, there was no norm with which to compare individual examples, no framework for organizing observations, no logical basis for prediction in new situations, and no basis for a consistent hydrodynamic interpretation. Yet gradually during the years 1950-1960, a relatively small but consistent set of sedimentary features began to be associated with turbidites. These are considered in the following list, and can now be taken as a set of descriptors for classical turbidites:

1) Sandstone beds had abrupt, sharp bases, and tended to grade upward into finer sand, silt and mud. Some of the mud was introduced into the basin by the turbidity current (it contained shallow benthonic forams), but the uppermost very fine mud contained bathyal or abyssal benthonic forams and represented the constant slow rain of mud onto the ocean floor.

2) On the undersurface (sole) of the sandstones there were abundant markings, now classified into three types: tool marks, carved into the underlying mud by rigid tools (sticks, stones) in the turbidity current; scour marks, cut into the underlying mud by fluid scour; and organic markings - trails and burrows - filled in by the turbidity current and thus preserved on the sole. The tool and scour markings give an accurate indication of local flow directions of the turbidity currents, and by now, many thousands have been measured and used to reconstruct paleoflow patterns in hundreds of turbidite basins.

3) Within the graded sandstone beds, many different sedimentary structures were recorded. By the late 1950s, some authors were proposing turbidite models, or ideal turbidites, based upon a generalization of these sedimentary structures and the sequence in which they occurred. This generalization is akin to the distillation process discussed in the first paper, and the final distillation and publication of the presently accepted model was done by Arnold Bouma in 1962. A version of the Bouma model is shown in Figure 1.

The Bouma Turbidite Facies Model

The Bouma sequence, or model (Figs. 1,2) can be considered as a very simple facies model that effectively carries out all of the four functions of facies models discussed in the first paper in this volume. I will examine these in turn, both to shed light upon turbidites in general, and to use turbidites as an illustration of a facies model in operation. I have described the model as very simple because it contains relatively few descriptive elements, and because it is narrowly focussed upon sandy and silty turbidites *only*. I shall later refer to these as "classical" turbidites.

1. The Bouma model as a NORM. The model (Fig. 1) as defined by Bouma consists of five divisions, A-E, which occur in a fixed sequence. Bouma did not give normalized thicknesses for the divisions, and this type of information is still unavailable. In Figure 3, I have sketched three individual turbidites which clearly contain some of the elements of the Bouma model, yet which obviously differ from the norm. They can be characterized as AE, BCE and CE beds. Without the model, we could ask no more questions about these three turbidites, but with the norm, we can ask *why* certain divisions of the sequence are missing. I will try and answer this rhetorical question later.

2. The Bouma model as a framework and guide for description. The model has served as the basis for description in a large number of studies, particularly in Canada, U.S.A. and Italy. With the framework provided by the model, one can quickly log a sequence of turbidites as AE/BCE/CE etc. (as in the three turbidites of Fig. 3), and then add to the

92

BOUMA DIVISIONS   INTERPRETATION

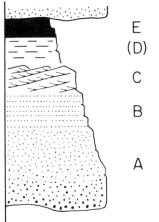

E
(D)
C
B
A

FINES IN TURBIDITY CURRENT, FOLLOWED BY PELAGIC SEDIMENTS		3
TRACTION IN	LOWER UPPER FLOW REGIME	2
RAPID DEPOSITION, ? QUICK BED		1

Figure 1
Five divisions of the Bouma model for turbidites: A—graded or massive sandstone; B—parallel laminated sandstone; C—ripple cross-laminated fine sandstone; D—faint parallel laminations of silt and mud, bracketed
to emphasize that in weathered or tectonized outcrops it cannot be separated from E— pelitic division, partly deposited by the turbidity current, partly hemipelagic. Interpretations of depositional process are grouped into three main phases, see text.

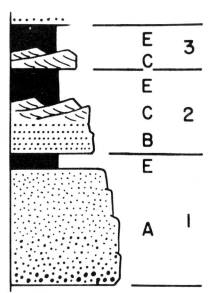

Figure 3
Hypothetical sequence of three turbidites, described as AE, BCE and CE in the Bouma model. See text.

5 CM

Figure 2
Complete "Bouma" turbidite (see Fig. 1), showing pelitic division E of lower bed (bottom left); graded division A, parallel laminated division B and ripple cross lamin-
ated division C. Divisions (D) and E were broken off this specimen, which is from the Côte Fréchette road cut, Lévis Formation (Cambrian), Quebec.

basic description any other features of note. With the model as a framework, one is not only aware of the features presented by any bed, but is also aware of any features embodied in the model but missing in a particular bed.

3. The model as a basis for hydrodynamic interpretation. The existence of the bouma model enables us to make one integrated interpretation of classical turbidites, rather than having to propose

different origins for each different type of bed. In Figure 1, the interpretation is considered in three parts. Division A contains no sedimentary structures except graded bedding. It represents very rapid settling of grains from suspension, possibly in such quantities and at such a rate that water is forcibly expelled upward, and momentarily, the grain/ water mixture becomes fluidized (or "quick"). The fluidization would destroy

any possible sedimentary structures. The second phase of deposition involves traction of grains on the bed. Flow velocities are lower, and the rate of deposition from suspension is much lower. By direct comparison with many experimental studies, division B represents the upper flow regime plane bed, and division C, the lower flow regime rippled bed. The third phase of deposition involves slow deposition of fines from the tail of the current. The origin of the delicate laminations in division (D) is not understood, and I prefer to place division (D) in brackets, implying that in all but the cleanest outcrops, (D) cannot be separated from E. In the uppermost part of division E, there may be some true pelagic mudstone with a deep water (bathyal or abyssal) benthonic fauna (forams in Tertiary and younger rocks).

4. The Bouma model as a predictor. Here, I shall show how the hydrodynamic interpretation of the model, together with departures from the norm, can be used on a predictive basis. Turbidite 1 (Fig. 3) begins with a thick sandy division (A), and was deposited from a high velocity current. Turbidite 2 (Fig. 3), by comparison with the norm, does not contain division A. It begins with Bouma division B, and was presumably deposited from a slower current. Turbidite 3 (Fig. 3) lacks divisions A and B, and presumably was deposited from an even slower current.

In a cautious way, we can now make some predictions based upon comparison with the norm, and upon the hydrodynamic interpretations. A sequence of many tens of turbidites in which all of the beds are thick and begin with division A (Fig. 4, and, for example, the Cambrian Charny Sandstones in the St. Romuald road cut near Levis, Quebec) probably represents an environment where all of the turbidity currents were fast-flowing during deposition. Such an environment was probably close to the source of the turbidity currents (proximal). By contrast (Fig. 5), a sequence of many tens of beds in which all the turbidites begin either with division B or C (Ordovician Cloridorme formation at Grande Vallee, Quebec) was deposited in an environment where all of the turbidity currents were flowing slowly during deposition. Such an environment was probably a long way from the source of the currents (distal). This conclusion will be slightly modified below.

This ideal proximal to distal scheme applies only to "classical" turbidites. In nature, variations in the size, sediment load, and velocity of individual currents will blur the proximal to distal distinctions, which is why I suggest taking the combined characteristics of a large number of beds before making environmental predictions. For example, if out of 250 beds, 70 per cent began with division A, the environment could be characterized as relatively proximal.

It follows from this application of the model that if one can work out the environment of deposition of a relatively large group of turbidites (let's say 300 beds - and a distal envrionment is indicated), and one knows the general paleoflow direction, one can make predictions as to what the same stratigraphic interval will look like closer to source and in a specific geographic direction. The reader is now referred to "A review of the geometry and facies organization of turbidites and turbidite-bearing basins" (Walker, 1970), and, if you are interested in the intimate details of lateral variability in classical turbidites, to an excellent paper by Enos (1965) on the Ordovician Cloridome Formation in Quebec.

It should be emphasized that for classical turbidites, the *descriptive* terminology is now *thick-bedded* and

Figure 4
Group of four parallel sided turbidites, AE, AE, AE and AE, suggesting that the beds are close to their source (proximal). Beds slightly overturned, top to right; Ordovician Cloridorme Formation at Grande Vallée, Quebec.

Figure 5
Very thin turbidite sandstones with thicker interbedded shales. Beds begin with Bouma divisions B and C, and suggest deposition far from their source (distal). Contrast with Figure 4. Ordovician Cloridorme Formation, Grande Vallée (near fish cannery), Quebec; stratigraphic top to left.

thin-bedded; the terms proximal and distal should only be used in an interpretive sense (see Walker, 1978).

Environments of Turbidite Deposition

Because a turbidite is simply the deposit of a turbidity current, turbidites can be found in any environment where turbidity or density currents operate. These environments include lakes and reservoirs, delta fronts, continental shelves, and most importantly, the deeper ocean basins. However, to be *preserved* and *recognized* as a turbidite, the features imposed on the bed by the current (ideally; sharp base with sole marks, graded bedding, Bouma divisions) must *not* be reworked by other types of currents. Small turbidites have been preserved in quiet water glacial lakes; thin prodeltaic turbidites can flow into water deep enough that agitation of the bottom by storms is very rare (say, less than one storm in 500 years), but to preserve a thick (hundreds or thousands of metres) turbidite sequence, the most likely environment is one that is consistently deep and quiet. Using present day morphological terms, these environments would include the continental rise (made up of coalescing submarine fans) and abyssal plains. It is important to emphasize that any sudden surge of sediment laden water can deposit a bed with all the characteristics of a classical turbidite. A levee break in a river, and a storm current transporting sediment out across the continental shelf (see paper 7 in this volume) would be two examples of this. Graded beds might be preserved in either situation, but the two environments would be characterized by the dominance of fluvial and shelf features, respectively. The presence of rare "turbidites" would indicate the possibility of density current activity, and would not condemn the entire sequences to deposition in great depths of water.

Other Facies Commonly Associated with Classical Turbidites

Classical turbidites can be characterized by three main features; first, the beds tend to be laterally extensive (hundreds of metres); second, they tend to be parallel sided and vary little in thickness laterally (hundreds of metres); and third, it is reasonable to use the Bouma model for this description and interpretation. However, along with classical turbidites there are other

coarse clastic facies also known to have been transported into very deep water (as defined by bathyal and abyssal benthonic forams in interbedded shales). These facies can be listed as:
1) massive sandstones
2) pebbly sandstones
3) clast supported conglomerates
4) chaotic matrix-supported pebbly sandstones and conglomerates.

This facies list stems initially from work of Emiliano Mutti and his colleagues in Italy, and an English language version is available (Walker and Mutti, 1973). I now believe that the classification of facies published by Walker and Mutti is unnecessarily subdivided (my opinion, not necessarily Mutti's), so I will stick to the simpler list above.

Massive sandstones. This facies (Fig. 6) consists of thick sandstone beds in which graded bedding is normally poorly developed. Most of the divisions of the Bouma sequence are missing, and interbedded shales tend to be very thin or absent. A typical sequence of beds

would be measured as A.A.A.A. using the Bouma model. However, I would consider this to be a misapplication of the model, because its function as a norm, predictor, framework and basis for hydrodynamic interpretation are all seriously weakened to the point of uselessness if the beds only show an A.A.A.A. sequence. The massive sandstones are commonly not so parallel sided as the classical turbidites; channelling is more common, and one flow may cut down and weld onto the previous one ("amalgamation") giving rise to a series of multiple sandstone beds.

The one common sedimentary structure found in the massive sandstones is termed "dish" structure (Fig. 7), and is indicative of abundant fluid escape during deposition of the sandstone. It indicates rapid deposition of a large amount of sand from a "fluidized flow" (akin to a flowing quicksand). This does *not* imply that the massive sandstone facies was transported all the way from source into the basin by a fluidized flow.

Figure 6
Massive sandstone facies. Note thickness of beds and absence of pelitic division of Bouma model. Stratigraphic top to left. Cambro-Ordovician Cap Enragé Formation near St-Simon, Quebec.

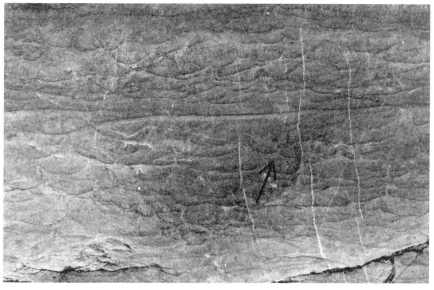

Figure 7
"Dish" structures, formed by rapid dewatering of a massive sandstone. Some of the dish edges curve upward into vertical dewatering pipes (arrow on photo). Ordovician Cap Enragé Formation, near St-Simon, Quebec.

Figure 8
Graded bed of pebbly sandstone, followed abruptly by a second bed without a pelitic division. St-Damase Formation (Ordovician) near Kamouraska, Quebec.

However, it does imply that a turbidity current, which normally maintains its sand load in suspension by fluid turbulence, can pass through a stage of fluidized flow during the final few seconds or minutes of flow immediately preceding deposition. The massive sandstone facies is prominent in the Cambrian Charny Formation around Quebec City and Lévis, and dish structures in massive sandstones are common in the Cambro-Ordovician Cap Enragé Formation near Rimouski, Quebec (Fig. 7).

Pebbly sandstones. The pebbly sandstone facies (Figs. 8, 9) cannot be described using the Bouma model, nor does it have much in common with the massive sandstone facies. Pebbly sandstones tend to be well graded (Fig. 8), and stratification is fairly abundant. It can either be a rather coarse, crude, horizontal stratification, or a well developed cross bedding of the trough, or planar-tabular (Fig. 9) type. At present, there is no "Bouma-like" model for the internal structures of pebbly sandstones; the sequence of structures, and their abundance and thickness has not yet been distilled into a general model. Pebbly sandstone beds are commonly channelled and laterally discontinuous, and interbedded shales are rare.

It is clear that with abundant channelling, and the presence of cross bedding

Figure 9
Pebbly sandstone facies, showing medium scale cross bedding. In isolation, this photograph could easily be confused with a photograph of fluvial gravels, but in fact is from the Cambro-Ordovician Cap Enragé Formation (near St-Simon, Quebec), and is interbedded with turbidites and graded pebbly sandstones.

96

in pebbly sandstones, this facies could easily be confused with a coarse fluvial facies. The differences are subtle and can be misleading to sedimentologists – the safest way to approach the interpretation of pebbly sandstones is to examine their context. If associated with, or interbedded with classical turbidites, the pebbly sandstone interpretation would be clear. Similarly, if associated with non-marine shales, root traces, caliche-like nodules, mud cracks, and other indicators of flood plain environments, the interpretation would also be clear. This facies highlights the fact that environmental interpretations *cannot* be based upon a "checklist" of features: the relative abundance and type of features, in their stratigraphic context, must always be the basis of interpretation.

Pebbly sandstones are particularly well exposed in the Cambro-Ordovician Cap Enragé Formation at St. Simon (near Rimouski, Quebec), where grading, stratification and cross bedding are prominent. The facies is also abundant in the Cambrian St. Damase Formation near Kamouraska, Quebec, and in the Cambrian St. Roch Formation at L'Islet Wharf (near St-Jean-Port-Joli, Quebec).

Clast supported conglomerates. Although volumetrically less abundant than classical turbidites, conglomerates are an important facies in deep water environments. They are abundant in California and Oregon, and are particularly well exposed at many localities in the Gaspé Peninsula. Sedimentologists have tended to ignore conglomerates, probably because without a facies model, there has been no framework to guide observations, and hence the feeling of "not being quite sure what to measure in the field". I have recently proposed some generalized "Bouma-like" models for conglomerates (Walker, 1975), but because the models are based upon less than thirty studies, they lack the universality and authority of the Bouma model for classical turbidites. The paper (Walker, 1975) discusses the models, their relationships, and how they were established. In Figure 10, it can be seen that the descriptors include the type of grading (normal (Fig. 11) or inverse), stratification (Fig. 11), and fabric; in different combinations they give rise to three models which are probably intergradational, and a fourth

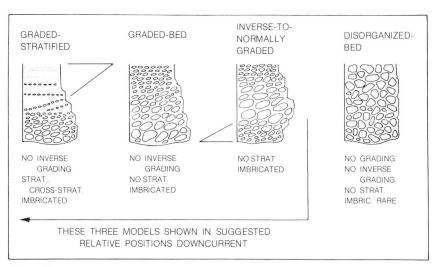

Figure 10
Four models for resedimented (deep water) conglomerates. The graded-stratified,
graded-bed, and inverse-to-normally graded models are probably intergradational.

Figure 11
Graded-stratified conglomerate, Cambro-Ordovician Cap Enragé Formation at Bic, Quebec. Basal conglomerate grades up into
stratified conglomerate, very coarse sandstone with crude "dish" structure (centre of photo) and finally into massive structureless sandstone (top left).

(disorganized-bed) characterized only by the absence of descriptors.

One of the most important features of conglomerates is the type of fabric they possess. In *fluvial* situations, where pebbles and cobbles are rolled on the bed, the long (a-) axis is usually transverse to flow direction, and the intermediate (b-) axis dips upstream, characterizing the imbrication. However, for most conglomerates associated with turbidites, the fabric is quite different: the

long axis is *parallel* to flow, and also dips upstream to define the imbrication (Fig. 12). This fabric is interpreted as indicating *no* bedload rolling of clasts. The only two reasonable alternatives involve mass movements (debris flows), or dispersion of the clasts in a fluid above the bed. Mass movements in which clasts are not free to move relative to each other do not produce abundant graded bedding, stratification, and cross-stratification, so I suggest the clasts were supported above the bed in

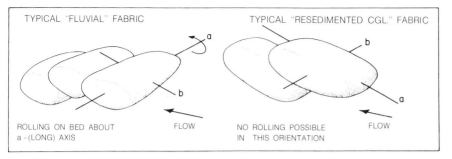

Figure 12
*Contrast between conglomerate fabric pro-
duced by rolling clasts on the bed (long axis
transverse to flow) with typical fabric in
resedimented conglomerates (no rolling, long
axis parallel to flow).*

a turbulent flow. The support mechanism may have been partly fluid turbulence, and partly clast collisions. Upon
deposition, the clasts immediately
stopped moving (no rolling), and the
fabric was "frozen" into the deposit.

In the absence of experimental work
on cobbles and boulders, the interpretation of the conglomerate models must be
based largely on theory. I suggest a
downcurrent trend from the inverse-to-
normally-graded model, into the graded-
stratified model. This trend does not
necessarily exist in any one bed: rather,
deposition from a particular current in
one of the three downstream positions in
Figure 10 will be of the type indicated in
the figure.

Clast supported conglomerates are
abundant in the Ordovician Grosses
Roches Formation and Cambro-
Ordovician Cap Enragé Formation,
Gaspé Peninsula, Quebec, and also
make up part of the Cambrian St. Roch
Formation east of Rivière-du-loup,
Quebec.

*Chaotic matrix-supported pebbly sand-
stones and conglomerates.* This facies
includes two different types of deposit.
First, there are conglomerates and
pebbly sandstones that have abundant
muddy matrix, and possibly show basal
inverse grading and preferred clast
alignment. They represent the deposits
of subaqueous debris flows. Because
the larger clasts in a debris flow are
maintained above the bed by the
strength of the debris flow matrix, the
deposit commonly has large blocks
projecting up above the top of the bed, or
even resting almost entirely on top of the

bed. The deposit shows no internal
evidence of slumping.

By contrast, the second type of
deposit commonly shows evidence of
slumping, and represents the mixing of
sediment within the depositional basin
by post-depositional slumping. The
deposits can range all the way from very
cohesive slumps involving many beds,
to very watery slumps generated by the
depostion of coarse sediment on top of
wet, poorly consolidated clays. The
latter process gives rise to the classical
pebbly mudstones.

Inasmuch as subaqueous debris
flows, and slumps, require greater
slopes than classical turbidity currents,
the chaotic facies is most abundant at
the foot of the slope into the basin, or in
the Inner Fan environment. Very few
examples have been described in
Canada. Large scale slumps are known
in Upper Ordovician turbidites in northeastern Newfoundland (Helwig, 1970),
and pebbly mudstones are known in
several units in western Newfoundland
(Stevens, 1970). The best described
debris flows are Devonian reef-margin
examples adjacent to the Ancient Wall,
Miette and Southesk-Cairn reef complexes in Alberta (Cook *et al.,* Srivastava
et al., 1972).

An Integrated Facies Model
for Turbidites and Associated
Coarse Clastic Rocks
The models discussed so far apply to
relatively closely defined facies, and do
not consider depositional environments.
Volumetrically, the turbidites and associated clastics are most abundant in
large submarine fans, which in many
areas have coalesced to form the

continental rise. Information on modern
fans is limited to short (1-5 m) cores,
surveys of surface morphology, and
relatively little subsurface geophysical
information. Ancient fans have been
proposed on the basis of paleocurrent
evidence, abundance of channels, and
distribution of facies. Two studies are
outstandingly important – Normark's
geophysical work and proposition of a
fan growth model based exclusively
upon recent sediment work, and Mutti
and Ghibaudo's fan model based exclusively on ancient sediments. These two
studies have been integrated into the
review by Walker and Mutti (1973). Here,
I will simply present the submarine fan –
abyssal plain model as it is currently
understood (Fig. 13), fit the various
facies into the various morphological
parts of the fan, and examine the
stratigraphic consequences of fan progradation.

Because of their generally parallel-
sided nature, the *classical turbidites* can
be assigned to the smooth areas of the
fan – the outer suprafan lobes and the
outer fan. The trend from proximal to
distal will develop most characteristically after the turbidites have flowed
beyond the confines of the braided
suprafan channels. The *massive sandstones* and *pebbly sandstones* are less
regularly bedded, and the common
presence of channelling suggests that
they be assigned to the braided suprafan
channels. As the channels become
plugged, and shift in position, a sand
body is gradually built up that consists of
coalesced channels but no overbank
deposits. In the absence of levees on the
suprafan, and with the lateral channel
shifting, any overbank fines that are
deposited are rapidly eroded again. In
nature, the gradual termination of the
suprafan channels is likely to result in a
very gradual facies change across the
suprafan lobes – some classical turbidites might be preserved in wide,
shallow channels, and some unusually
large pebbly sandstone flows may spill
out onto the smooth area of the
suprafan.

Similarly, there is likely to be a similar
facies change toward the feeder channel, from pebbly sandstones into *conglomerates* (assuming that such coarse
clasts were available in the source
area). Conglomerates are probably
restricted to channels, mainly the inner
fan channel, but also as coarse lags in

98

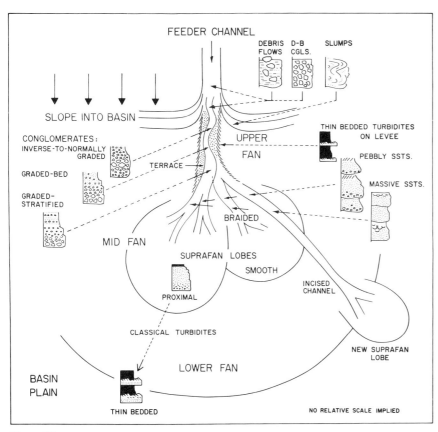

Figure 13
*Submarine fan environmental model. See text
for full discussion. D-B is disorganized-bed
conglomerate model.*

the bottoms of some suprafan channels.
The gradual downfan change from
inverse-to-normally graded types to
graded-stratified types is suggested in
Figure 13, but this change is tentative
and is indicated only by theory, not by
direct observation. The bottom of the
feeder channel and the foot of the slope
are the most likely environments for
slumping and debris flows because of
the steeper gradients. The disorganized-
bed (D-B in Fig. 13) conglomerates
might also be assigned here.

The inner fan levees are built up by
flows which fill the channel and spill
onto the levees and the area behind the
levees. Sediment consists only of the
finest suspended material (silt and clay)
but there may be sufficient current
strength to ripple the silt and produce a
turbidite that would be described as CE
in the Bouma model. Hence although a
thick sequence of CE, BCE and C(D)E
beds may define a distal environment,
silty CE beds could also indicate levee or
back-levee environments on the inner
fan (a *proximal* environment by any
definition). Again, I emphasize that one

cannot use a checklist to define environments – in this case, the abundance of
CE beds and their facies relationships
(with conglomerates, or with basin plain
muds) must be considered before an
interpretation can be made. Problems
of thin-bedded turbidites have been
discussed by Mutti (1977) and Nelson
et al (1977).

**Stratigraphic Aspects
of Fan Progradation**
By comparison with a deltaic situation,
we can reasonably assume that submarine fan progradation would result in a
stratigraphic sequence passing from
outer fan, through mid fan, into inner fan
deposits upwards in the succession
(Fig. 14). Progradation in the outer fan
area would result in the deposition of a
sequence classical turbidites that became more proximal in aspect upwards.
This type of sequence is now termed
"thickening- and coarsening-upward".

The progradation of individual suprafan lobes might also be expected to
result in thickening- and coarsening-
upward sequences, but these may not

be restricted to classical turbidites. The
smooth, outer suprafan lobes would be
represented by classical turbidites, but
these would pass upward into massive
and pebbly sandstones as the braided
portion of the suprafan prograded. The
stratigraphically higher suprafan lobe
sequences might therefore contain
more massive and pebbly sandstones,
and fewer classical turbidites.

The result of steady fan progradation
so far would be one thickening- and
coarsening-upward sequence of classical turbidites (outer fan), overlain by
several thickening- and coarsening-
upward sequences of classical turbidites, massive, and pebbly sandstones,
representing several superimposed
suprafan lobes that shifted laterally and
built on top of each other during mid-fan
progradation. The inner fan deposits
would probably consist of one deep
channel fill (Fig. 15), conglomeratic if
coarse material were available at the
source, and laterally equivalent to
mudstones deposited on the channel
levees and in the low areas behind the
levees. It is possible during progradation, even in a generally aggrading
situation, that the inner fan channel
could cut into one of the braided
suprafan lobes.

Channel fill sequences, both in the
inner fan and braided suprafan channels, may consist of "thinning- and
fining-upward sequences" (Fig. 16).
Mutti and his Italian colleagues have
suggested that these sequences result
from progressive channel abandonment, depositing thinner and finer beds
from smaller and smaller flows in the
channels. Thus an inner fan channel
might have a conglomeratic basal fill,
and pass upward into finer conglomerates, and massive and pebbly sandstones (see Walker, 1977).

There are at least two alternative
stratigraphic records of submarine fans,
other than the steady progradation
discussed above. First, if supply for the
fan is cut off at source (or diverted
elsewhere), the fan will be abandoned,
and will be covered by a rather uniform
layer of hemipelagic mud. The previously active channels will also be mud-filled.
Abandoned mud-filled channels are
known in the stratigraphic record, and
include the Mississippi submarine channel (abandoned by post-Pleistocene rise
of sea level), the Rosedale Channel
(Late Miocene, Great Valley of Califor-

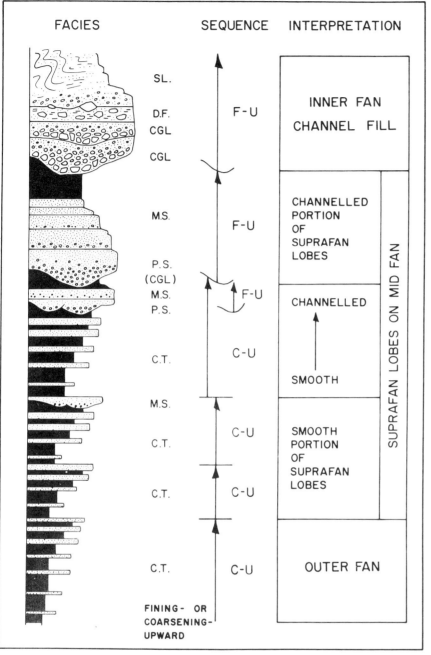

FACIES   SEQUENCE   INTERPRETATION

SL.

D.F.
CGL   F-U   INNER FAN
CGL   CHANNEL FILL

M.S.   F-U   CHANNELLED
                      PORTION
                      OF
                      SUPRAFAN
P.S.   LOBES
(CGL)
M.S.   F-U   CHANNELLED
P.S.

C.T.   C-U   SUPRAFAN LOBES ON MID FAN
                      SMOOTH

M.S.
           C-U   SMOOTH
C.T.   PORTION
                      OF
                      SUPRAFAN
           C-U   LOBES

C.T.

C.T.   C-U   OUTER FAN

FINING- OR
COARSENING-
UPWARD

Figure 14
Hypothetical submarine fan stratigraphic sequence produced by fan progradation. C.T., classical turbidite; M.S., massive sandstone; P.S., pebbly sandstone; D.F., debris flow. Arrows show thickening- and coarsening-upward sequences (C-U) and thinning- and fining-upward sequences (F-U). See text for details.

nia) and the Yoakum Channel (Middle Eocene, Texas Gulf Coast).

Second, if the sediment supply increases considerably, or the gradient of the slope into the basin increases (tectonically?), the fan channel may be incised across the entire fan, and all sediment transported much farther into the basin. This is the situation in the modern La Jolla Fan (California), which

has been entirely by-passed, with most of the coarser sediment (sand and coarser) being transported much farther into the San Diego Trough. A possible ancient example is the Cambrian St. Roch Formation at L'Islet Wharf (near St-Jean-Port-Joli), Quebec, where a thinning- and fining-upwards sequence of conglomerates and pebbly sandstones rests in a channel (Fig. 17). The

channel cuts into a thick sequence of relatively thinly bedded turbidites (beds commonly begin with Bouma B and C divisions) that appear more distal than proximal. The juxtaposition of conglomerates in a channel, cutting into relatively distal turbidites, suggests an envrionment such as that labelled "incised channel" in Figure 13.

Limitations of the Fan Model
The fan model presented here is based upon data from geophysical surveys of relatively small modern fans such as La Jolla, San Lucas, and the many other fans of the Southern California Borderland. The model may not apply so well to some larger fans (Monterey and Astoria, off northern California-Oregon-Washington; the Bengal Fan) because they are characterized by major channels which cross the entire length of the fan - in the case of the Bengal Fan, the channels are over 1000 km long.

However, the fan model as presented seems to be a useful framework for considering many small to medium scale ancient basins. It cannot be applied to the long (hundreds of km) exogeosynclinal troughs in which the paleoflow pattern is dominantly parallel to the tectonic strike. Examples of turbidites in such troughs include the M. Ordovician Cloridorme Formation (Gaspé Peninsula) and its time equivalent in the Central Appalachians, the Martinsburg Formation. The deposits consist dominantly of classical turbidites hundreds of metres thick, but showing no consistent proximal to distal change along the length of the trough in the downflow direction. It is commonly suggested that turbidity currents flowed downslope toward the trough axis, perhaps constructing fans at the trough margin. However, at the trough axis the flows turned and continued to flow *parallel* to the trough axis. The marginal fans were presumably destroyed by subsequent tectonics, and the absence of consistent proximal to distal changes along the trough axis is probably due to input from a whole series of fans along the trough margin. Thus any consistent changes developing from one source would be masked by input from adjacent sources up and down the trough. At present, there is no facies model that acts as a good predictor in this type of turbidite basin.

Figure 15
Portion of large channel cutting into shales. Channel fill consists of disorganized-bed conglomerates and lenticular sandstones, with an overall thinning – and fining-upward sequence. Ordovician Grosses Roches, Quebec, Appalachians. Figure far left for scale.

Figure 16
Example of a thinning- and fining-upward sequence (see Figure 14) from the Cambro-Ordovician Cap Enragé Formation near St-Simon. The conglomerate (lower right) *contains large boulders which die out upward (toward top left). Centre of sequence is a pebble conglomerate, passing into pebbly sandstones (centre left) and finally into massive sandstones (near water's edge).*

Examples of Turbidites and Associated Coarse Clastics

The papers listed below do *not* constitute a general set of readings with respect to an introduction to the turbidite concept. Rather, they are significant contributions mostly to Canadian geology, either because they discuss turbidites and their importance to specific problems or regional geology, or because they are important contributions to a general understanding of turbidites.

1. Precambrian turbidites

Walker, R.G. and F.J. Pettijohn, 1971, Archean sedimentation: analysis of the Minnitaki Basin, northwestern Ontario, Canada: Geol. Soc. Am. Bull., v. 82, p. 2099-2130.

Henderson, J.B., 1972, Sedimentology of Archean turbidites at Yellowknife, Northwest Territories: Can. Jour. Earth Sci., v. 9, p. 882-902.

Turner, C.C. and R.G. Walker, 1973, Sedimentology, stratigraphy and crustal evolution of the Archean greenstone belt near Sioux Lookout, Ontario: Can. Jour. Earth Sci., v. 10, p. 817-845.

Rousell, D.H., 1972, The Chelmsford Formation of the Sudbury Basin – a Precambrian turbidite; *in* J.V. Guy-Bray, ed., New Developments in Sudbury Geology: Geol. Assoc. Can. Spec. Paper 10, p. 79-91.

Cantin, R. and R.G. Walker, 1972, Was the Sudbury Basin circular during deposition of the Chelmsford Formation?, *in* J.V. Guy-Bray, ed., New Developments in Sudbury Geology: Geol. Assoc. can. Spec. Paper 10, p. 93-101.

2. Appalachian area

Enos, P., 1969, Anatomy of a flysch: Jour. Sed. Petrol., v. 39, p. 680-723. (Note: this is the classic paper on the Cloridorme Formation.)

Parkash, B., 1970, Downcurrent changes in sedimentary structures in Ordovician turbidite greywackes: Jour. Sed. Petrol., v. 40, p. 572-590.

Parkash, B. and G.V. Middleton, 1970, Downcurrent textural changes in Ordovician turbidite greywackes: Sedimentology, v. 14, p. 259-293. (Note: these two papers by Parkash are detailed studies of the Cloridorme Formation.)

Skipper, K., 1971, Antidune cross-stratification in a turbidite sequence. Cloridorme Formation, Gaspé, Quebec: Sedimentology, v. 17, p. 51-68. (See also discussion of this paper, Sedimentology, v. 18, p. 135-138.)

Skipper, K. and G.V. Middleton, 1975, The sedimentary structures and depositional mechanics of certain Ordovician turbidites, Cloridorme Formation, Gaspe, Quebec: Can. Jour. Earth Sci., v. 12, p. 1934-1952.

Hubert, C., J. Lajoie and M.A. Leonard, 1970, Deep sea sediments in the Lower Paleozoic Quebec Supergroup, *in* J. Lajoie, ed., Flysch Sdimentology in North America: Geol. Assoc. Can. Spec. Paper 7, p. 103-125. (Note: the main areas discussed in the paper are L'Islet Wharf, and the Cap Enragé Formation in the Bic

Figure 17
*Channel in Cambrian St. Roch Formation at
L'Islet Wharf, Quebec. Stratigraphic top to
right. Channel cuts into classical turbidites,
and consists of at least two main portions—*
*foreground (with geologist for scale), and cliff
at top right. Note the graded-stratified
conglomerate filling lower part of channel,
and passing up into massive sandstone
(lower right).*

- St. Fabien area. See also Rocheleau
and Lajoie, and Davies and Walker,
below.)

Rocheleau, M. and J. Lajoie, 1974,
Sedimentary structures in resedimented
conglomerate of the Cambrian flysch,
L'Islet, Quebec Appalachians: Jour. Sed.
Petrol., v. 44, p. 826-836.

Davies, I.C. and R.G. Walker, 1974,
Transport and deposition of resediment-
ed conglomerates: the Cap Enragé
Formation, Cambro-Ordovician, Gaspé,
Quebec: Jour. Sed. Petrol., v. 44, p.
1200-1216.

Hendry, H.E., 1973, Sedimentation of
deep water conglomerates in Lower
Ordovician rocks of Quebec - composite bedding produced by progressive
liquefaction of sediment?: Jour. Sed.
Petrol., v. 43, p. 125-136.

Schenk, P.E., 1970, Regional variation of
the flysch-like Meguma Group (Lower
Paleozoic) of Nova Scotia, compared to
recent sedimentation off the Scotian
Shelf, *in* J. Lajoie, ed., Flysch Sedimentology in North America: Geol. Assoc.
Can. Spec. Paper 7, p. 127-153.

Stevens, R.K., 1970, Cambro-Ordovician
flysch sedimentation and tectonics in
west Newfoundland and their possible
bearing on a Proto-Atlantic ocean, *in* J.
Lajoie, ed., Flysch Sedimentology in

North America: Geol. Assoc. Can. Spec.
Paper 7, p. 165-177.

Horne, G.S. and J. Helwig, 1969, Ordovi-
cian stratigraphy of Notre Dame Bay,
Newfoundland *in* M. Kay, ed., North
Atlantic - Geology and Continental Drift:
Am. Assoc. Petrol. Geol. Mem. 12,
p. 388-407.

Osborne, F., 1956, Geology near Quebec
City: Nat. Can., v. 83, p. 157-223.

Hiscott, R.N., 1979, Clastic sills and
dykes associated with deep-water
sandstones, Tourelle Formation, Ordovician, Quebec: Jour. Sed. Petrol., v. 49,
p. 1-9.
Emphasizes the importance of clastic
sills in turbidites sequences, and shows
how easily they can be misinterpreted as
turbidites.

3. Campus, University of Montreal
Lajoie, J., 1972, Slump fold axis orienta-
tions: an indication of paleoslope?: Jour.
Sed. Petrol., v. 42, p. 584-586.

4. Canadian Arctic
Trettin, H.P., 1970, Ordovician-Silurian
flysch sedimentation in the axial trough
of the Franklinian geosyncline, northeastern Ellesmere Island, Arctic Canada, *in* J. Lajoie, ed., Flysch Sedimentology in North America: Geol. Assoc. Can.
Spec. Paper 7, p. 13-35.

5. Western Canada
Danner, W.R., 1970, Western Cordilleran
flysch sedimentation, southwestern
British Columbia, Canada, and northwestern Washington and central Oregon, U.S.A., *in* J. Lajoie, Flysch Sedimentology in North America: Geol. Assoc.
Can. Spec. Paper 7, p. 37-51.

Cook, H.E., P.N. McDaniel, E. Mountjoy
and L.C. Pray, 1972, Allochthonous
carbonate debris flows at Devonian
bank ("reef") margins, Alberta, Canada:
Bull. Can. Petrol. Geol., v. 20, p. 439-497.

Srivastava, P., C.W. Stearn, and E.W.
Mountjoy, 1972, A Devonian megabreccia at the margin of the Ancient Wall
carbonate complex, Alberta: Bull. Can.
Petrol. Geol., v. 20, p. 412-438.

6. Field Guidebooks
Hubert, C.M., 1969, ed., Flysch sedi-
ments in parts of the Cambro-
Ordovician sequence of the Quebec
Appalachians: Geol. Assoc. Can.,
Guidebook for field trip 1, Montreal, 38 p.

Riva, J., 1972, Geology of the environs of
Quebec City: Montreal, Internatl. Geol.
Cong., Guidebook B-19, 53 p.

St. Julien, P., C. Hubert, W.B. Skidmore
and J. Beland, 1972, Appalachian
structure and stratigraphy, Quebec:
Montreal, Internatl. Geol. Cong., Guidebook A-56, 99 p.

Harris, I.M., ed., 1975, Ancient sediments
of Nova Scotia, Eastern Section, Soc.
Econ. Paleont. Min., Guidebook. Also *in*
Maritime Sediments, v. 11, numbers
1, 2 and 3.

Poole, W.H. and J. Rodgers, 1972,
Appalachian geotectonic elements of
the Atlantic Provinces and southern
Quebec: Montreal, Internatl. Geol.
Congr., Guidebook A-63, 200p.

7. Turbidite Reservoirs
A list of papers discussing oil and gas in
turbidite reservoirs is given by Walker,
1978. For the reader who wishes to
consult a few papers, I suggest the
following:

MacPherson, B.A., 1978, Sedimentation
and trapping mechanism in Upper
Miocene Stevens and older turbidite
fans of Southeastern San Joaquin
Valley, California: Amer. Assoc. Petrol.
Geol. Bull. v. 62, p. 2243-2274.
A comprehensive paper that is particularly valuable for the discussion of the
larger scale geometry of turbidite sands.

Also important is the discussion of lateral wedging out, and the superposition of sand bodies (controlled by interbedded shales).

Dickas, A.B., and J.L. Payne, 1967, Upper Paleocene buried channel in Sacramento Valley, California: Amer. Assoc. Petrol. Geol. Bull., v. 51, p. 873-882.
Discusses a large, deep abandoned submarine channel filled with fines. The fines act as an up-dip seal for gas in surrounding rocks. See Walker (1978) for further discussion of large mud-filled channels.

Weagant, F.E., 1972, Grimes gas field, Sacramento Valley, California: in Stratigraphic Oil and Gas Fields: Amer. Assoc. Petrol. Geol., Memoir 16, p. 428-439.
The gas is trapped in small turbidite lobes separated by thick shale blankets. The paper contains good isopach maps of lobes, and a cross-section showing the overall superposition of lobes.

Hsu, K.J., 1977, Studies of Ventura field, California, 1: facies geometry and genesis of lower Pliocene turbidites. Amer. Assoc. Petrol. Geol. Bull., v. 61, p. 137-168.
The first release of information of Shell Oil Company work in the Ventura Basin in the 1950's. The paper contains good descriptions of the rocks, and very interesting E-log correlations and isopach maps. No modern interpretation was suggested in terms of submarine fan models (although Walker, 1978, attempted to read the data in terms of submarine fans). Hsu himself prefers to emphasive longitudinal sediment transport in the basin, downplaying fans.

Selected References - Basic Reading
This list is intentionally very brief. It is intended to serve as basic reading for those wishing to read further in various aspects of turbidites and associated coarse clastics in their basinal setting.

1. Turbidites in basins - facies and facies associations
Walker, R.G., 1970, Review of the geometry and facies organization of turbidites and turbidite-bearing basins, in J. Lajoie, ed., Flysch Sedimentology in North America: Geol. Assoc. Can. Spec. Paper 7, p. 219-251.
This paper discusses at length the various turbidite and associated facies, but predates the Normark-Mutti fan model. It contains an extensive reference list.

Walker, R.G. and E. Mutti, 1973, Turbidite facies and facies associations, in G.V. Middleton and A.H. Bouma, eds., Turbidites and deep water sedimentation: Pacific Section, Soc. Econ. Paleont. Min., Short Course Notes (Los Angeles), p. 119-157.
An extended discussion of the facies and models discussed in the present article.

Walker, R.G., 1978, Deep-water sandstone facies and ancient submarine fans: models for exploration for stratigraphic traps: Amer. Assoc. Petrol. Geol. Bull., v. 62, p. 932-966.
A review of the submarine fan model, emphasizing the separation of individual lobes by mud blankets, and suggesting how the model can be used for exploration.

2. Modern submarine fans
Normark, W. R., 1974, Submarine canyons and fan valleys: factors affecting growth patterns of deep sea fans, in R.H. Dott, Jr. and R.H. Shaver, eds., Modern and Ancient Geosynclinal Sedimentation: Soc. Econ. Paleont. Min. Spec. Publ. 19, p. 56-68.
An updated version of Normark's original (1970) discussion of fan growth.

Nelson, C.H. and L.D. Kulm, 1973, Submarine fans and deep-sea channels, in G.V. Middleton and A.H. Bouma, eds., Turbidites and Deep Water Sedimentation: Pacific Section, Soc. Econ. Paleont. Min. Short Course Notes (Los Angeles), p. 39-78.
Although emphasizing the N.W. Pacific, this review paper, with abundant references, is a good overall summary of fan morphology and sedimentation.

Normark, W.R., 1978, Fan valleys, channels and depositional lobes on modern submarine fans: characters for recognition of sandy turbidite environments: Amer. Assoc. Petrol. Geol. Bull., v. 62, p. 912-931.
Normark's most recent and comprehensive discussion of modern submarine fans.

3. Modern and Ancient fans - comparison
Nelson, C.H. and T.H. Nilsen, 1974, Depositional trends of modern and ancient deep sea fans, in R.H. Dott, Jr. and R.H. Shaver, eds., Modern and Ancient Geosynclinal Sedimentation: Soc. Econ. Paleont. Min. Spec. Paper 19, p. 69-91.
Good comparison of modern and ancient fans, showing how information from both sources can be dovetailed ("distilled") together.

4. Processes - turbidity currents and associated sediment gravity flows
Middleton, G.V. and M.A. Hampton, 1976, Subaqueous sediment transport and deposition by sediment gravity flows, in D.J. Stanley and D.J.P. Swift, eds., Marine Sediment Transport and Environmental Management: New York, Wiley Interscience.
All you need to know about turbidity currents, and associated processes. Non-mathematical.

5. History and philosophy of the turbidity current concept
Walker, R.G., 1973, Mopping-up the turbidite mess, in R. N. Ginsburg, ed., Evolving Concepts in Sedimentology: Baltimore, Johns Hopkins Press, p. 1-37.
Detailed history, with philosophical commentary, on the evolution of the turbidity current concept. This paper will not help you find oil, however!

Other references cited in this article
Bouma, A.H., 1962, Sedimentology of Some Flysch Deposits: Amsterdam, Elsevier Publ. Co., 168 p. Cited only as the first documentation of the now-accepted turbidite model.

Kuenen, P.H. and C.I. Migliorini, 1950, Turbidity currents as a cause of graded bedding: Jour. Geol., v. 58, p. 91-127. Cited for historical reasons, as the first paper that directed geologists' attention to the possibility of high density turbidity current deposits in the geological record. This paper represents one of the most important foundation stones of modern (post World War II) sedimentology.

Mutti, E., 1977, Distinctive thin-bedded turbidite facies and related depositional environments in the Eocene Hecho Group (South-central Pyrenees, Spain): Sedimentology, v. 24, p. 107-131.
A general discussion of thin-bedded turbidites emphasizing a variety of possible depositional environments, with examples from the Hecho Group in Spain.

Nelson, C.H., E. Mutti, and F. Ricci-Lucchi, 1977, Upper Cretaceous resedimented conglomerates at Wheeler Gorge, California: description and field guide: a discussion: Jour. Sediment. Petrol., v. 47, p. 926-928.
A discussion of a paper by Walker with respect to thin-bedded turbidites. I have subsequently looked at these rocks (January, 1979) and now suggest that the turbidites *below* the conglomerates are distal, but that the turbidites *above* the conglomerates represent a thin-bedded facies in some other environment - possibly levee as suggested by Nelson *et al.* This discussion emphasizes the problems of interpreting the thin-bedded facies.

Walker, R.G., 1977, Deposition of upper Mesozoic resedimented conglomerates and associated turbidites in southwestern Oregon: Geol. Soc. Amer. Bull., v. 88, p. 273-285.
Description and interpretation of thick graded conglomerates in thinning- and fining-upward sequences.

Hiscott, R.N., 1979, Clastic sills and dykes associated with deep-water sandstones, Tourelle Formation, Ordovician, Quebec: Jour. Sed. Petrol., v. 49, p. 1-9.
Emphasizes the importance of clastic sills in turbidites sequences, and shows how easily they can be misinterpreted as turbidites.

Walker, R.G., 1975, Generalized facies models for resedimented conglomerates of turbidite association: Geol. Soc. Am. Bull., v. 86, p. 737-748.
This is the most recent paper on resedimented conglomerates - it shows how Bouma-like models were set up for different types of conglomerates.

Helwig, J., 1970, Slump folds and early structures, northeastern Newfoundland Appalachians: Jour. Geol., v. 78, p. 172-187.

MS received November 24, 1975. Revised April, 1979.

Reprinted from Geoscience Canada, Vol. 3, No.1, p. 25-36.

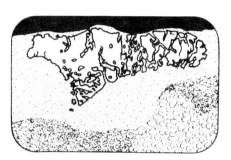

Facies Models 9. Introduction to Carbonate Facies Models

Noel P. James
Department of Geology
Memorial University of Newfoundland
St. John's, Newfoundland A1B 3X5

Introduction
This article is a general introduction to facies models in carbonate sedimentary rocks, within the larger *Facies Models* series. Here I would like to set the stage for these articles by outlining the inherent differences between siliciclastic and carbonate sedimentation (Table I).

Carbonate Sediments are Born, Not Made
This deceptively simple phrase encapsulates the main theme of the differences between the two sediment types. Siliciclastic sediments, made primarily by the disintegration of parent rock, are transported to the environment of deposition and once there the patterns of texture and fabric are impressed upon the sediment by the hydraulic regimen. The signature of siliciclastic facies is thus in sedimentary

Table I
Differences between siliciclastic and carbonate sediments.

Carbonate Sediments	Siliclastic Sediments
The majority of sediments occur in shallow, tropical environments	Climate is no constraint, sediments occur worldwide and at all depths
The majority of sediments are marine	Sediments are both terrestrial and marine
The grain size of sediments generally reflects the size of organism skeletons and calcified hard parts	The grain size of sediments reflects the hydraulic energy in the environment
The presence of lime mud often indicates the prolific growth of organisms whose calcified portions are mud size crystallites	The presence of mud indicates settling out from suspension
Shallow water lime sand bodies result primarily from localized physicochemical or biological fixation of carbonate	Shallow water sand bodies result from the interaction of currents and waves
Localized buildups of sediments without accompanying change in hydraulic regimen alter the character of surrounding sedimentary environments	Changes in the sedimentary environments are generally brought about by widespread changes in the hydraulic regimen
Sediments are commonly cemented on the sea floor	Sediments remain unconsolidated in the environment of deposition and on the sea floor
Periodic exposure of sediments during deposition results in intensive diagenesis, especially cementation and recrystallization	Periodic exposure of sediments during deposition leaves deposits relatively unaffected
The signature of different sedimentary facies is obliterated during low grade metamorphism	The signature of sedimentary facies survives low-grade metamorphism

structures and grain size variations. Carbonate sediments, on the other hand, are born in or close to the environment of depositon. Thus, in addition to the purely physical sedimentary parameters used in the analysis of non-carbonate sediments, the composition of the sedimentary particles themselves, either precipitated out of seawater (e.g., ooids) or formed by organisms (e.g., corals and clams), is equally important in characterizing the depositional environment.

Variations of Carbonate Producing Organisms with Time
To interpret ancient sedimentary sequences and construct facies models we rely heavily upon observations in modern environments of deposition. This approach works and is seen to work because the basic composition of most sedimentary particles has remained the same through time; a quartz sand grain or an ooid is the same in the Pleistocene, Permian or Precambrian. Because organisms have changed with time it is difficult, at first glance, to compare

modern and specific ancient carbonate facies. I think, though, that carbonate secreting organisms in the rock record, when viewed as sediment producers, do have living equivalents in modern oceans, although they may not even be in the same phyla. This is because, despite the numerous groups of organisms with hard parts, there are only two ways in which these hard parts are arranged: 1) as whole, rigid skeletons (foraminifers, snails, corals), and 2) as numerous individual segments held together in life by organic matter (trilobites, clams, fish). Table II lists the more important *carbonate* producing and binding organisms and their fossil equivalents.

Zones of Carbonate Accumulation

Because the precipitation of carbonate is easiest in warm, shallow seawater, most carbonate sedimentation takes place on continental shelves or banks in the tropics. Although the majority of sediments produced in this 'carbonate factory' remain in the source area, some are transported landward and some are transported basinward (Fig. 1). Thus, there are three different zones of accumulation: 1) the subtidal, open shelf and shelf margin, characterized by in place accumulations of lime sands, lime muds and reefs; 2) the shoreline, where sediments are transported from the open shelf onto beaches and tidal flats; and 3) the slope and basin, where shelf-edge sediments are transported seaward, often by mass movements, and redeposited at depth. In the basins, especially in post-Jurassic time, the fallout of calcareous zooplankton and phytoplankton has also contributed significantly to carbonate sediments.

The reader who wishes a detailed account of different carbonate sedimentary facies through time is referred to a recent outstanding documentation by Wilson (1975).

As each of the three zones of accumulation have distinctive sedimentary environments and produce differing sedimentary facies they will form a framework for the subsequent articles on carbonate facies models. The shoreline and slope to basin facies models are most like siliclastic facies models because sediments are transported from one area and deposited in another. Roger Walker (1976) has already alluded to slope to

Table II
The sedimentary aspect of modern carbonate producing and binding organims and their counterparts in the fossil record.

Modern Organism	Ancient Counterpart	Sedimentary Aspect
Corals	Archaeocyathids, Corals, Stromatoporoids, Bryozoa, Rudistid bivalves, Hydrozoans.	The large components, often in place, of reefs and mounds.
Bivalves	Bivalves, Brachiopods, Cephalopods, Trilobites and other arthropods.	Remain whole or break apart into several pieces to form sand and gravel-size particles.
Gastropods, Benthic Foraminifers	Gastropods, Tintinids, Tentaculitids, Salterellids, Benthic Foraminifers, Brachiopods.	Whole skeletons that form sand and gravel-size particles.
Codiacean algae - *Halimeda*, sponges	Crinoids and other pelmatozoans, Sponges.	Spontaneously disintegrate upon death into many sand-size particles.
Planktonic foraminifers	Planktonic foraminifers, Coccoliths (post-Jurassic).	Medium sand-size and smaller particles in basinal deposits.
Encrusting foraminifers and coralline algae	Coralline algae, Phylloid algae, Renalcids, Encrusting Foraminifers.	Encrust on or inside hard substrates, build up thick deposits or fall off upon death to form lime sand particles.
Codiacean algae - *Penicillus*	Codiacean algae - *Penicillus*-like forms.	Spontaneously disintegrate upon death to form lime mud.
Blue-green algae	Blue-green algae (especially in Pre-Ordovician).	Trap and bind fine-grained sediments to form mats and stromatolites.

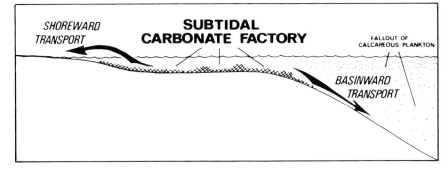

Figure 1
A sketch illustrating the main zones of carbonate accumulation.

basin facies in the context of carbonate turbidites and debris flows. At the other end of the spectrum, reefs and reef-like deposits are the most unlike siliclastic facies as they are predominantly accumulations of biologically produced, in place, carbonate.

I have chosen to begin the discussion of carbonate facies models with a description of the nearshore zone of accumulation, reflected in the rock record by shallowing-upward sequences. They are most like the siliclastic models described previously in this series by Walker, are reasonably well understood and more or less independent of variations in carbonate-producing organisms with time.

General References

The reference list on this topic is relatively short because there have recently appeared several excellent texts on carbonate sediments and facies. From these the reader can easily gain access to most of the pertinent literature on any specific aspect.

Bathurst, R. G. C., 1975, Carbonate Sediments and their Diagenesis: Developments in Sedimentology, No. 12, Elsevier, 2nd Ed., 658 p.
    This book is the most complete reference on the topic of modern and ancient carbonates. Chapters 1 and 2 detail the petrography and occurrence of modern and ancient carbonate particles, Chapters 3 and 4 summarize several different and well-studied environments of carbonate deposition. The book does not cover ancient sedimentary rock sequences.

Milliman, J. D., 1974, Marine Carbonates: Springer-Verlag, 375 p.
    This book is devoted wholly to modern carbonate sediments. The first half of the book is an exhaustive documentation of different carbonate particles, the second half is a discussion of modern environments of carbonate deposition.

Wilson, J. L., 1975, Carbonate Facies in Geologic History: Springer-Verlag, 471 p.
    Chapters 1, 2 and 12 of this book are an excellent summary of the principles and stratigraphic aspects of carbonate sedimentation. The bulk of the text is a detailed review of carbonate sedimentary facies at different times in geologic history.

Horowitz, A. S. and P. E. Potter, 1971, Introductory Petrography of Fossils: Springer-Verlag, 302 p.
    Chapter 2 is a concise introduction to carbonate sedimentology and the remainder of the book is devoted to the recognition of various skeletal particles in thin section.

Ham, W. E., ed., 1962, Classification of Carbonate Rocks, a Symposium: Amer. Assoc. Petrol. Geol., Memoir 1, Tulsa, Okla., 279 p.
    This symposium contains several papers, notably those by W. E. Ham and L.C. Pray, M. W. Leighton and C. Pendexter, R. L. Folk, R. J. Dunham, which, by attempting to classify sedimentary carbonates, outline succinctly the important factors governing carbonate sedimentation.

Ginsburg, R. N. and N. P. James, 1974, Holocene carbonate sediments of Continence Shelves: in C. A. Burke and C. L. Drake, eds., The Geology of Continental Margins: Springer-Verlag, p. 137-157.
    A short article summarizing the sedimentology of eight different well-studied areas of carbonate sedimentation in the modern ocean.

Ginsburg, R. N., R. M. Lloyd, K. W. Stockman, and J. S. McCallum, 1963, Shallow Water Carbonate Sediments: in M. N. Hill, ed., The Sea, Vol. 3, p. 554-578.
    This article illustrates how the architecture of modern marine carbonate skeletons governs the grain-size of the resultant sediments.

Folk, R. L. and R. Robles, 1964, Carbonate Sands of Isla Perez, Alacran Reef, Yocatan: Jour. Geol., v. 72, p. 255-292.
    A classic study illustrating how two different skeletal organisms, corals and the codiacean alga *Halimeda*, break down under different conditions into specific grain sizes.

MS received May 3, 1977.
Revised January, 1979.
Originally appeared in Geoscience Canada, Volume 4, Number 3, p. 123-125.

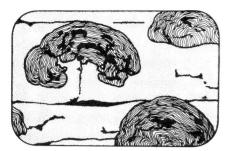

Facies Models 10. Shallowing-Upward Sequences in Carbonates

Noel P. James
Department of Geology
Memorial University of Newfoundland
St. John's, Newfoundland A1B 3X5

Introduction

Perhaps the most commonly encountered carbonate deposits are laterally persistant, evenly bedded limestones and dolomites of apparent shallow water origin, as demonstrated by abundant fossil mud cracks and stromatolites. These deposits, which occur most commonly on the continents and in relatively undeformed portions of mountain belts, are not only important sources of paleontological and sedimentological information, but are also common host rocks for hydrocarbons and metallic ores (particularly lead and zinc). As such, it is critical that we be able to determine, as precisely as possible, the environment in which each of the interbedded sediments was deposited.

A quantum jump in our understanding of these deposits occurred when modern carbonate tidal flats were examined in detail, notably by Robert Ginsburg and his colleagues in Florida and the Bahamas about 20 years ago. It was quickly realized that there were a host of sedimentary structures and textures on these flats that would allow a much more precise definition of environments of deposition than was possible before: these findings were quickly applied to fossil sequences (Fischer, 1964; Laporte, 1967; Aitken, 1966, Roehl, 1967). This application in turn generated two different lines of

investigation: 1) description of other areas of modern tidal flat deposition, in particular the southern shore of the Persian Gulf where evaporites are common and Shark Bay, Western Australia, where a great variety of modern stromatolites are forming; and 2) documentation of different styles of tidal flat deposits in the geologic record.

Despite the great number of studies which record carbonate sequences with evidence of periodic exposure, there has not been, to date, a synthesis of the various sequences. In siliciclastic deposits, it is commonly the sequence of rock types that defines the environment of deposition. In carbonate stratigraphy, however, the actual sequence is of less importance, possibly because sedimentary environments commonly can be defined accurately on the basis of grain type, fossil fauna and structures.

The Model

Carbonate sediments characteristically accumulate at rates much greater than the rate of subsidence of the shelf or platform upon which they are deposited. This is because carbonate sediments are produced mainly in the environment of deposition - especially in shallow

water where conditions for the biological and physicochemical fixation of carbonate are optimum. As a result, carbonate accumulations repeatedly build up to sea level and above, resulting in a characteristic sequence of deposits, in which each unit is deposited in progressively shallower water. This shallowing-upward sequence commonly is repeated many times in a succession of shallow water deposits (Fig. 1).

Readers will recognize that such a shallowing upward sequence also may be termed a 'regressive sequence'. This term has led to much confusion in the past, because it has been used to describe deposits associated with a high rate of sediment production and accumulation under relatively static sea level - sea bottom conditions. So I have abandoned the term 'regressive' altogether in favor of a rock-descriptive term, albeit interpretive, the shallowing-upward sequence, which others have called shoaling-upward (even though all such deposits are not shoals).

1) The model as a norm. The ideal carbonate shallowing-upward sequence comprises four units,

Figure 1
Bedded carbonates ranging in age from Middle to Late Cambrian near Fortress Lake, B.C.: Arrows mark the top of large-scale shallowing-upward sequences (L - Lyell FM., S - Sullivan Fm., W - Waterfoul Fm., A - Arctomys Fm., E - Eldon and Pica Fms.).

Striping of the Waterfoul Fm. is caused by repetitive smaller scale shallowing-upward sequences between subtidal-intertidal limestones (dark) and supratidal dolomites (light). Photo courtesy J. D. Aitken.

illustrated in Figure 2. The basal unit, which is generally thin, records the initial transgression over pre-existing deposits and so is commonly a high energy deposit. The bulk of the sequence which may be of diverse lithologies consists of normal marine carbonate, as discussed below. The upper part of the sequence consists of two units: the intertidal unit within the normal range of tides; the other a supratidal unit, deposited in the area covered only by abnormal, windblown or storm tides. Each of these units exhibits the characteristic criteria of subaerial exposure.

The thread that binds all such sequences together is the presence of the distinctive intertidal unit, which, once recognized, allows one to interpret the surrounding lithologies in some kind of logical sequence (Fig. 3).

2) *The model as a predictor.* First-order variation on the basic model revolves around the two main types of intertidal environment: 1) quiet, low-energy situations, commonly referred to as tidal flats, and 2) agitated, high-energy situations, or quite simply, beaches. Second-order variation involves the kind of subtidal units below and of supratidal units above; the subtidal reflects the type of marine environment adjacent to the

tidal flat and the supratidal reflects the adjacent terrestrial environment, in particular the climate (Fig. 4).

For purposes of discussion I will begin with those sequences that contain low-energy intertidal units (tidal-flats) because they exhibit the greatest variety of distinctive features and consequently are well documented, both in modern and ancient settings. To place the observed features in context we should first examine modern carbonate tidal flats.

Sequences with a Low-Energy Intertidal Unit

Modern Tidal Flats

The main elements of a modern carbonate tidal flat system as exemplified by the narrow shelf and embayments of Shark Bay, Western Australia, the southern coast of the Persian Gulf and the wide platform of the Bahama Banks are shown in Figure 5. A characteristic of most modern examples is that they occur in protected locations: protected that is from the open ocean waves and swells, yet still affected by tides and severe storms. This unique setting is commonly afforded by the presence of a semi-

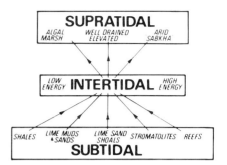

Figure 4
A flow diagram indicating the various possible environmental transitions present in a carbonate shallowing-upward sequence.

protective barrier composed of lime sand shoals, locally associated with reefs and/or islands. The barrier commonly is dissected by tidal channels through which flow high velocity tidal currents. A shallow muddy lagoon lies in the lee of this barrier. The lagoon may be enormous as in the case of the Bahamas, relatively narrow and elongate as in the Persian Gulf, or very small as in the pocket embayments of Shark Bay. In such an arrangement, tidal flats are present as: 1) small flats atop and on the lee side of the emergent sand shoals of the barrier, and 2) large flats along the shoreline of the shallow lagoon (Fig. 5). Thus tidal flats occur in association with two separate carbonate accumulations, high energy sand bodies and low energy lime muds. A third type of association which is less common in modern situations is the association with reefs, especially the interior of large reef complexes.

Intertidal environments. The intertidal zone, especially along rocky coasts and beaches is commonly a gradual transition from sea to land without much noticeable variation. On wide gradually sloping tidal flats, this zone can be the familiar gradual transition or a complex area of many subenvironments. At one end of the spectrum the flats have few, very shallow, short tidal creeks (Fig. 6). At the other end of the spectrum the flats are dissected by many tidal creeks flanked by levees. Slight depressions between the creeks are occupied by tidal ponds (which fill and partially empty during each rise and fall of the tide), and the whole complex is fronted by small beach ridges or erosional steps (Fig. 6). Perhaps in this case it would be better to refer to the whole zone as the "pond and

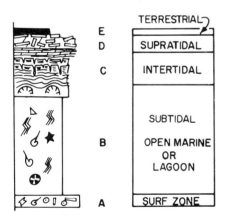

Figure 2
Five divisions of the shallowing-upward model for carbonates: A - lithoclast rich lime conglomerate or sand, B - fossiliferous limestone, C - stromatolitic, mud-cracked cryptalgal limestone or dolomite, D - well laminated dolomite or limestone, flat-pebble breccia, E - shale or calcrete, bracketed to emphasize that the unit is often missing - see text. Symbols used throughout are from Ginsburg (1975).

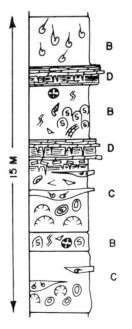

Figure 3
Actual sequence of several shallowing-upward sequences from the Manlius Fm., New York State (From Laporte, 1975).

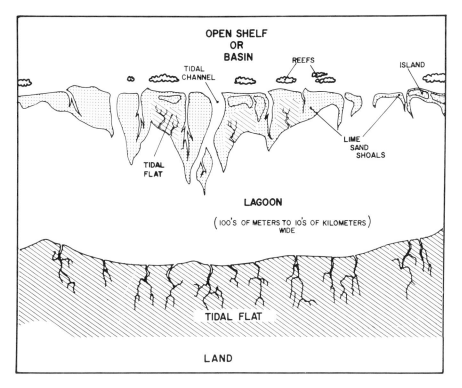

Figure 5
Plan view of the geometry of a modern tidal flat complex. Note that tidal flats can be present both adjacent to the land or in the lee of lime sand shoals.

creek belt" because some of the areas are dry most of the time (levees and beaches) whereas others are continuously submerged (ponds and creeks). These complications have led some workers (e.g., Ginsburg and Hardie, 1975) to despair of conventional terms and instead to relate different zones to the per cent of time that they are exposed rather than to their position.

On some tidal flats, in the Bahamas for example where there are many tidal creeks and noticeable relief between levee and tidal pond (about 1 m), the true intertidal zone which lies between the two may comprise only 60 to 70 per cent of the intertidal environment. In other areas such as the Persian Gulf, where there are fewer creeks and less relief, almost the whole flat is truly intertidal. The important point to grasp is that numerous environments may exist in very close proximity, not only perpendicular to the shoreline but parallel to it as well, so that in the geologic record rapid, local lithological variations are to be expected, both vertically and laterally, rather than a smooth succession of progressively shallower environments.

The tidal flat wedge is built up of fine grained sediments brought onto the flats from the adjacent offshore marine zone

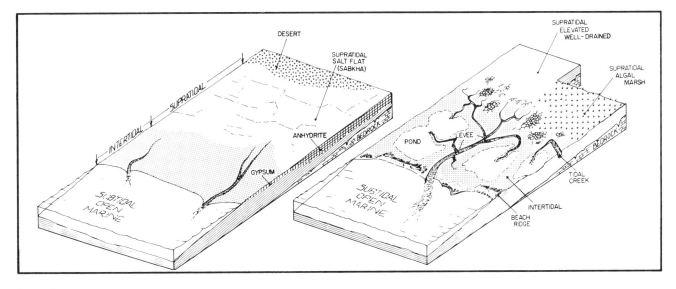

Figure 6
Block diagrams showing the major morphological elements of a tidal flat; left - a hypersaline tidal flat with few channels and bordering a very arid desert (similar to the modern Persian Gulf), right - a normal marine tidal flat with many channels and ponds and bordering an elevated well-drained area of low swamp algal marsh in a humid climate (similar to the modern Bahamas).

by storms rather than by daily tides. Large storms such as hurricanes which flood the flat with sheets of water white with suspended sediment are particularly effective. Shinn, Lloyd and Ginsburg (1969) have suggested that *the tidal flat is a river delta turned wrong-side out* with the sea as the "river" supplying sediment to the channeled flats as the "delta".

Sediments of the intertidal zone are characterized by three distinctive features, not found elsewhere: 1) algal mats, 2) irregular to even laminations (cryptalgal laminites), with fenestral porosity, and 3) desiccation features.

The algal mats, gelatinous to leathery sheets of blue-green algae growing on top of the sediment surface are widely regarded as the signature of intertidal deposition; they may occur throughout the intertidal zone but their precise distribution is controlled by climate and the presence or absence of other organisms. The upper limit is controlled by climate; in arid areas they cannot grow above the high intertidal into the supratidal zone, whereas in areas of high rainfall where the supratidal zone is moist or flooded for days at a time, mats are prolific. The lower limit is more variable and appears to be controlled by the presence of gastropods that eat algae. In areas of normal salinity, mats are prevented from developing below the middle intertidal zone because they are browsed by gastropods; in areas of hypersalinity (deadly for gastropods), mats grow down into the subtidal zone. In addition algal mats will colonize only a temporarily or permanently stable bottom, and will not grow on shifting sand.

Although the algal mats may themselves vanish with time, evidence of their presence during deposition remains because of the peculiar pores that they help to create, generally referred to as 'laminoid fenestrae'. These are irregular, elongate to mostly sub-horizontal sheet-like cavities (loferites or birds-eyes of some workers) with no obvious support and much larger than can be explained by grain packing. They are simply due to the fact that the mats are covered with sediment and eventually rot away as they are buried, leaving voids as well as holes due to entrapped gas and shrinkage.

Another structure recording the presence of blue-green algal mats are

the finely laminated carbonates (Fig. 7), ranging from stratiform and slightly crenulated to the familiar arched domes of stomatolites. These have been called cryptalgal (hidden, algal) laminations by Aitken (1967) in reference to the fact that the influence of algae in the rock-forming process is more commonly inferred than observed.

Lower intertidal zone. Much of the subtidal character remains evident in sediments from this part of the environment, and the deposits are commonly well burrowed and bioturbated. In hypersaline areas, however, the surface of the sediment is veneered with a thick algal mat, frequently broken into desiccation polygons. Beneath the mat grains are blackened due to reducing conditions and altered by boring algae to peloids of lime mud.

Tidal ponds and the creeks that drain them on hypersaline tidal flats support the most prolific growth of algal mats anywhere on the flat. The algal mat flourishes in water depths greater than those in the immediate offshore area because of relatively elevated salinites in the ponds. On tidal flats where the salinity is closer to normal, marine tidal ponds are populated by a restricted but prolific fauna of foraminifers and gastropods and the gastropods prevent

the growth of algal mats. Similarly, if tidal creeks are common in such areas, the channels are devoid of mats but do contain concentrations of the pond fauna, which may accumulate as bars of skeletal lime sand. As the channels migrate these skeletal sands commonly form a basal lag deposit.

Middle and upper intertidal zone. Sediments here are commonly light-grey to light-brown (oxidizing conditions), have good fenestral porosity (the variable growth of algal mats), are graded (episodic storm deposition) and are broken into desiccation polygons' (prolonged exposure). There is generally good growth of algal mats throughout: in the lower parts thick leathery mats are separated into desiccation polygons a few cms to a metre in a diameter with cracks filled by lime mud in the lower parts (Fig. 8); in the central parts, thinner, leathery mats have surfaces that are puffed up into blisters and convoluted into crenulated forms; and in the upper parts, shriveled, crinkled and split mats are found. Bedding generally is irregular especially in the upper zones with mats alternating with graded storm layers.

In some settings sediment in the upper intertidal zone dries out to form chips of lime mud while in others the

Figure 7
Cryptalgal laminites that have been mud-cracked. The intertidal unit of a shallowing-upward sequence in the Petit Jardin Fm. (Upper Cambrian) on the south shore of the Port-au-Port Peninsula, Nfld. (Photo courtesy R. Levesque).

sediment below the mats is lithified to a depth of as much as 10 cms.

Although sediments commonly are laminated throughout the intertidal environment, they are also riddled with small-scale tubules produced by insects and worms, larger tubes produced by crabs and other crustaceans. Sediments also may be penetrated by the prolific shallow roots systems of salt tolerant plants.

Supratidal environment. In all situations (including channel levees) this area is characterized by long periods of exposure. This is reflected by the lithification of storm deposited sediments in the form of surface crusts, several cms thick, and which in turn are fractured into irregular polygons. These polygons may be pushed up by the force of crystallization (or by plant roots) to form 'teepees', or dislodged completely to form pavements of flat-pebble breccia. Clasts are commonly cemented on modern tidal flats by cryptocrystalline aragonite or calcite, and characteristically contain considerable (25 to 50%) fine crystalline dolomite.

If the creek levees in the intertidal zone have built up above normal high tide level, they consist of hard, finely to very finely laminated sediment,

extremely regular, and composed of alternating layers of sediment and thin algal mats with excellent fenestral porosity.

The landward parts of the supratidal zone may grade into various terrestrial environments, the end members of which are: 1) areas of elevated, pre-existing bedrock and no sedimentation in which the surface of the rock is characterized by intensive subaerial diagenesis, and the development of caliche (calcrete crusts); 2) areas of contemporaneous sedimentation which grade between: a) low-lying environments in regions of high rainfall, occupied by algal marshes, b) low-lying environments in arid, desert regions, characterized by evaporite formation, and c) well drained zones, often slightly elevated and with little deposition.

Algal marshes, flooded by fresh water during the rainy season are an ideal environment for the growth of algal mats and these mats are periodically buried by layers of sediment swept in during particularly intense storms: thus the preserved record is one of thick algal mats alternating with storm layers. With progressive aridity the supratidal zone dries out. If the chlorinity of the groundwaters remains constantly above 39º/oo, cementation, particularly by

aragonite, is common. Cementation is most common if there is minor but consistent input of fresh water from inland to dilute the hypersaline groundwaters somewhat. If the chlorinity of the groundwaters remains constantly above 65º/oo then authigenic evaporites precipitate within the sediment below ground level. In this setting (called a supratidal sabkha, or salt flat in the Middle East) dolomitization is also common in the subsurface, saline brine pools occur at the surface and terrigenous wind-blown sand is common in the sediment

In relatively well-drained zones the supratidal environment is a deflation surface, occasionally cut by the upper reaches of tidal creeks, sometimes damp from rising capillary waters and covered by a thin film of algal mat. Scoured and rippled sediment is common and clasts are sometimes encrusted with algae to form oncolites.

Common Sequences with a Low-Energy Intertidal Unit

a) *Muddy and grainy sequences.* These sequences developed either by progadation of the wide continental tidal flat or by shoaling of the lime sand bodies that formed the barrier offshore (Fig. 9). The climate in the region of deposition was generally too wet or the groundwater table too low or diluted by fresh water to permit precipitation of evaporites.

The muddy sequences, those in which skeletal lime muds or muddy lime sands are the main subtidal unit, are well developed today in well-drained areas of Shark Bay where salinities are too high to permit browsing of the algal mats by gastropods, and in the tidal creek and pond belt of the Bahamas. These sequences are generally regarded as the 'classic' tidal flat sequences. The basal unit, if present, records the initial incursion of the sea onto land and as such is commonly coarse-grained, composed of clasts, etc., all diagnostic of surf-zone deposition. The subtidal unit is characteristically a bioturbated lime wackestone to packstone with a normal and diverse marine fauna, commonly containing stromatolites in deposits older than middle Paleozoic. In Precambrian and lower Paleozoic deposits the characteristic tidal flat features such as desiccation polygons,

Figure 8
A bedding plane of mud-cracked polygons with the edges of each polygon curled up, likely because the algal mats in the polygons shrivelled upon exposure and drying out;

Near the top of a shallowing-upward sequence in the East Arm Fm. (Upper Cambrian), Bonne Bay, Nfld.

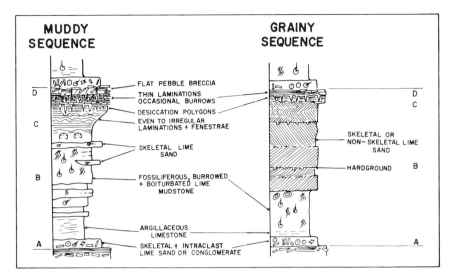

Figure 9
Two hypothetical sequences with a low energy, tidal flat unit developed on a low energy subtidal unit (left) and a high energy lime sand unit (right).

Figure 10
Shallowing-upward sequences comprising lower intertidal-subtidal limestones (L) overlain by supratidal dolomites (D –

Cryptalgal laminites, sandy in part) in the Lyell Fm. at Takakkaw Falls, Yoho National Park, B.C. (photo courtesy J. D. Aitken).

well-laminated sediments and fenestrae will occur at the base of the intertidal zone (Figs. 10, 11). In deposits younger than middle Paleozoic, the prolific browsing and burrowing activity in the lower intertidal zone (unless the water mass was hypersaline) have homogenized the sediment, so that the signature of intertidal deposition is recorded only within the mid and upper intertidal sediments.

If the tidal flat were extensively channelled, the migration of channels back and forth may also have destroyed some of the subtidal character, forming instead a partial fining-upward sequence (much like that of a river), with a basal skeletal lime sand.

Where fenestrae are present they show a zonation: horizontal to laminated in the lower intertidal environments (smooth mat), irregular and in some

cases vertical in the middle and upper intertidal environments (pustular, shriveled and crinkled mats). Desiccation polygons are most common near the top, apparently coincident with cementation. The supratidal zone is characterized by very evenly laminated deposits or flat pebble breccias.

Readers interested in the finer details of such sequences are referred to studies by Laporte (1967) and Fischer (1964), the latter outlining and documenting a similar facies sequence but in reverse order, forming a deepening-upward sequence.

The same style of shoaling sequence also may be present off-shore from the low-energy tidal flat, on the lime-sand shoals. Here low energy tidal flats developed in the lee of the leading edge of the shoal once beach ridges were developed or currents had swept sand together to form islands. This will be reflected in the sequence as a sudden change from obvious high energy deposition to low energy intertidal deposition. The subtidal unit is generally well-sorted, oolitic, pelletoidal or skeletal lime sand (pelmatozoans are particularly common in the Paleozoic), with a few containing oncolites. Bedding is characteristically planar, with herringbone cross-laminations, large at the base and becoming smaller upwards, and individual bedding planes commonly covered with small-scale ripples. Early cementation is characteristic, and so deposits contain many intraclasts of cemented lime sand, and bored surfaces. Once the shoals, or parts of the shoals are inactive they may be burrowed and much of the original cross-bedding may be destroyed.

The intertidal to supratidal units are similar to those described above, but are generally relatively thin. If the shoal is exposed for a long time caliche and soil profiles commonly develop, reflected by brown irregular laminations, breccias, and thin shale zones.

b) *Stromatolite and reef sequences.* One common variation on the model is the development of shoaling-upward sequences in association with abundant stromatolites in the lower Paleozoic/Precambrian and with reefs in the Phanerozoic in general.

In Shark Bay, Western Australia, where all environments are hypersaline and so stromatolites abound, the interrelationship between stromatolite morphology and environment nas only recently been documented (Hoffman, 1976). In the intertidal zone columnar to club-shaped forms, up to one metre high are found rimming headlands. In relatively high energy, exposed environments the relief of the columns is proportional to the intensity of wave action. These grade laterally away from the headlands to the lower energy bights where the stromatolites are more prolate and elongate, oriented normal to the shoreline. In tidal pools digitate columnar structures abound.

These growth forms are the result of active sediment movement; algal mats only grow on stabilized substrate, thus columns are nucleated upon pieces of rock, etc.; growth is localized there and does not occur on the surrounding shifting sands. Early lithification of the numerous superimposed layers of mat and sediment turns the structures into resistant limestone. Moving sand continuously scours the bases of the stromatolites. The mounds or pillars are largest in subtidal or lower intertidal environments and decrease in synoptic relief upwards, finally merging with stratiform mats in upper intertidal zones, above the zone of active sediment movement.

The resulting model sequence, summarized in Figure 12, is integrated from the Shark Bay example and the summary sequence of 200 or more shoaling sequences present in the Rocknest Formation of middle Precambrian age near Great Slave Lake (Hoffman, 1976). In the intertidal zone deposits reflect higher energy than normal, indicating a more exposed shoreline. These sediments underlie and surround the domal to columnar stromatolites, which in turn grade up into more stratiform stromatolites, and finally into very evenly bedded structures. The supratidal unit of this sequence will be characterized by both desiccation polygons and flat-pebble breccias as well as occurrences of delicate branching stromatolites, forming in supratidal ponds. Care should be taken in delineating this sequence because stromatolites that are similar to those in the intertidal zone also occur in the subtidal (Playford and Cockbain, 1976).

Shallowing-upward sequences are also common as the last stage of sedimentation in large bioherms, as numerous successions within the large back-reef or lagoonal areas of reef complexes, and as 'caps' on widespread biostromes. In this type of sequence the shoaling upward is first reflected in the subtidal unit itself, generally as a transition from large massive hemispherical colonial metazoans of the reef facies, to the more delicate, stick-like forms that are common in the shallow protected locations. These stick-like skeletons may be swept together on beaches at the edge of the tidal flat. As a result, the intertidal unit commonly contains a conglomerate within it, or at the base. The upper part of the sequence is otherwise similar to the others described. For a more detailed description of "reefy" sequences see the two studies by Havard and Oldershaw (1976) and Read (1973).

Figure 11
Numerous shallowing-upward sequences comprising thick subtidal oolite lime sands and thin intertidal-supratidal cryptalgal laminites with fenestrate porosity; Petite Jardin Fm., Port-au-Port Peninsula, Nfld.

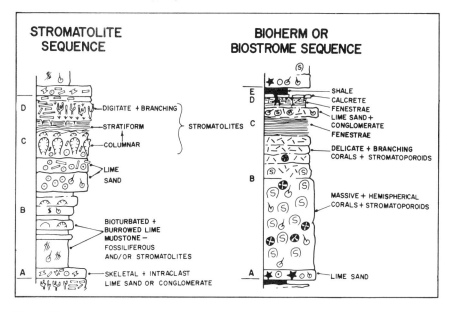

Figure 12
Two hypothetical sequences with a low energy intertidal unit developed in conjunction with stromatolites (left) and on top of a skeletal metazoan bioherm or biostrome (right).

116

c) *Carbonate-evaporite sequences*
(also see Kendall, Facies Models No. 13
and 14). The other major variation on the
model proposed at the beginning of this
article is at the opposite end of the
environmental spectrum, in the
supratidal zone, in this case emergent in
a very arid environment and flushed by
hypersaline groundwaters. The
hypersalinity of the groundwaters and
attendant high evaporation results in the
formation of *authigenic* evaporites
which in turn raises the $Mg++/Ca++$
ratio of the groundwaters and induces
dolomitization of the sediment. The
processes occur within the sediment,
above the water table in the intertidal
zone and both above as well as below
the water table in the supratidal zone. If
the water compositions are barely within
the field of gypsum precipitation, and
there are fluctuations due to brackish
flow of groundwater from the mainland,
evaporites will occur in the form of
isolated masses or crystals in the upper
part of the sequence. If the groundwater
compositions are continuously well
within the field of gypsum precipitation,
growth of evaporite minerals takes place
as a mush of gypsum crystals in the
intertidal zone or as layers of anhydrite
nodules, as complex masses with a
characteristic chickenwire texture, and
as layers contorted into enterolithic
(intestine-like) shapes (Fig. 13). The
important point, which is often ignored, is
the growth of the evaporites *within*
the sediment, as a diagenetic overprint
on depositional facies of various
environments. As evaporite growth is
porphyroblastic, the host sediment
commonly is displaced to
intercrystalline areas and earlier fabrics
are destroyed. Accompanying
dolomitization commonly is intense, with
sediments of the intertidal and much of
the subtidal zones affected.

Evaporites, however, are very soluble
when exposed to percolating meteoric
waters of low salinity and have a
tendency to vanish from the record.
Dissolution of the evaporites affects the
sequence in several ways, but the most
important is the formation of collapse
breccias (Fig. 14). This collapse occurs
when the evaporites dissolve leaving no
support for the overlying sediments,
which subside into the void created by
evaporite removal. Thus the top of the
sequence is a breccia of marine
limestone from the overlying sequence
with a mixture of terrigenous sand if a

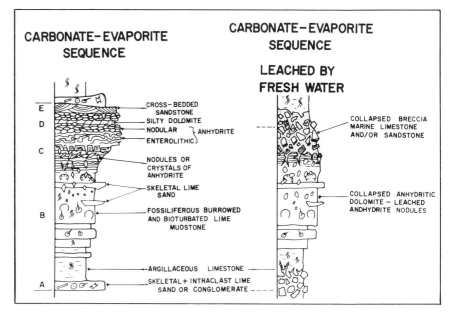

Figure 13
Two hypothetical sequences with a low-
energy intertidal unit and a supratidal unit
developed under arid conditions; on the right
the evaporites have been dissolved by
percolating fresh waters.

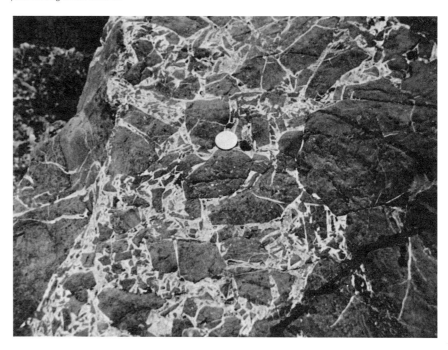

Figure 14
A collapse breccia of subtidal lime mudstone
clasts in white calcite; caused by the solution
of anhydrite at the top of a shallowing-upward
sequence in the Shunda Fm. (Mississippian)
at Cadomin, Alberta (photo courtesy
R. W. Macqueen).

terrigenous facies capped the original sequence (Fig. 13). Isolated anhydrite crystals in lower parts of the sequence may be leached out, forming vugs which may be subsequently filled with quartz or chalcedony (usually length-slow). The dolomite, at least in the upper part, is commonly altered to calcite, in the reverse of the dolomitization process (so-called "dedolomitization").

Sequences with a High Energy Intertidal Unit

In contrast to the low-energy intertidal (the tidal flat) the higher energy beach zone is not commonly recognized in the rock record. This may be partly because it resembles many subtidal grainstone deposits and so is not obviously distinctive. Also, it is relatively narrow compared to the tidal flat, and has a lower preservation potential; and finally, the beach deposits lack the distinctive sedimentary features of the tidal flat. These very reasons illustrate the value of the concept of a shoaling-upward sequence as a guide. Once the potential for such a sequence is recognized in the geologic record, then one can concentrate on the search for subtle features that characterize beach deposition, which otherwise might go unnoticed.

Modern carbonate beaches. The beach is characterized by two zones: 1) *the lower foreshore*, that zone unusually below the zone of wave swash, and 2) *the upper foreshore*, the zone of wave swash. Sediments of the lower foreshore are coarse grained, poorly sorted, have a matrix of lime mud (if it is available), and are characterized by small and large-scale festoon cross-bedding, oriented parallel to the shoreline and generally attributed to longshore drift. The upper foreshore comprises thick-bedded, internally laminated, very well-sorted lime sands and gravels in planar cross-bedded accretionary beds that dip gently seaward (generally less than 15 degrees). Sediments in the upper foreshore zone may have many open-space structures, the equivalent of the fenestrae of muddy intertidal sediment, called keystone vugs (Dunham, 1969) or microcaverns (Purser, 1972). These are due to gas escape and in the geological record are partly to completely filled with cement.

As on the tidal flat, periodic exposure of beach deposits leads to cementation and partial subaerial diagenesis. The textures thus created are difficult to recognize in the field but are important keys to recognizing the beach environment. The two most important of these diagenetic phenomena are beachrock and calcrete.

Beachrock is composed of seaward-dipping beds of lime sand and gravel that are generally cross-laminated and occur in the lower intertidal to middle intertidal environment. It is formed by the precipitation of carbonate cement out of seawater or mixed seawater and rainwater. The beds of limestone may be up to one metre thick, are commonly jointed at right angles to the beach and are encrusted and/or bored by numerous intertidal organisms. Lithification disappears seaward and rarely extends higher than the intertidal zone. The partly cemented beds may be broken up and redeposited as conglomerates, made up of cemented sand clasts. In the upper parts of the intertidal zone cementation takes place in intergranular voids partly filled with air: the cements, as a result, are often stalactitic (more extensively developed on the undersides of grains).

If exposed for long periods of time and if located in an environment where there is at least periodic rainfall, the lime sands will begin to undergo subaerial diagenesis (see Bathurst, 1975, for an extended discussion of subaerial diagenesis). In addition the upper metre or so of such subaerially exposed deposits develop calcrete or caliche horizons which have many features that closely resemble those produced by laminar to laterally-linked stromatolites and oncolites. These features are discussed in detail by James (1972) and Read (1976).

The supratidal unit in these sequences may be any of the ones described above, although calcrete (caliche) is very common. Beaches may act as small barriers protecting supratidal ponds and flats so that the cap in such sequences will be thin beds of lime mud (often dolomitized) with all of the associated supratidal features. One variation not found elsewhere occurs where the high energy surf zone of the overlying sequence erodes the top of the sequence down to the cemented portions, resulting in truncation layers or hardgrounds that separate sequences.

Shallowing upward sequence with a high-energy intertidal unit. The lower two units of this type of sequence are similar to those described in the preceding sections on sequences with low-energy intertidal units (Fig. 15). In this sequence, however, characteristic subtidal carbonates grade upward into coarse-grained lime sands with all the characteristics of the lower and upper foreshore described above (Fig. 9). The supratidal unit, in the form of a thin shale (soil), may be present, but more commonly the supratidal is represented not by a deposit but by intensive diagenesis of the upper unit (cementation, dissolution, calcrete formation and microkarst). This is in many ways similar to the diagenetic overprint of other facies by supratidal evaporite formation.

Summary

In the past there has been a natural tendency to use obvious sedimentary structures (e.g., mud cracks, stromatolites) to infer that parts of a carbonate sedimentary sequence had been periodically exposed. Individual structures, however, often have counterparts in other sedimentary environments (e.g., syneresis cracks, subtidal stromatolites) resulting, in many cases, in questionable paleoenvironmental interpretations. With all the data now available on carbonate strandline deposition we can frequently use what have become

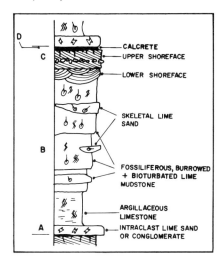

Figure 15
A hypothetical sequence with a high-energy intertidal unit: a beach, developed, in this case, adjacent to a low energy subtidal environment.

natural associations of sedimentary features in a vertical succession and define, with precision, specific strandline facies and their interrelationships.

While this is true for low-energy shoreline sequences, it is much less so for high-energy shoreline sequences. To bring all aspects of this type of facies model to comparable levels of understanding much more data is needed on exposed or high-energy intertidal environments, not from the modern, but from the rock record. In addition, the time is ripe to test whether or not the diagenetic features which result from periodic subaerial exposure (cementation, microkarst, calcrete) can be commonly recognized in ancient sequences.

Acknowledgements

Roger Macqueen, Bob Stevens and David Kobluk kindly criticized early versions of this manuscript and kept me honest.

References

Basic References on Shallowing-Upward Carbonate Sequences

Ginsburg, R. N., ed., 1975, Tidal Deposits: Springer-Verlag, 428 p.

The best all around, up-to-date reference on siliclastic and carbonate tidal deposits.

Merriam, D. F., ed., 1964, Symposium on Cyclic Sedimentation: State Geol. Survey Kansas Bull., 2 vols., 636 p.

40 papers many of which are still basic references, on cyclic sedimentary sequences.

Lucia, F. J., 1972, Recognition of evaporite-carbonate shoreline sedimentation: in K. J. Rigby, and K. Hamblin, eds., Recognition of Ancient Sedimentary Environment: Soc. Econ. Paleontol. Mineral. Spec. Publ. 16, p. 160-192.

A well-written summary with many samples.

Hoffman, P., 1973, Recent and Ancient Algal Stromatolites: Seventy years of Pedagogic cross-pollination: in R. N. Ginsburg, Evolving Concepts in Sedimentology: Johns Hopkins Press, p. 178-191.

A very readable essay on the evolution of our understanding of stromatolites and related sediments.

Irwin, M. L., 1965, General theory of epeiric clear water sedimentation: Amer. Assoc. Petrol. Geol. Bull., v. 49, p. 445-459.

The first integrated synthesis of shallow water carbonate sequences and their meaning.

Shaw, A. B., 1965, Time in Stratigraphy: McGraw-Hill, p. 1-71.

A pedantic, but thought-provoking analysis of epeiric sea carbonate sedimentation.

Fischer, A. G., 1964, The Lofer cyclotherms of the Alpine Triassic: in D. F. Merriam, ed., Symposium on Cyclic Sedimentation: State Geol. Survey Kansas Bull. 169, v. 1, p. 107-149.

A regressive sequence of shallow water carbonates, superbly documented and illustrated.

Laporte, L., 1967, Carbonate deposition near mean sea-level and resultant facies mosaic: Manlius Formation (Lower Devonian) of New York State: Amer. Assoc. Petrol. Geol. Bull., v. 51, p. 73-101.

An extremely clear, well-written analysis of a shallowing-upward sequence.

Bosellini, A. and L. A. Hardie, 1973, Depositional theme of a marginal evaporite: Sedimentology, v. 20, p. 5-27.

An analysis which concentrates on the sedimentology of, not evaporites in, a series of shallowing-upward sequences.

Modern Carbonate Tidal Flats

Bathurst, R. G. C., 1975, Carbonate Sediments and Their Diagenesis: Developments in Sedimentology No. 12, Elsevier, p. 178-209, 217-230, 517-543.

Shinn, E. A., R. M. Lloyd, and R. N. Ginsburg, 1969, Anatomy of a modern carbonate tidal flat, Andros Island, Bahamas: Jour. Sed. Petrology, v. 39, p. 1202-1228.

Hardie, L.A. (ed.), 1977, Sedimentation on the Modern Carbonate Tidal Flats of Northwest Andros Island, Bahamas: The Johns Hopkins Press, Baltimore Md., 202 p.

Garrett, P., 1970, Phanerozoic stromatolites: non-competitive ecological restriction by grazing and burrowing animals: Science, v. 169, p. 171-173.

Logan, B. W., G. R. Davies, J. F. Read, and D. Cebulski, 1970, Carbonate sedimentation and environments, Shark Bay, Western Australia: Amer. Assoc. Petrol. Geol. Memoir 13, 223 p.

Logan, B. W. et al., 1974, Evolution and diagenesis of Quaternary carbonate sequences, Shark Bay, Western Australia: Amer. Assoc. Petrol. Geol. Memoir 22, 358 p.

Hoffman, P., 1976, Stomatolite morphogenesis in Shark Bay, Western Australia: in M. R. Walter, Stomatolites, Elsevier, p. 261-273.

Kinsman, D. J. J., 1966, Gypsum and anhydrite of Recent Age, Trucial Coast, Persian Gulf: Second Symposium on Salt, Northern Ohio Geol. Soc. Cleveland, p. 1302-1326.

Kendall, C. St. G. C. and P. A. d'E. Skipwith, 1968, Recent Algal mats of a Persian Gulf Lagoon: Jour. Sed. Petrology, v. 38, p. 1040-1058.

Kendall, C. St. G. C. and P. A. d'E. Skipwith, 1969, Holocene shallow-water carbonate and evaporite sediments of Khor al Bazam, Abu Dhabi, Southwest Persian Gulf: Amer. Assoc. Petrol. Geol. Bull., v. 53, p. 841-869.

Purser, B. H., ed., 1973, The Persian Gulf: Springer-Verlag, 471 p.

Modern Carbonate Sand Bodies

Imbrie, J. and H. Buchanan, 1965, Sedimentary structures in modern carbonate sands of the Bahamas: in Primary Sedimentary Structures and Their Hydrodynamic Interpretation: Soc. Econ. Paleontol. Mineral., Spec. Publ. 12, p. 149-173.

Ball, M. M., 1967, Carbonate sand bodies of Florida and the Bahamas: Jour. Sed. Petrology, v. 37, p. 556-591.

Hine, A.C., 1977, Lilly Bank, Bahamas: History of an active oolitic sand shoal: Jour. Sed. Petrology, v. 47, p. 1554-1583.

Low-Energy Intertidal

The papers listed below are studies in which sequences or partial sequences have been well documented. I have omitted papers that simply mention that the rocks are deposited in very shallow environments.

1. Predominately Muddy or Shaley Sequences

Aitken, J. D., 1966, Middle Cambrian to Middle Ordovician cyclic sedimentation, Southern Rocky Mountains of Alberta: Can. Petrol. Geol. Bull., v. 14, p. 405-441.

Macqueen, R. W. and E. W. Bamber, 1968, Stratigraphy and facies relationships of the Upper Mississippian Mount Head Formation, Rocky Mountains and Foothills, Southwestern Alberta: Can. Petrol. Geol. Bull., v. 16, p. 225-287.

Roehl, P. O., 1967, Stony Mountain (Ordovician) and Interlake (Silurian) Facies analogs of recent low-energy marine and subaerial carbonates, Bahamas: Amer. Assoc., Petrol. Geol. Bull., v. 51, p. 1979-2032.

Mukherji, K. K., 1969, Supratidal carbonate rocks in the Black River (Middle Ordovician) Group of Southwestern Ontario, Canada: Jour. Sed. Petrol., v. 39, p. 1530-1545.

Trettin, H.P., 1975, Investigations of Lower Paleozoic geology, Foxe Basin, Northeastern Melville Peninsula, and parts of Northeastern and Central Baffin Island: Geol. Survey Canada Bull. 251, 177 p.

Kahle, C.F. and J.C. Floyd, 1971, Stratigraphic and environmental significance of sedimentary structures in Cayugan (Silurian) Tidal Flat carbonates, Northwestern Ohio: Geol. Soc. America Bull. 82, p. 2071-2098.

Walker, K.R., 1972, Community ecology of the Middle Ordovician Black River Group of New York State: Geol. Soc. America Bull. 83, p. 2499-2524.

Assereto, R. and C.G.St. C., Kendall, 1971, Megapolygons in Ladinian limestones of Triassic of Southern Alps evidence of deformation by penecontemperaneous dessication and cementation: Jour. Sed. Petrology, 43, 715-723.

Reinhardt, J. and L.A., Hardie, 1976, Selected Examples of Carbonate Sedimentation, Lower Paleozoic of Maryland: Maryland Geological Survey, Guidebook No. 5, Baltimore Md., 53 p.

2. Grainy Sequences

Smith, D.L., 1972, Stratigraphy and Carbonate Petrology of the Mississippian Lodgepole Formation in Central Montana, *summarized in,* J.L., Wilson, 1975, Carbonate Facies in Geologic History, Springer-Verlag, p. 283-285.

Lohmann, K.C., 1976, Lower Dresbachian (Upper Cambrian) platform to deep-shelf transition in Eastern Nevada and Western Utah: an evaluation through lithologic cycle correlation, *in* R.A., Robison, and A.J. Rowell, (ed.), Paleontology and Depositional Environments: Cambrian of Western North America, Geological Studies, Brigham Young University, v. 23, p. 111-122.

3. Reef or Stromatolite-rich Sequences

Hoffman, P., 1976, Environmental diversity of middle Precambrian stromatolites: *in* M. R. Walter, Stromatolites: Elsevier, p. 599-613.

Donaldson, J. A., 1966, Marion Lake map area, Quebec-Newfoundland. Geol. Survey Canada Memoir 338, 85 p.

Havard, C. and A. Oldershaw, 1976, Early diagenesis in back-reef sedimentary cycles, Snipe Lake, reef complex, Alberta: Can. Petrol. Geol. Bull., v. 24, p. 27-70.

Mountjoy, E. W., 1975. Intertidal and supratidal deposits within isolated Upper Devonian Buildups: Alberta: *in* R. N. Ginsburg, ed., Tidal Deposits: Springer-Verlag, p. 387-397.

Read, J.F., 1973, Carbonate cycles, Pillara Formation (Devonian) Canning Basin, Western Australia: Can. Bull. Petrol. Geol. Bull., v. 21, p. 38-51.

4. Carbonate-Evaporite Sequences

Wilson, J. L., 1967, Carbonate-evaporite cycles in lower Duperow Formation of Williston Basin: Can. Petrol. Geol. Bull., v. 15, p. 230-312.

Schenk, P. E., 1969, Carbonate-sulfate-redbed facies and cyclic sedimentation of the Windsorian Stage (Middle Carboniferous), Maritime Provinces: Can. Jour. Earth Sci., v. 6, p. 1037-1066.

Fuller, J. G. C. M. and J. W. Porter, 1969, Evaporite Formations with petroleum reservoirs in Devonian and Mississippian of Alberta, Saskatchewan and North Dakota: Amer. Assoc. Petrol. Geol. Bull., v. 53, p. 909-927.

Shearman, D. J. and J. G. Fuller, 1969, Anhydrite diagenesis, calcitization, and organic laminites, Winnipegosis Formation, Middle Devonian, Saskatchewan: Can. Petrol. Geol. Bull., v. 17, p. 496-525.
    For an alternate interpretation see N. C. Wardlaw, and G. E. Reinson, 1971, Amer. Assoc. Petrol. Geol. Bull., v. 55, p. 1759-1787.

Mossop, G. D., 1973, Lower Ordovician evaporites of the Baumann Fiord Formation, Ellesmere Island: *in* Rept. of Activities, Part A, Geol. Survey Canada Paper 73-1, p. 264-267.

Mossop. G. D., 1973, Anhydrite-carbonate cycles of the Ordovician Baumann Fiord Formation, Ellesmere Island, Arctic Canada: A Geologic History: Univ. London, Ph.D. Thesis, 231 p.

Wood, G.V., and M.J. Wolfe, 1969, Sabkha cycles in the Arab-Darb Formation of the Trucial Coast of Arabia: Sedimentology, v. 12, p. 165-191.

Meissner, F.F., 1972, Cyclic sedimentation in Middle Permian strata of the Permian Basin, West Texas and New Mexico, *in* Elam, J.C. and S. Chuber, S. (eds.), Cyclic Sedimentation in the Permian Basin, 2nd edition, West Texas Geological Soc., Midland, Texas, p. 203-232.

High-Energy Intertidal

There appear to be few well described beach sequences in carbonate successions.

Purser, B. H., 1972, Subdivision et interpretation des sequences carbonates: Mém. Bur. Rech. Géol. Min., v. 77, p. 679-698.

Inden, R. F., 1974, Lithofacies and depositional model for a Trinity Cretaceous sequence: *in* B. F. Perkins, ed., Geoscience and Man, vol. VIII. Aspects of Trinity Geology, Louisiana State Univ., p. 37-53.

References Cited in Text

Aitken, J. D., 1967, Classification and environmental significance of cryptalgal limestones and dolomites, with illustrations from the Cambrian and Ordovician of southwestern Alberta: Jour. Sed. Petrology, v. 37, p. 1163-1178.

Dunham, R. J., 1969, Early vadose silt in Townsend mound (reef), New Mexico: *in* G. M. Friedman, ed., Depositional Environments in carbonate rocks: a symposium: Soc. Econ. Paleontol. Mineral., Spec. Publ. 14, p. 139-181.

Playford, P. E. and A. E. Cockbain, 1976, Modern algal stomatolites at Hamelin Pool, a hypersaline barred basin in Shark Bay, Western Australia: *in* M. R. Walter, ed., Stromatolites, Developments in Sedimentology: Elsevier 389-413.

James, N.P., 1972, Holocene and Pleistocene calcareous crust (caliche) profiles; criteria for subaerial exposure: Jour. Sed. Petrology, v. 42, p. 817-836.

Read, J.F., 1976, Calcretes and their distinction from stromatolites: *in* M.R. Walter, ed., Stromatolites, Developments in Sedimentology, No. 20: Elsevier, Amsterdam, p. 55-71.

Ms received May 3, 1977.
Revised January, 1979.
Reprinted from Geoscience Canada, Vol. 4, No. 3, p. 126-136.

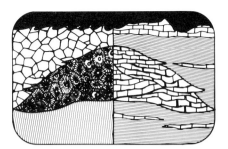

Facies Models 11. Reefs

Noel P. James
Department of Geology
Memorial University of Newfoundland
St. John's, Newfoundland A1B 3X5

Introduction

A reef, rising above the sea floor, is an entity of its own making – a sedimentary system within itself. The numerous, large calcium carbonate secreting organisms stand upon the remains of their ancestors and are surrounded and often buried by the skeletal remains of the many small organisms that once lived on, beneath and between them.

Because they are built by organisms, fossil reefs are storehouses of paleontological information and modern reefs are natural laboratories for the study of benthic marine ecology. This, together with the fact that fossil reefs buried in the subsurface contain a disproportionately large amount of our oil and gas reserves compared to other types of sedimentary deposits, has resulted in their being studied in detail by paleontologists and sedimentologists, perhaps more intensely than any other single sedimentary deposit, yet from two very different viewpoints. This article is an integration of these two viewpoints and I shall concentrate less on the trinity of back-reef, reef and fore-reef, familiar to most readers, and more on the complex facies of the reef proper.

The Organism/Sediment Mosaic

Reef facies are best differentiated on the basis of several independent criteria including: 1) the relationship between, and relative abundance of large skeletons and sediments, i.e., the type of reef limestone; 2) the diversity of reef building species; and 3) the growth form of the reef builders.

Types of reef limestone. The present state of any thriving reef is a delicate balance between the upward growth of large skeletal metazoans, the continuing destruction of these same organisms by a host of rasping, boring and grazing organisms, and the prolific sediment production by rapidly growing, short-lived, attached calcareous benthos (Fig. 1).

The large skeletal metazoans (e.g., corals) generally remain in place after death, except when they are so weakened by bioeroders that they are toppled by storms. The irregular shape and growth habit of these reef-builders results in the formation of roofed-over cavities inside the reef that may be inhabited by smaller, attached calcareous benthos, and may be partly to completely filled with fine-grained "internal" sediment. Encrusting organisms grow over dead surfaces and aid in stabilizing the structure. Branching reef-builders frequently remain in place but just as commonly are fragmented into sticks and rods by storms to form skeletal conglomerates around the reef.

Most reef sediment is produced by the post-mortem disintegration of organisms that are segmented (crinoids, calcareous green algae) or non-segmented (bivalves, brachiopods, foraminifers, etc.) and that grow in the many nooks and crannies between the larger skeletal metazoa. The remainder of the sediment is produced by the various taxa that erode the reef: boring organisms (worms, sponges, bivalves) produce lime mud; rasping organisms that

graze the surface of the reef (echinoids, fish) produce copious quantities of lime sand and silt. These sediments are deposited around the reefs as an apron of sediment and also filter into the growth cavities to form internal sediment, which is characteristically geopetal.

Many different classifications have been proposed for the resulting reef carbonates but the most descriptive and widely accepted is a modification of Dunham's (1960) classification of lime sand and mudrocks proposed by Embry and Klovan (1971) at the University of Calgary (Fig. 3). They recognize two kinds of reef limestone, allochthonous and autochthonous. The allochthonous limestones are the same as the finer grained sediments, but with two categories added to encompass large particles. If more than 10 per cent of the particles in the rock are larger than two mm and they are matrix supported it is a *Floatstone;* if the rock is clast supported it is a *Rudstone.* The autochthonous limestones are more interpretative; *Framestones* contain in place, massive fossils that formed the supporting framework; *Bindstones* contain in place, tabular or lamellar fossils that encrusted or bound the sediment together during deposition; *Bafflestones* contain in place, stalked fossils that trapped sediment by baffling.

Diversity amongst reef-building metazoans. Very diverse faunas, in terms of both growth form and taxa, occur when a community is well established and conditions for growth are optimum, i.e., nutrients are in good supply, chemical

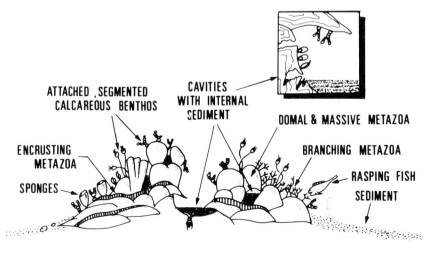

Figure 1
A sketch illustrating the different aspects of the organism/sediment mosaic that is a reef.

122

Figure 2
*A patch reef of Lower Cambrian age exposed
in sea cliffs along the northern shore of the
Straits of Belle Isle, Southern Labrador.*

BAFFLESTONE

BINDSTONE

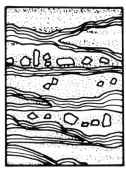

FRAMESTONE

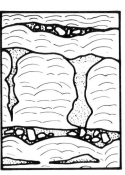

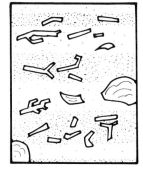

FLOATSTONE

RUDSTONE

REEF

LIMESTONE

Figure 3
*An interpretative sketch of the different types
of reef limestone recognized by Embry and
Klovan (1971).*

and physical stresses are low. In such optimum environments the division of biomass amongst various species is due mainly to complex biological controls.

In contrast, low diversity environments commonly fall into three general categories: 1) unpredictable environments; 2) new environments (fauna moving into a new environment); and 3) severe environments (high chemical and physical stress). Among the factors most likely to stress modern and fossil reef-building communities are: a) temperature and salinity fluctuations – most modern and likely most ancient reef-builders grow or grew best in tropical sea water of normal salinity; b) intense waves and swell – the skeletons of most reef-builders will be broken or toppled by strong wave surge; c) low light penetration – in modern reef-building organisms rapid calcification takes place because symbiots, which are light dependent, take over some of the bodily functions of the host; d) heavy sedimentation – all reef-builders are sedentary filter-feeders or micro-predators and water filled with fine-grained sediments would clog the feeding apparatus.

The growth form of reef-building metazoans The relationship between organism shape and environment is one of the oldest and most controversial topics in biology and paleobiology. In terms of reef-building metazoans, however, many observations of the interrelationship between organisms and surrounding sediments from the rock record combined with studies of modern coral distribution on tropical reefs allow us to make some generalizations about form and environment that are very useful in reef facies analysis (Table I).

The Model
The reef can generally be sudivided into three facies (Figs. 4, 5):

1) Reef-core facies – massive, unbedded, frequently nodular and lenticular carbonate comprising skeletons of reef-building organisms and a matrix of lime mud.

2) Reef-flank facies – bedded lime conglomerates and lime sands of reef-derived material dipping and thinning away from the core.

3) Inter-reef facies – normal shallow-water, subtidal limestones, unrelated to

Table I
The growth form of reef-building metazoa and the types of environments in which they most commonly occur.

Growth Form	Environment	
	Wave Energy	Sedimentation
Delicate, branching	low	high
Thin, delicate, plate-like	low	low
Globular, bulbous columnar	moderate	high
Robust, dendroid branching	moderate-high	moderate
Hemispherical, domal, irregular, massive	moderate-high	low
*Encrusting	intense	low
*Tabular	moderate	low

*Enscrusting vs. tabular forms are very difficult, if not often impossible to differentiate in the rock record, yet they indicate very different reef environments.

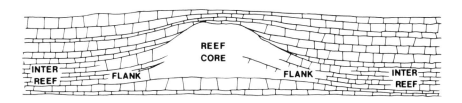

Figure 4
A sketch illustrating the three major reef facies in cross-section.

Figure 5
Reef and reef-flank deposits (R) ca. 100 meters thick (Peechee Formation) of Upper Devonian age in the Flathead Range, southern Rocky Mountains, Alberta (Photograph courtesy B. Pratt).

reef formation, or fine-grained siliciclastic sediments.

A useful, non-generic term for such a structure is "bioherm". For a thoughtful discussion of this and other reef terminology the interested reader is referred to essays by Dunham (1970) and Heckel (1974).

The key to understanding reef facies is unravelling the complex series of lithologies that comprise the reef core.

It has long been recognized that there is an ecological succession in many Paleozoic reefs (Lowenstam, 1959), i.e., the replacement of one community of reef-building organisms by another as the reef grew. A recent synthesis by Walker and Alberstadt (1975) of reefs ranging in age from Early Ordovician to Late Cretaceous suggests that a similar community succession is present in reefs throughout the Paleozoic and Mesozoic. Application of this concept to Oligocene reefs (Frost, 1977) which are dominated by scleractinian corals (the reef-builders in today's oceans) allows us now to equate ancient reef community succession with observations on modern reef communities with some measure of confidence.

In most cases four separate stages of reef growth can be recognized, and these stages along with the types of limestone, relative diversity of organisms and growth form of reef-builders in each, are summarized in Figure 6.

Pioneer (stabilization) stage. This first stage is most commonly a series of shoals or other accumulations of skeletal lime sand composed of pelmatozoan or echinoderm debris in the Paleozoic and Mesozoic, and plates of calcareous green algae in the Cenozoic. The surfaces of these sediment piles are colonized by algae (calcareous green), plants (sea grasses) and/or animals (pelmatozoans) that send down roots or holdfasts to bind and stabilize the substrate. Once stabilized, scattered branching algae, bryozoans, corals, soft sponges and other metazoans begin to grow between the stabilizers.

Colonization stage. This unit is relatively thin when compared to the reef structure as a whole, and reflects the initial colonization by reef-building metazoans. The rock is generally characterized by few species, sometimes massive or lamellar forms but more commonly

124

STAGE	TYPE OF LIMESTONE	SPECIES DIVERSITY	SHAPE OF REEF BUILDERS
DOMINATION	bindstone to framestone	low to moderate	laminate encrusting
DIVERSIFICATION	framestone (bindstone) mudstone to wackestone matrix	high	domal massive lamellar branching encrusting
COLONIZATION	bafflestone to floatstone (bindstone) with a mudstone to wackestone matrix	low	branching lamellar encrusting
STABILIZATION	grainstone to rudstone (packstone to wackestone)	low	skeletal debris

Figure 6
A sketch of the four divisions of the reef-core facies with a tabulation of the most common types of limestone, relative species diversity and shape of reef-builders found in each stage.

Figure 7
*A small patch of domal shaped corals (*Diploria *sp.) in cross-section on a cliff exposure of Late Pleistocene reef limestone, Barbados, W. I.*

thickets of branching forms, often monospecific (Fig. 8). In Cenozoic reefs the one characteristic common to all corals in this stage of reef growth is that they are able to get rid of sediment and clean their polyps, and so are able to grow in areas of high sedimentation. The branching growth form creates many smaller subenvironments or niches in which numerous other attached and encrusting organisms can

live – forming the first stage of the reef ecosystem. Stromatactis (cavity filling of laminated fibrous calcite and sediment) is common in rocks representing this stage.

Diversification stage. This stage usually provides the bulk of the reef mass and is the point at which most pronounced upward-building towards sea level occurs and easily definable, lateral facies

develop. The number of major reef-building taxa is usually more than doubled, and the greatest variety in growth habit is encountered (Fig. 15). With this increase in form and diversity of framework and binding taxa, comes increased nestling space, i.e., surfaces, cavities, nooks and crannies, leading to an increase in diversity of debris-producing organisms.

Domination stage. The change to this stage of reef growth is commonly abrupt. The most common lithology is a limestone dominated by only a few taxa with only one growth habit, generally encrusted to laminated. Most reefs show the effects of surf at this stage, in the form of beds of rudstone.

The reason for this ecologic succession is at present a topic of much debate. Some workers feel that the control is extrinsic and reflects a progressive replacement of deep-water communities by shallower water ones as the reef grows to sea level and into more turbulent water – yet there is often abundant evidence that the first two stages are developed in shallow water. Other workers feel that the control is intrinsic and reflects a natural succession as the organisms gradually alter the substratum and change the energy flow pathways as the community develops – yet there is abundant evidence of increasing water turbulence as the structure grows.

Superimposed reefs. Reef structures in the rock record are often impressive because of their size, not only laterally but vertically. Careful examination of stratigraphically thick reefs, however, often reveals that they are not a single structure, but a series of superimposed or stacked reefs that grew on top of one another in more or less the same place. Individual episodes of reef growth are commonly separated by periods of exposure, reflected in the rock by intensive diagenesis, calcrete horizons, or shales (paleosols). When the ocean floods one of these surfaces that has been exposed, reef growth begins at the diversification stage because there is already a hard, often elevated, substrate present.

Figure 8
An accumulation of branching corals (Porites
porites) *and bivalves in the colonization stage
of a late Pleistocene reef, Barbados, W. I.*

The Model as a Framework or Guide

The reef facies model is predicated on
the assumption that a full spectrum of
reef-building organisms are present, as
we see in the tropical oceans today, but
such was not the case for much of the
Phanerozoic. The critical element that is
often missing, and without which the four
stages of development in the reef core
cannot occur, is the presence of skeletal
metazoa that secrete large robust,
branching, hemispherical or tabular
skeletons. Without them the reef cannot
exist in the zone of constant turbulence,
usually wave induced, because smaller
and more delicate forms would be
broken and swept away (unless submarine cementation is very rapid, pervasive
and near-surface). This zone of turbulence is the optimum area for growth and
diversity because sediment is constantly removed, water is clear and nutrients
are constantly swept past the sessile
organisms. Such large skeletal metazoa
were, however, present only at certain
times during the Phanerozoic (Fig. 9),
and each period has its own specialized
group of frame-builders: a) Middle and
Upper Ordovician - bryozoa, stromatoporoids, tabulate corals; b) Silurian and
Devonian - stromatoporoids, tabulate
corals; c) Late Triassic - corals, stromatoporoids; d) Late Jurassic - corals,

stromatoporoids; e) Upper Cretaceous -
rudist bivalves; f) Oligocene, Miocene
(?), Plio-Pleistocene - scleractinian corals.

What then of the rest of the Phanerozoic record - were there no reefs? While
there were certainly periods when no
reefs at all formed, these periods were
generally short and represent either
climatic/tectonic crises or the complete
lack of any reef builders, even small
ones (e.g., Middle and Upper Cambrian).
During most of the Phanerozoic there
were structures that some workers call
reefs, some call mounds, some call
banks: they lack many of the characteristics we ascribe to reefs, yet were
clearly rich in skeletal organisms and
had relief above the sea floor. The origin
of these structures, which I have called
reef mounds, has probably caused more
discussion than any other topic in the
literature on reefs (Heckel, 1974). When
viewed against the backdrop of the
general reef facies model, however, I
think of them as half-reefs or incomplete
reefs because they represent only
stages one and two of the model. These
structures did not develop the other
upper two stages either because the
environment was not condusive to the
growth of large skeletal metazoa or
because these larger metazoa simply

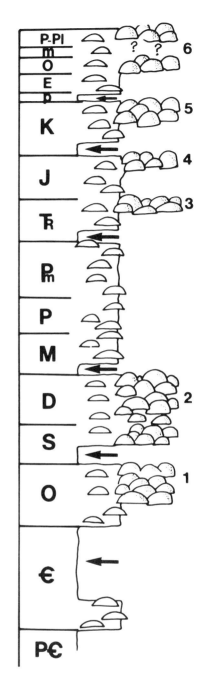

△ **Reef Mound**

Reef

Figure 9
*An idealized stratigraphic column represent-
ing the Phanerozoic and illustrating times
when there were no reefs (arrows), times
when there were only reef mounds and times
when there were both reefs and reef mounds;
numbers indicate different associations of
reef-building taxa discussed in the text.*

did not exist at the time when the structure formed.

Reef mounds are, as the name suggests, flat lenses to steep conical piles with slopes up to 40 degrees consisting of poorly sorted bioclastic lime mud with minor amounts of organic boundstone. With this composition they clearly formed in quiet water environments and from the rock record appear to occur in three preferred locations: 1) arranged just downslope on gently-dipping platform margins (Fig. 13); 2) in deep basins; and 3) spread widely in tranquil reef lagoons or wide shelf areas. When viewed in section, reef mounds display a similar facies sequence in each case (Wilson, 1975) (Fig.11).

Reef Mound Core Facies.
Stage 1: Basal bioclastic lime mudstone to wakestone pile - very muddy sediment with much bioclastic debris but no baffling or binding organisms.
Stage 2: Lime mudstone or bafflestone core - the thickest part of the mound, consisting of delicate to dendroid forms with upright growth habits in a lime mudstone matrix. The limestone is frequently brecciated, suggesting partial early lithification, dewatering and slumping, and contains stromatactis. Each geologic age has its own special fauna that forms this stage: a) Lower Cambrian - archaeocyathids; b) Lower Ordovician - sponges and algae; c) Middle Ordovician, Late Ordovician, Silurian, Early Carboniferous (Mississippian) - bryozoa; d) Late Carboniferous (Pennsylvanian) and Early Permian - platy algae; e) Late Triassic - large fasciculate dendroid corals; f) Late Jurassic - lithistid sponges; g) Cretaceous - rudist bivalves.
Stage 3: Mound cap - a thin layer of encrusting or lamellar forms, occasional domal or hemispherical forms, or winnowed lime sands.

Reef Mound Flank Facies.
These massive, commonly well-bedded carbonates comprise extensive accumulations of archaeocyathid, pelmatozoan, fenestrate bryozoan, small rudist, dendroid coral, stromatoporoid, branching red algae or tabular foraminifer debris and chunks of wholly to partly lithified lime mudstone. Volumetrically these flank beds may be greater than the core itself and almost bury it (Fig. 16).

Figure 10
A small patch reef built by bryozoa, corals, and stromatoporoids, Long Point Formation (Middle Ordovician), Port-au-Port Peninsula, Newfoundland.

REEF MOUND

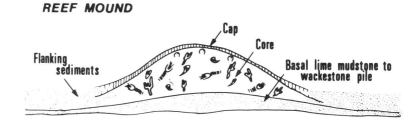

Figure 11
Cross-section through a hypothetical reef mound illustrating the geometry of the different facies.

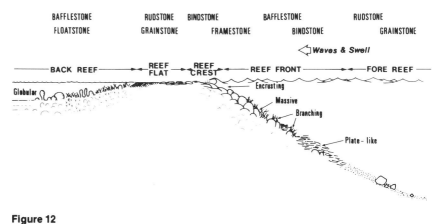

Figure 12
Cross-section through a hypothetical, zoned, marginal reef illustrating the different reef zones, spectrum of different limestones produced in each zone and environment of different reef-building forms.

Figure 13
Massive reef limestone (right) of the Nansen Formation (Permo-Pennsylvanian) extending downward and basinward into dark, argillaceous limestones of the Hare Fiord Formation

(left); arrows point out small reef mounds developed on the seaward slopes of the reef front, western side of Blind Fiord, Ellesmere Island, N. W. T.

PATCH REEFS

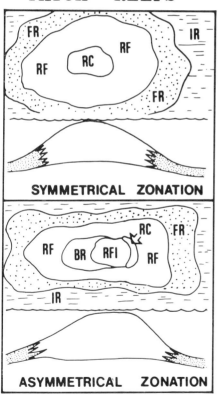

BARRIER REEF

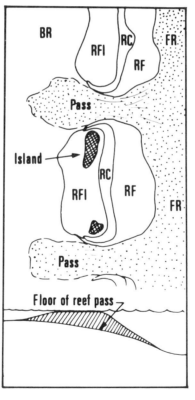

Figure 14
A sketch of different patch reef types and a barrier reef, both in plan view and in section; IR = inter-reef; FR = fore-reef (reef flank in symmetrically zoned reefs); RF = reef front (reef margin or fore-reef of some workers); RC = reef crest; RFl = reef flat; BR = back-reef.

Although the origin of most of these mounds can clearly be related to some combination of baffling and encrusting by organisms, localized prolific production of carbonate sediment, and possible shaping by currents and storms, those found in rocks of Mississippian age (Tournaisian-Visean) are particularly puzzling. Commonly called Waulsortian mud mounds (from the name of a village in Belgium) these structures are just as large as most reef mounds and have sides as steep but possess *no* major large organisms, only crinoids and bryozoa as tiny fragments which make up no more than 20 per cent of the rock – the rest is lime mud.

In summary (Fig. 9), there are times when the model is inapplicable because there are no reefs at all, there are times when only reef mounds form, and there are times when both reef mounds and reefs occur, but in different environments.

The Model as a Basis for Hydrodynamic Interpretation

Once a reef has reached the colonization stage, and especially the diversification stage, the structure is frequently high enough above the surrounding sea floor to affect water circulation and thus to alter sedimentation patterns. At this point not only are the surrounding sedimentary environments altered but the reef itself develops a zonation of different organism/sediment associations, because its margins reach from shallow to deep water.

Modern reefs are best developed and most successful on the windward sides of shelves, islands and platforms where wind and swell are consistent and onshore. The asymmetry of many ancient reefs and distribution of sediment facies suggests that this was so in the past as well. The reason for the preferential development of reefs on the windward side is by no means established but sedimentation is likely the most important. Shallow water reef-building species characteristically produce abundant fine sediment, yet the major reef-builders, because they are filter feeders and micropredators, are intolerant of fine sediment. The open ocean, windward locations are the only ones in which fine sediment is continuously swept away.

Growth of reefs into the zone of onshore waves and swell forms a natural

Figure 15
The diversification stage of an Upper Devonian reef, comprising domal stromatoporoids, and domal to branching tabulate corals, Blue Fiord Formation, south side of Eids Fiord, Ellesmere Island, N. W. T.

Figure 16
Flank beds (above helicopter) dipping off a reef mound (R) of Permo-Pennsylvanian age (Nansen Formation), eastern side Blind Fiord, Ellesmere Island, N. W. T.

breakwater and so creates a relatively quiet environment in the lee of the reef. Frequently, this restriction significantly changes water circulation on the shelf, platform or lagoon behind the reef. In such a marginal location, the symmetrical reef facies model comprising a reef-core facies surrounded on all sides by reef-flank facies is no longer discernable. Instead facies are more asymmetrically distributed with the reef-core facies flanked on the windward side by the fore-reef facies and on the leeward side by the platform facies (often called the back-reef facies).

Reef Core Facies. The massive bedded limestones of the reef core commonly illustrate several different lithologies which develop in one of the following four zones (Fig. 12).

Reef crest zone: This is the highest part of the reef at any stage in its growth, and if in shallow water, it is that part of the reef top that receives most of the wind and wave energy. The composition of the reef crest depends upon the degree of wind strength and swell. In areas where wind and swell are intense only those organisms that can encrust, generally in sheet-like forms, are able to survive. When wave and swell intensity are only moderate to strong, encrusting forms still dominate but are commonly also bladed or possess short, stubby branches. In localities where wave energy is moderate, hemispherical to massive forms occur with scattered clumps of branching reef-builders, although the community is still of low diversity. The lithologies formed in these three cases would range from bindstones to framestones.

Reef front zone: This zone extends from the surf zone to an indeterminate depth, commonly less than 100 metres, where the zone of abundant skeletal growth grades into sediments of the fore-reef zone. Direct analogy between modern reefs, especially Caribbean reefs, and ancient reefs is difficult because today the sea floor from the surf zone to a depth of 12 metres or so is commonly dominated by the robust branching form *Acropora palmata* a species which developed only recently, in the late Pleistocene. Such branching forms are rarely found in ancient reefs and instead the most abundant forms are massive, laminar to hemispherical skeletons, forming framestones and sometimes bindstones.

The main part of this zone supports a diverse fauna with reef-builders ranging in shape from hemispherical to branching to columnar to dendroid to sheet-like. Accessory organisms and various niche dwellers such as brachiopods, bivalves, coralline algae, crinoids and green segmented calcareous algae (Halimeda), are common. On modern reefs where the reef-builders are corals this zone commonly extends to a depth of 30 metres or so. The most common rock type formed in this zone would still be framestone but the variety of growth forms also leads to the formation of many bindstones and bafflestones as well.

Below 30 metres or so wave intensity is almost non-existent and light is very attenuated. The response of many reef-building metazoans is to increase their surface area, by having only a small basal attachment and a large, but delicate, plate-like shape. Rock types from this zone look like bindstones, but binding plays no role in the formation of these rocks and perhaps another term is needed.

The deepest zone of growth of coral and green calcareous algae on modern coral reefs is around 70 metres. The lower limit may depend upon many factors, perhaps one of the most important being sedimentation, especially in shale basins which border many reefs, so that this lower limit should be used with caution in the interpretation of fossil reefs.

Sediments on the reef front are of two types: 1) internal sediments within the reef structure, generally lime mud giving the rocks a lime mudstone to wackestone matrix; 2) coarse sands and gravels in channels running seaward between the reefs. These latter deposits have rarely been recognized in ancient reefs.

As a result of numerous observations on modern reefs it appears that most of the sediment generated on the upper part of the reef front and on the reef crest is transported episodically by storms up and over the top and accumulates in the lee of the reef crest. Sediments on the intermediate and lower regions of the reef front, however, are transported down to the fore-reef zone. Shallow-water material is contributed to the fore-reef zone only when it is channelled by way of passes through the reef.

Reef flat zone: The reef flat varies from a pavement of cemented, large skeletal debris with scattered rubble and coral-line algae nodules in areas of intense waves and swell, to shoals of well-washed lime sand in areas of moderate wave energy. Sand shoals may also be present in the lee of the reef pavement. Vagaries of wave refraction may sweep the sands into cays and islands. These obstructions in turn create small protected environments very near the reef crest. Water over this zone is shallow (only a few metres deep at most) and scattered clumps of reef-building metazoans are common. The resulting rock types range from clean skeletal lime grainstones to rudstones.

Back reef zone: In the lee of the reef flat conditions are relatively tranquil and much of the mud formed on the reef front comes out of suspension. This, coupled with the prolific growth of mud and sand-producing bottom fauna such as crinoids, calcareous green algae, brachiopods, ostracodes, etc., commonly results in mud-rich lithologies. The two most common growth habits of reef-builders in these environments are stubby, dendroid forms, often bushy and knobby, and/or large globular forms that extend above the substrate to withstand both frequent agitation and quiet muddy periods.

The rock types characteristic of this environment are bafflestones or float-stones to occasional framestones with a skeletal wackestone to packstone matrix. In some reefs there are beds of nothing but disarticulated branches in lime mud (e.g., Amphipora limestone of the Upper Devonian), but there is little evidence of much transport.

Fore-Reef Facies. This facies consists of thin to thick and massively bedded skeletal lime grainstones to lime packstones which are composed of whole or fragmented skeletal debris, blocks of reef limestone and skeletons of reef-builders, and which grade basinward into shales or lime muds. In contrast to the reef facies these beds are rarely dolomitized.

Platform Facies. The most abundant limestones are thin-bedded, skeletal-rich, often bioturbated, lime wackestones to packstones. Evaporites are commonly interbedded with carbonates if the reef has severely restricted water circulation.

The Model as a Predictor
If we know the age of a sequence of carbonate rocks and we have some idea of the paleotectonic setting then we can predict, from limited data, the types of reefs we might expect to be present in a shelf or platform setting.

Times When a Complete Spectrum of Reef Builders was Present. The edge of the shelf or platform is occupied by a marginal reef (Fig. 14). The reef is well-zoned if the front is steep and wave action intense, but zonation is weak if the front slopes gradually seaward and the seas are relatively quiet. The linear reef is cut by passes through which platform sediments are funnelled basinward.

Patch reefs on the platform in the lee of the barrier reef range from circular to elliptical to irregular in plan and are sometimes large enough to enclose a lagoon themselves. Each reef is zoned with respect to depth. If the wind waves on the platform are small and conditions are tranquil the zonation is symmetrical; if the wind waves are strong, zonation is asymmetrical and resembles that of a barrier reef (Fig. 14).

Reef mounds occur on the inner, shallow parts of the platform, in areas of normal salinity but turbid water. Reef mounds also occur at depth, in front of the barrier reef down on the reef front or fore-reef.

Patch reefs or reef mounds commonly form a very widespread lithofacies compared to the barrier reef. The stratigraphic thickness of these reefs is dependent upon the rate of subsidence: if subsidence rate is low, reefs are thin; if subsidence rate is high, reefs may be spectacular in their thickness.

Times When Only Delicate, Branching and Encrusting Metazoa Prevail. The margin of the shelf or platform is normally a complex of oolitic or skeletal (generally crinoidal) sand shoals and islands. The only reef structures are reef mounds which occur below the zone of active waves down on the seaward slopes of the shelf or platform, and if conditions are relatively tranquil behind the barrier, on the shelf itself. Mounds may display a zonation, with ocean-facing sides in shallow water armoured with accumulations of fragmented and winnowed skeletal debris.

130

Stromatolite Reefs. During the Precambrian and earliest Paleozoic, prior to the appearance of herbivorous metazoa, stromatolites formed impressive structures. These stromatolite complexes clearly had relief above the sea floor and in terms of morphology were surprisingly similar to later skeletal reefs. One such impressive variety of stromatolites and associated sediments in a platform- to-basin transition has been documented by Hoffman (1974) from the Pethei Group of Lower Proterozoic age (ca. 1,800 Ma) at Great Slave Lake, N.W.T.

In this sequence the edge of the platform is generally occupied by a narrow zone, the "mound and channel belt" similar to barrier reefs that would form later. Large elongate stromatolites with up to three metres relief are separated by channels filled with crossbedded and megarippled coarse-grained sands and conglomerates composed of stromatolite fragments and clasts of oolitic grainstone. This belt separates platform facies (a complex array of laminar to columnar stromatolites and intervening ooid, intraclast and oncolitic limestones) from a slope-to-basin facies (lime mudstone-shale rhythmites with bedded slump breccias and siliclastic mudstones containing poorly laminated small calcareous columnar stromatolites).

Summary
The purpose of this article has been to marry the sedimentological and paleontological approaches to the study of reefs into a single facies model, useful to both disciplines. This model is an integration of data from two very different sources: from the modern sea floor, predominantly in the horizontal dimension; and from the rock record, predominantly in the vertical dimension as recorded in mountain exposures, quarries and drill core.

To alter and refine this model more information is needed from two areas. We must learn more about the succession of organisms and sediments that underlie the living surface of modern reefs, by drilling into these reefs. We must learn more about reefs from those parts of the stratigraphic record where reefs are known to occur, but have been little studied - the Precambrian, the Lower Paleozoic and Cenozoic.

The trend in the past has been to compare specific fossil reefs with modern reefs. This comparative approach should now be used on fossil reefs, to compare and contrast the sedimentology and paleoecology of reefs formed by different groups of organisms at different times in geologic history.

Acknowledgements
Bob Stevens and David Kobluk kindly read earlier versions of this manuscript and offered *many* helpful suggestions for its improvement.

Bibliography

Examples of Fossil Reefs
The following papers are examples of different reef types from the fossil record.

Precambrian
Hoffman, P., 1974, Shallow and deepwater stromatolites in lower Proterozoic platform-to-basin facies change, Great Slave Lake, Canada: Amer. Assoc. Petrol. Geol., Bull., v. 58, p. 856-867.

Aitken, J.D., 1977, New data on correlation of the Little Dal Formation and a revision of the Proterozoic map-unit "H-5": Geol. Survey Canada Paper 77-1A, p. 131-135.

Cambrian
James, N.P. and D.R. Kobluk, 1978, Lower Cambrian patch reefs and associated sediments, southern Labrador, Canada: Sedimentology, v. 25, p. 1-32.

Ordovician
Pitcher, M., 1961, Evolution of Chazyan (Ordovician) reefs of eastern United States and Canada: Can. Petrol. Geol. Bull., v. 12, p. 632-691.

Toomey, D.F. and R.M. Finks, 1969, The paleoecology of Chazyan (lower Middle Ordovician) "reefs" or "mounds" and Middle Ordovician (Chazyan) mounds, southern Quebec, Canada: New York Assoc. Guidebook to Field Excursions, College Arts, Sciences, Plattsburg, N.Y., a summary report, p. 121-134.

Toomey, D.F., 1970, An unhurried look at a Lower Ordovician mound horizon, Southern Franklin Mountains, West Texas: Jour. Sed. Petrology, v. 40, p. 1318-1335.

Ross, R.J., V. Jaanuson, and I. Friedman, 1975, Lithology and origin of Middle Ordovician Calcareous mudmound at Meiklejohn Peak, Southern Nevada. U. S. Geol. Survey Prof. Paper No. 871, 45p.

Silurian
Lowenstam, H.A., 1950, Niagaran reefs in the Great Lakes area: Jour. Geology, v. 58, p. 430-487.

Scoffin, T.P., 1971, The conditions of growth of the Wenlock reefs of Shropshire, England: Sedimentology, v. 17, p. 173-219.

Manten, A.A., 1971, Silurian reefs of Gotland: Developments in Sedimentology, No. 13, Elsevier, Amsterdam, 539 p.

Heckel, P.H. and D. O'Brien, eds., 1975, Silurian reefs of Great Lakes Region of North America: Amer. Assoc. Petrol. Geol. Reprint Ser. No. 14, 243 p.

Fisher, J.H., ed., 1977, Reefs and Evaporites - Concepts and Depositional Models: Amer. Assoc. Petrol. Geol. Studies in Geology No. 5, 196 p.

Shaver, R.H., *et al*, 1978, The search for a Silurian reef model: Great Lakes Area: Spec. Rept. No. 15, Indiana Geol. Survey, 36 p.

Devonian
Davies, G.R., ed., 1975, Devonian Reef Complexes of Canada, I and II: Can. Soc. Petrol. Geol. Reprint Ser., No. 1, 229 p. and 246 p.

Krebs, W., 1971, Devonian reef limestones in the eastern Rhenish Shiefergebirge, *in* G. Muller, ed., Sedimentology of Parts of Central Europe, Guidebook, 8th Internatl. Sed. Congress, Heidelberg, p. 45-81.

Krebs, W. and E.W. Mountjoy, 1972, Comparison of central European and western Canadian Devonian reef complexes: 24th Internatl. Geol. Cong., sect. 6, p. 294-309.

Playford, P.E., and D.C. Lowry, 1966, Devonian reef complexes of the Canning Basin, Western Australia: Geol. Survey Western Australia, Bull. 118, 50 p.

Elloy, R., 1972, Réflexions sur quelques environments récifaux du paléozoique: Bull. Centre Rech. Pau-SNPA, v. 6, p. 1-105.

Jansa, L.F. and N.R. Fischbuch, 1974, Evolution of a Middle and Upper Devonian sequence from a clastic coastal plain-deltaic complex into overlying carbonate reef complex and banks, Sturgeon-Mitsue area, Alberta: Geol. Bull., v. 58, p. 787-799.

Klovan, J.E., 1974, Development of western Canadian Devonian reefs and comparison with Holocene analogues: Amer. Assoc. Petrol. Geol. Bull., v. 58, p. 787-799.

Mississippian
Lees, A., 1964, The structure and origin of the Waulsortian (Lower Carboniferous) "Reefs" of west-central Eire: Phil Trans. Roy. Soc. London, Ser. B. No. 740, p. 485-531.

Cotter, E., 1965, Waulsortian-type carbonate banks in the Mississippian Lodgepole Formation of central Montana: Jour. Geol., v. 73, p. 881-888.

Pennsylvanian

Heckel, P.H. and J.M. Cocke, 1969, Phylloid algal mound complexes in out-cropping Upper Pennsylvanian rocks of mid-continent: Amer. Assoc. Petrol. Geol. Bull., v. 53, p. 1085-1074.

Toomey, D.F. and H.D. Winland, 1973, Rock and biotic facies associated with a Middle Pennsylvanian (Desmoinesian) algal buildup, Neca Lucia Field, Nolan County, Texas: Amer. Assoc. Petrol. Geol. Bull., v. 57, p. 1053-1074.

Toomey, D.F., J.L. Wilson, and R. Rezak, 1977, Evolution of Yucca Mound Complex, Late Pennsylvanian Phylloid-algal buildup, Sacremento Mountains, New Mexico: Amer. Assoc. Petroleum Geologists Bull., v. 61, p. 2115-2135.

Permian

Newell, N.D., et al., 1953, The Permian Reef Complex of the Guadalupe Mountains Region Texas and New Mexico: San Francisco, Freeman and Co., 236 p.

Malek-Aslani, M., 1970, Lower Wolfcamp reef in Kemnitz Field, Lea County, New Mexico: Amer. Assoc. Petrol. Geol. Bull., v. 54, p. 2317-2335.

Davies, G.R., 1970, A Permian hydrozoan mound, Yukon Territory: Can. Jour. Earth Sci., v. 8, p. 973-988.

Dunham, R.J., 1972, Guide for study and discussion for individual reinterpretation of the sedimentation and diagenesis of the Permian Capitan Geologic Reef and associated rocks, New Mexico and Texas: Soc. Econ. Paleo. and Min. Permian Basin Section Publ. 72-14, 235 p.

Hileman, M.E. and S.J. Mazzulo, 1977, Upper Guadalupian Facies, Permian Reef Complex, Guadalupe Mountains New Mexico and Texas: Soc. Econ. Paleo. and Min. Permian Basin Section, Publ. 77-16, 508 p.

Triassic

Leonardi, P., 1967, Le Dolomiti: Geologic dei monti tra Isarco e Piave: v. 1 and 2, 1010 p. Nat. Res. Council, Rome.

Zankl, H., 1971, Upper Triassic carbonate facies in the Northern Limestone Alps, in G. Muller, ed., Sedimentology of Parts of Central Europe, Guidebook 8th Internatl. Sed. Cong. Heidelberg, p. 147-185.

Bosellini, A. and D. Rossi, 1974, Triassic Carbonate buildups of the Dolomites, Northern Italy: Soc. Econ. Paleontol. and Mineral. Spec. Publ. 18, p. 209-233.

Jurassic

Rutten, M.D., 1956, The Jurassic reefs on the Yonne (southeastern Paris Basin): Amer. Jour. Sci., v. 254, p. 363-371.

Gwinner, M.P., 1968, Carbonate rocks of the Upper Jurassic in S. W. Germany in G. Muller, ed., Sedimentology of Parts of Central Europe, 8th Internatl. Sedimentological Cong., Heidelberg, p. 193-207.

Eliuk, L.S., 1979, Abenaki Formation, Nova Scotia Shelf Canada: a depositional and diagenetic model for Mesozoic Carbonate Platform: Can. Petrol. Geology Bull., v. 24 (in prep.).

Cretaceous

Kauffman, E.G. and N.F. Sohl, 1974, Structure and Evolution of Antillean Cretaceous Rudist Frameworks: Verhandl. Naturf. Ges. Basel, v. 84, p. 399-467.

Perkins, B.F., 1974, Paleoecology of a rudist reef complex in the Comanche Cretaceous Glen Rose Limestone of Central Texas: in B.F. Perkins, ed., Aspects of Trinity Division Geology, Geoscience and Man VIII, Louisiana State University, p. 131-173.

Enos, P., 1974, Reefs, platforms, and basins of Middle Cretaceous in northeast Mexico: Amer. Assoc. Petrol. Geol. Bull., v. 58, p. 800-809.

Bebout, D.G., and R.G. Loucks, 1974, Stuart City Trend, Lower Cretaceous, South Texas: Beaueau Econ. Geol. Rept. No. 78, Austin Texas, 80 p.

Philip, J., 1972, Paleoecologie des formations a rudistes du Cretace Superior - l'example du sud-est de la France: Paleogogr., Paleoclimat. Paleoecol., v. 12, p. 205-222.

Cenozoic

Forman, M.J. and S.O. Schlanger, 1957, Tertiary reefs and associated limestone facies from Louisiana and Guam: Jour. Geology, v. 65, p. 611-627.

Frost, S.H., 1977, Cenozoic reef systems of Caribbean - prospects for paleoecological synthesis: in Studies in Geology: Amer. Assoc. Petrol. Geol., No. 4, p. 93-110.

Pleistocene

James, N.P., C.S. Stearn, and R.S. Harrison, 1977, Field guidebook to modern and Pleistocene reef carbonates, Barbados, W. I.: 3rd Internatl. Coral Reef Symp., Univ. of Miami, Fisher Island, Miami, Florida, 30 p.

Mesolella, K.J., H.A. Sealy, and R.K. Matthews, 1970, Facies geometries within Pleistocene reefs of Barbados, W. I.: Amer. Assoc. Petrol. Geol. Bull., v. 54, p. 1899-1917.

Stanley, S.M., 1966, Paleoecology and diagenesis of Key Largo limestone, Florida: Amer. Assoc. Petrol. Geol. Bull., v. 50, p. 1927-1947.

General Articles Reviewing Fossil Reefs

Wilson, J. L., 1975, Carbonate Facies in Geologic history: Springer-Verlag, N.Y., Heidelberg, Berlin, 471 p.
The best overall reference, especially chapters II, IV to VIII, XI and XIII.

Dunham, R. J., 1970, Stratigraphic reefs versus ecologic reefs: Amer. Assoc. Petrol. Geol. Bull., v. 54, p. 1931-1932.
A thoughtful essay on what is, and is not, a reef - from the point of view of a sedimentologist.

Heckel, P. H., 1974, Carbonate buildups in the geologic record: a review: in L. F. Laporte, ed., Reefs in Time and Space: Soc. Econ. Paleont. Mineral Spec. Publ. 18, p. 90-155.
A detailed account of reefs or buildups, their classification, zonation with geologic time and general models for reef formation - more than you ever wanted to know about fossil reefs and an excellent set of references.

Laporte, L. F., ed., 1974, Reefs in Time and Space: Soc. Econ. Paleont. Mineral Spec. Publ. 18, 256 p.
Seven papers dealing with various aspects of modern and fossil reefs.

Walker, K. R. and L. P. Alberstadt, 1975, Ecological succession as an aspect of structure in fossil communities: Paleobiol., v. 1, p. 238-257.
The first half of this paper outlines succinctly, the main theme of ecological succession in fossil reefs.

Copper, P., 1974, Structure and development of early Paleozoic reefs: Proc. Second Internatl. Coral Reef Symp., v. 6, p. 365-386.
An essay on the different types of Paleozoic reefs.

General Articles Reviewing Modern Reefs

Bathurst, R. G. C., 1975, Carbonate Sediments and Their Diagenesis: Amsterdam, Elsevier Publ. Co., 658 p.
Chapter 3 and 4 nicely summarize specific modern reef environments.

Darwin, C., 1842, Structure and Distribution of Coral Reefs: reprinted by Univ. of California Press, from 1851 edition, 214 p.
The beginning of modern coral reef research by a young biologist.

Frost, S. H., M. P. Wiss, and J. B. Saunders, 1977, Reefs and related carbonates - ecology and sedimentology: in Studies In Geology: Amer. Assoc. Petrol. Geol., no. 4, 421 p.
Twenty-eight papers with copious references, mainly on modern reefs.

Ginsburg, R. N. and N. P. James, 1974, Spectrum of Holocene reef-building communities in the western Atlantic: in A. M. Ziegler et al., eds., Principles of Benthic Community Analysis (notes for a short course): Univ. of Miami, Fisher Island Station, p. 7.1 – 7.22.
A succinct summary of the major reef types of the Atlantic-Caribbean area.

Stoddart, D. R., 1969, Ecology and morphology of recent coral reefs: Biol. Rev., v. 44, p. 433-498.
A superb summary of modern coral reefs – an excellent place to start.

Taylor, D. E., ed., 1977, Proceedings of Third International Coral Reef Symposium, Miami Florida. 2. Geology: Fisher Island Station, Miami Beach, Florida, 628 p.
Many short papers and guideboods with up to date research on reefs.

Different Types of Modern Reefs

Pacific Atolls
Emery, K. O., J. I. Tracey, and H. S. Ladd, 1954, Geology of Bikini and nearby atolls: U.S. Geol. Survey, Prof. Paper 260-A, 265 p.

Marginal Reefs
Goreau, T. F., 1959, The ecology of Jamaican coral reefs. I. Species, composition and zonation: Ecol., v. 40, p. 67-90.

Goreau, T. F. and N. I. Goreau, 1973, The ecology of Jamaican coral reefs. II. Geomorphology, zonation and sedimentary phases: Bull. Mar. Sci., v. 23, p. 399-464.

James, N. P., R. N. Ginsburg, D. S. Marszalek, and P. W. Choquette, 1976, Facies and fabric specificity of early subsea cementation in shallow Belize (Br. Honduras) reefs: Jour. Sed. Petrol., v. 46, p. 523-544.

Maxwell, W. G. H., 1968, Atlas of the Great Barrier Reef: Amsterdam, Elsevier Publ. Co., 258 p.

Pinnacle Reefs
Korniker, L. A. and D. W. Boyd, 1962, Shallow water geology and environment of Alacran reef complex, Campeche Bank, Mexico: Amer. Assoc. Petrol. Geol. Bul., v. 46, p. 640-673.

Logan, B. W., J. I. Garding, W. M. Ahr, J. D. Williams, and R. G. Snead, 1969, Carbonate sediments and reefs, Yucatan shelf, Mexico: Amer. Assoc. Petrol. Geol. Mem. 11, p. 1-196.

Patch Reefs
Garrett, P., D. L. Smith, A. O. Wilson and D. Patriquin, 1971, Physiography, ecology and sediments of two Bermuda patch reefs: Jour. Geol., v. 79, p. 647-668.

Maiklem, W. R., 1970, The Capricorn Reef complex, Great Barrier Reef, Australia: Jour. Sed. Petrol., v. 38, p. 785-798.

Reef Mounds
Turmel, R. and R. Swanson, 1976, The development of Rodriguez Bank, a Holocene mudbank in the Florida Reef Tract: Jour. Sed. Petrol., v. 46, p. 497-519.

Algal Reefs
Adey, W. and R. Burke, 1976, Holocene bioherms (algal ridges and bank-barrier reefs) of the eastern Caribbean: Geol. Soc. Amer. Bull., v. 87, p. 95-109.

Ginsburg, R. N. and J. H. Schroeder, 1973, Growth and Submarine fossilization of algae cup reefs, Bermuda: Sedimentology, v. 20, p. 575-614.

Other References Cited

Embry, A. F. and J. E. Klovan, 1971, A Late Devonian reef tract on northeastern Banks island, N. W. T.: Bull. Can. Petrol. Geol., v. 19, p. 730-781.

Frost, S. H. 1977, Ecologic controls of Caribbean and Mediterranean Oligiocene reef coral communities: in D. L. Taylor, ed., Proceedings of Third International Coral Reef Symposium: Miami, Florida, p. 367-375.

Lowenstam, H. A., 1950, Niagaran reefs in the Great Lakes area: Jour. Geol., v. 58 p. 430-487.

MS received November 28, 1977.
Revised January, 1979.
Reprinted from Geoscience Canada, V. 5, No. 1, p. 16-26.

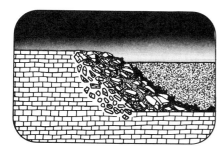

Facies Models 12. Carbonate Slopes

I.A. McIlreath
Petro-Canada,
P. O. Box 2844
Calgary, Alberta T2P 2M7

N.P. James
Department of Geology
Memorial University of Newfoundland
St. John's, Newfoundland A1B 3X5

Introduction

To any geologist who has seen them in the field the sediments that comprise the slope facies of many carbonate complexes are often the most staggering and long remembered of all. The sheer size of the enormous limestone blocks chaotically intercalated with delicately laminated lime mudstones tests our understanding of sediment genesis and deposition as almost no other deposits.

While the deposits themselves are intriguing they are also useful as the only remaining clues as to the nature and composition of a now dolomitized or tectonically obliterated platform margin. Furthermore, the very presence of this debris is an excellent indicator of a nearby carbonate platform or reef complex and this principle has been successfully used to locate reefs in the subsurface. The lime sands of these deposits, where intercalated with organic-rich basinal sediments, can be reservoirs.

We cannot interpret these deposits with the same level of confidence as shallow water carbonate sediments because: 1) modern slope deposits are not easily accessible for field study, although the limited use of small research submersibles and

seismology is slowly changing this; 2) ancient slope deposits commonly occur in orogenic belts, where facies and tectonic relationships are so complicated that these deposits are often mistaken for tectonic breccias or mélanges; 3) slope sediments in the subsurface generally have not been serious exploration targets as long as adjacent platform margins remained the primary objective; 4) slope sediments are formed in a series of environments that transect major pressure, temperature and oxygen-level boundaries in the ocean and the precise effects of these physicochemical parameters on the sediments are poorly known.

As a result our present facies models are based on the rock record with additions from the modern. In addition, our understanding of sediment emplacement is based upon mechanisms determined for siliciclastic deposits. It has been easy to apply this comparative approach because many similarities do exist, but there are also differences which have been ignored for too long.

In the second article in this facies models series, Walker (1976) outlined the attributes of a turbidite model and then integrated all associated siliciclastic lithofacies that encompass the slope-to-basin transition into an overall, larger scale, submarine fan model. Similarly our approach in this article will be to outline the major aspects of the slope facies in carbonate sedimentary sequences, first by examining the major sediment types and their modes of emplacement, and second by relating these to general facies models, which are very much dependent upon the nature of the adjacent margin.

Carbonate Slope Sedimentation

The slope facies is a transitional one between the rapid and active production of calcium carbonate in shallow water and the slow gentle rain of fine-grained pelagic sediments in the basin. The platform-to-basin transition may in places be abrupt, in the form of a steep cliff, but more commonly is a gently inclined slope decreasing in grade with depth and merging imperceptibly into basinal deposits at some distances, which may be 100's of km from the actual margin. Because the

environment as a whole is an incline, short periods of gravity-induced catastrophic sedimentation alternate with long periods of relatively quiet pelagic sedimentation, or to paraphrase Ager (1973, p. 100), "long periods of boredom alternating with short periods of terror".

Pelagic Carbonates. Pelagic carbonates are those sediments deposited in the open sea and derived from the skeletons of planktonic microorganisms which inhabit the overlying water column. Such deposits include ooze and chalk and consist primarily of the skeletons of various nannofossil groups, especially coccoliths, the tests of planktonic and sometimes benthic foraminifers. Macrofossils such as pteropods, pelecypods, echinoderms and, in older units, ammonites are present as accessory components. An excellent summary of such deposits can be found in Hsu and Jenkyns (1974) and Scholle (1977).

True pelagic carbonates are apparently not known from the early Paleozoic and are first recognized from rocks of Upper Silurian age (Tucker, 1974). Planktonic foraminifers and coccoliths appear to have evolved in the Jurassic and during post-Jurassic time pelagic carbonate has increased to the point that in the last 100 Ma it comprises about 67 per cent of world-wide carbonate deposition (Hay *et al.,* 1976).

Most chalks accumulate at a rate of between one and 30 cm per year. The sedimentary structures and colours depend upon the degree of circulation and oxygenation. Dark colours and preserved laminations reflect stagnation; lighter colours, more burrows and fewer preserved sedimentary structures reflect stronger bottom circulation.

The water depth of pelagic carbonate deposition ranges from less than 100 m to greater than 4500 m. The limiting factors for such accumulations are the relative rates of sedimentation of carbonate versus non-carbonate components, physical erosion and chemical dissolution. Chemical dissolution is particularly important in carbonate slope facies because the environment passes, with depth, through several important increasing pressure and decreasing temperature boundaries. Aragonite components, such as pteropods and benthic foraminifers, may be selectively removed by dissolution in water as shallow as 500 m (the aragonite compensation depth) while calcite components are completely dissolved at the carbonate compensation depth, between 4,000 and 5,000 m in today's oceans. Much less is known about the removal or recrystallization of Mg-calcite. This progressive removal by dissolution results in a residual sediment composed largely of siliceous skeletons, red hemipelagic clays and wind-blown silt. Dissolution also takes place in the oxygen minimum layer, that zone just below the thermocline in the modern ocean where, due to the increased metabolism of aerobic organisms and the lack of oxygen replacement, the oxygen level is often reduced to less than 0.2 ml/1. The higher levels of CO_2 associated with the oxygen deficiency lead to an increase in the $CaCO_3$ solubility. Where this zone impinges on the sea floor sedimentary carbonate is removed and the resulting sediment is enriched in organic matter, contains more opaline silica and where the oxygen values are too low even for burrowing organisms, the sediment is dark and laminated.

In some areas of the modern ocean the production of siliceous plankton (silicioflagellates, diatoms and radiolaria) exceeds that of calcareous nanno-and microplankton. During the Paleozoic, when pelagic carbonate was reduced or absent, siliceous sediment was much more widespread in deep-water areas.

Hemipelagic Slope Sediments. Sediments that make up the fine-grained pelagic component of most slope deposits come not only from the water column but from the adjacent platform as well (Wilson, 1969). While the contribution at any one time from the water column is more or less constant, that portion derived from the platform is episodic. Most often storms stir up the wide, shallow, mud-floored areas of the shelf and the milk-white water streams out across the shelf margin to settle in deep water. A less voluminous but more regular transfer process exists at such near-vertical

shelf-to-deep-oceanic-basin transitions as St. Croix, Virgin Islands where warm sediment-rich shelf waters 'float' over the cooler basinal waters by tidal exchange. These fine-grained, shallow-water derived slope sediments have been called 'peri-platform ooze' by Schlager and James (1978) because they occur as an apron around the platform and because they are significantly different in their mineralogy and composition from the wholly pelagic sediments of the open sea.

In the Precambrian and Paleozoic most pelagic slope carbonates may well have been almost wholly peri-platform ooze.

The resultant hemipelagic slope deposits are monotonous, uniform dark grey, fine-grained lime mudstones, generally thin-bedded with flat planar contacts and internal microlaminations (Fig. 1). Mudstone beds are often separated by partings into very thin beds of similar mudstone or beds of shale, forming characteristic 'rhythmites' or 'ribbon limestones'. The original depositional textures and fabrics are often modified by sedimentary boudinage, while differential compaction and/or cementation frequently

transforms the evenly-bedded sediments into a nodular limestone. The irregular nodules may, in some cases, be so packed together to form a jig-saw puzzle resembling an *in situ* breccia.

Peri-platform Talus. Directly seaward of the shallow water reefs or lime-sand shoals that form a platform margin, there is commonly a debris apron of limestone blocks (Fig. 2), skeletons of reef building metazoa, lime sand and muds. These accumulations are the result of rock-fall and sand-streams from shallow water and, as illustrated in Figure 3, are very common along the seaward margins of modern reef complexes (James and Ginsburg, in press; Land and Moore, 1977). The blocks themselves may be multigeneration in composition because the reefs, sand shoals and other deposits at the platform margin are characteristically susceptible to early lithification, either by submarine cementation, or if there are slight fluctuations in sea-level, by complex subaerial diagenesis. In addition, parts of the talus wedge are commonly cemented on the sea floor (James and Ginsburg, in press; Land and Moore, 1977). The

Figure 1
Peri-platform ooze; evenly-bedded, grey lime mudstone with thin interbeds of argillaceous lime mudstone, Cooks Brook Formation (Middle Cambrian), Humber Arm, Western Newfoundland.

lithified portions of these limestones become hard and brittle, and so are particularly susceptible to fracturing and fragmentation.

Large passes through a reef (see James, 1978) also act as conduits, funnelling back-reef sediments into this zone so that, along strike, areas of chaotic breccia may alternate with fans of lime sand. The latter sediment is also commonly cemented, forming numerous hardgrounds.

Examination of sediment dispersal seaward of the platform in areas with low to intermediate slopes (up to 30 degrees) indicates that this talus does not travel any significant distance away from the margin by day-to-day processes.

Lime Breccias. These deposits, which have been called debris flows, submarine mass flows, mass breccia flows, breccia and megabreccia beds, rudite sheets, or olistostromes (in the non-tectonic sense) are certainly the most impressive parts of the slope sequence. They originate in two very different areas, high up on the slope in shallow water or from lower down the slope profile.

Figure 2
Peri-platform talus; a block of shallow-water reef limestone (approximately 30m high) enclosed in thin-bedded, dark grey, peri-platform lime mudstones. Block occurs approximately 250m down slope from the toe of a near-vertical, 200m high platform margin. Note vertical orientation of bedding within the block, Cathedral Formation (Middle Cambrian), north face Mt. Stephen, British Columbia.

A. Breccias Derived from Shallow Water. These breccias are generally exposed in discontinuous to laterally extensive sheets, channels with lenticular cross sections or irregular masses. They stand out as resistant masses of light-coloured carbonate against a background of dark-coloured, well-bedded limestone and shale (Figs. 4 and 5). They are characterized by blocks of all sizes and shapes, but often equi-dimensional and somewhat rounded. Some of the blocks are so enormous that they have been mistaken for bioherms (see Mountjoy et al., 1972). One exceptional clast in the Cow Head Group (Cambro-Ordovician) at Lower Head, Newfoundland is 0.2 km x 50 m in size, with surrounding blocks often 30 x 15 m in dimension (Kindle and Whittington, 1958). The breccias commonly have a matrix of lime mud, lime sand or argillaceous lime mud.

The deposits are bedded, with a planar to undulating basal contact accentuated by differential compaction and an irregular to hummocky upper contact. The nature of the bedding contacts often cannot be determined accurately because the bedding planes are stylolitic, and so any original bedding-plane features

are often destroyed. Davies (1977) made the interesting observation that the common occurrence of crinoids, bryozoa and ammonites at the upper surface of Permo-Pennsylvanian deposits on Ellesmere Island may represent an indigenous fauna inhabiting the 'reef-like' upper surface of the deposit.

The polymict nature of the clasts reflects the complexity of the source area; the platform margin consisting of partly lithified reefs and/or lime-sand shoals, down slope (yet still shallow) reef mounds, or peri-platform talus. Well-sorted and well-bedded lime sands which can be differentially submarine cemented (Fig. 6), individual colonies of reef builders, multigeneration reef rock, limestones with subaerial karst features, tidal flat lithologies and even cemented talus that has been refractured to give breccia clasts within breccia, are all to be expected.

The fabrics of such coarse clastic deposits have been discussed by Walker (1976) and they range from mainly chaotic to imbricated to horizontal to wave form and are rarely graded or even reverse graded. They range from clast-supported to most commonly matrix-supported, with the matrix ranging from shale to argillace-

Figure 3
Peri-platform talus; looking across the steeply dipping fore-reef slope at a depth of 130m seaward of the Belize barrier reef complex. The slope is composed of blocks of limestone, plates of coral and lime sand composed of the plates of the green alga Halimeda; the small block at the center (arrow) is about one metre high.

Figure 4
Lime breccias; light grey, shallow-water reef-derived limestone breccias occurring in a 'channel' – a (approximately 8m thick), sheets – b (approximately 2m thick), and irregular masses – c (up to 12m thick), enclosed in thin-bedded, dark grey, peri-platform lime mudstones, Cathedral Formation (Middle Cambrian), southface Mt. Field, British Columbia.

Figure 5
A sequence illustrating two different types of carbonate slope deposits; debris flows with large limestone clasts (right) and thin-bedded, graded calcarenites (the thin, grey limestone beds), interbedded with black fissile shale. This overturned sequence (top at lower left) of Middle Ordovician age occurs at Cape Cormorant, Port-au-Port Peninsula, Western Newfoundland.

flows in which granular solids such as boulders, pebbles and sand are more or less "floated" during transport by the yield strength of the matrix composed of interstitial fluid and fine sediment. Buoyancy of the fluid matrix also contributes to the support. Since not all such deposits have a clay mineral matrix the transport mechanism is thought to be a combination of debris flow and grain flow (Middleton and Hampton, 1973). A major problem in this regard is that almost all experimental work to date has been done on clay-water mixtures; none of the experiments has been carried out on sediments with a clay-lime or lime mud matrix.

B. Breccias Derived from the Slope. The evenly-bedded calcilutites or lime muds of the slope facies are often prone to downslope creep. Individual beds can be seen to neck or wedge out, or whole intervals will move downslope within a series of slump folds (Fig. 7). Dislocation and movement of large masses of slope material downslope leads to the formation of breccias or submarine glide masses composed of numerous tabular clasts of slope limestone that have been bent or fractured, that are poorly-sorted and that exhibit random to subparallel orientations, often resembling shallow-water 'flat pebble conglomerates' (Fig. 8). Enormous blocks of bedded slope sediments, often internally folded, are caught up in the breccias.

The source of these breccias is thought to be the large 'intraformational truncation surfaces' (Fig. 9) or "cut-and-fill structures" (Wilson, 1969) which are sharp concave-up discontinuity surfaces that truncate underlying beds and are overlain by a downslope thickening wedge of sediment with an angular relationship on the truncated beds. In these deposits, reduction of shear stress occurs by displacement of coherent masses along discrete shear planes and not usually by deformation within the mass as occurs in slumps.

The tabular clasts of slope material clearly indicate that the slope sediments were partly consolidated very early, probably by submarine cementation. Cementation may have been

ous lime mud to lime mudstone with occasional lime sand. As Hopkins (1977) points out, however, what is often taken to be lime mud in outcrop turns out to be peloid lime sand in thin section, so that sand-sized matrix may be more common than supposed.

The exact mechanisms by which these sediments are transported are not yet clear. Submarine debris flows (Hampton, 1972) are sediment gravity

Figure 6
Thinly-bedded, nodular foreslope sequence comprising cemented nodules in compacted calcarenite (N) and a laterally continuous bed of cemented calcarenite *(S). These calcarenites form the predominant foreslope facies below the Miette and Ancient Wall buildups (Upper Devonian), Alberta. Photo courtesy of J.C. Hopkins.*

Figure 8
Slope-derived breccia; clasts of partly lithified peri-platform ooze (see Fig. 1) that have been eroded and transported as a clast-supported breccia, Cooks Brook Formation (Middle Cambrian), Humber Arm, Western Newfoundland.

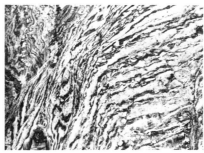

Figure 7
Extensive syn-sedimentary distortion of bedding developed by creep in thin to very thin-bedded, upper basinal slope, peri-platform lime mudstones, Eldon Formation (Middle Cambrian), Wapta Mountain, British Columbia.

similar to that in shallow-water with lithified and unlithified layers reflecting times of slow and rapid sedimentation respectively. If cementation took place below the thermocline, dissolution of aragonite and possible precipitation of calcite may have caused the same effect in layers of different original composition. Alternatively, if the lime mudstone is interlaminated with shale, cementation of the carbonate may have taken place while the shale remained soft.

Deposits of the two end members, one originating high on the slope and the other down on the proper slope are sometimes found intermixed in extensive breccia masses (Fig. 10). Such deposits are similar to what Schlager and Schlager (1973) term marl-flaser breccia, characterized by a chaotic fabric of plastically deformed, dark grey, argillaceous lime mudstone lithoclasts separating irregular lenses of subangular limestone and other lithoclasts, with the deformed marls forming the flaser fabric. There are thought to be shallow-water derived breccia flows that incorporated lime mudstone clasts from the floor of the slope environment as they moved basinward and they may grade downslope, as do many other breccias, into turbidites.

Graded Calcarenites. A large proportion of any slope sequence is commonly size-graded beds of clastic textured limestone, mainly of sand size, interpreted to be the carbonate equivalent of siliciclastic turbidites (Fig. 5). They are envisaged to be deposited from turbidity currents that formed by the sudden surge-type release of dense fluid rather than from a steady state flow such as described recently by Harms (1974). These

Figure 11
A bed of light grey calcarenite comprised of a graded lower portion, planar laminated middle unit, and the upper portion having climbing ripples (A, B and C Bouma subdivisions respectively), capping a lime breccia, Sekwi Formation (Lower Cambrian), Cariboo Pass, Mackenzie Mountains. Photo courtesy of F.F. Krause.

Figure 9
Large intraformational truncation surface in argillaceous and cherty limestones of the Hare Fiord Formation (Permo-Pennsylvanian), north side of Svartfjeld Peninsula, Ellesmere Island. Note smooth, *curved concave-up (listric) geometry of the truncation surface and the lack of macro-scale deformation of beds below or above truncation surface. Shadow at lower left center is of heliocopter: width of view 150m. Photo courtesy of G.R. Davies.*

Figure 10
A large deformed clast of well-bedded peri-platform ooze that was eroded, transported and redeposited as part of a debris flow, Cow Head Group (Middle Ordovician), Cow Head, Western Newfoundland.

deposits have also been called allodapic limestones (Meischner, 1964). Such sediments are well-bedded and characteristically have sharp planar bases that can be coplanar with, or locally scour and truncate underlying slope beds. Sole marks and load structures are usually absent although in some cases they may be obliterated because of stylolitization and solution along bedding contacts. Calcareous turbidites can exhibit all five of the typical ABCDE divisions of the Bouma

sequence but most commonly it is the A, and sometimes the B and C divisions that characterize the deposits (Fig. 11). The particles in the basal parts of division A are often cobble size and larger and the more common grain types are lithoclasts, skeletal debris and ooids, the petrology of which indicates a shallow water origin.

The most obvious sources for these units are the unstable accumulations of lime sand and gravel that build up near the platform margin and are occasionally set into motion. It is also possible that they are the distal parts of carbonate debris flows. Davies (1977) has suggested a third origin, the indigenous fauna, especially pelmatozoans, that live on the slopes and produce abundant skeletal material that may be easily remobilized.

Post-Paleozoic graded calcarenites derived from sediments further down the slope profile can be virtually indistinguishable compositionally from pelagic limestone. These calcarenites are generally rich in pelagic components such as coccoliths and foraminifers but may also contain lesser amounts of pteropods, sponge spicules, radiolarians, and coarser-grained skeletal debris (especially pelmatozoans). The sediments are size-sorted and may be mixed with clastic terrigenous or volcaniclastic sediment if they have travelled great distances. Although the sedimentary structures such as horizontal laminations, convolutions, occasional channels, flute and groove casts and trace fossils may be present, the A and B divisions of the Bouma sequence are commonly missing and they generally start with the C or D divisions.

Non-graded Calcarenites. Massive to cross-bedded and ripple-marked calcarenites are an enigmatic type of deposit found in many slope sequences. These deposits are fine- to coarse-grained wackestones to grainstones with occasional large clasts or fossils. Individual beds have sharp bases and vary in geometry from lenticular to irregular masses. The fabric may be random or grains may be aligned parallel to the paleoslope. The grains in these deposits are variable, ranging in composition from shallow-water derived particles to pelagic grains.

At present we envisage these deposits as resulting from one of three depositional mechanisms, liquified flow, grain flow, or reworking of pre-existing sediments by bottom currents. Perhaps the massive deposits having an apparent lack of sedimentary structures are nothing more than the product of downslope mass movement of well-sorted lime sands produced at a rapid rate near the platform margin.

Sedimentary structures in the cross-bedded deposits indicate some sort of bottom currents, often running parallel to the slope (contour currents). Well-sorted, rippled lime sands, sometimes with large scale bed forms, and composed of ooid sand occur in the deeper parts of the slopes around the margins of the Tongue of the Ocean, Bahamas (NJP, pers. obs.) and are also common on the slopes along the western parts of the Bahama Banks (Mullins and Neumann, in press) where currents flow along and parallel to the slope at speeds of 50 cm/sec and more (although such currents are high and not characteristic of today's oceans). These currents may rework pre-existing pelagic slope deposits, leaving only the larger foraminifers

and pteropods together with lithoclasts of cemented pelagics to form a deep-water grainstone. They may also winnow the upper parts of turbidites, removing the finer layers and leaving a sequence composed only of shallow-water clasts, and divisions A and B of the Bouma sequence, capped by a cross-bedded lime sand.

Such clean, well-sorted sands are commonly sites of submarine cementation and hardground formation. In such areas precipitation of cement may lead to displacive expansion of grain-to-grain distance, resulting in fracturing and the formation of *in situ* breccias.

Facies Models
With the information presently available we cannot integrate the spectrum of carbonate slope deposits into one simple model as Walker (1976) has done for siliciclastic deposits. Rather, the style of carbonate slope sedimentation is equally a function of the abruptness of the margin to basin transition, and the nature of the shallow portion of the margin. Where the margin itself is a facies transition with a gradual slope profile, then the sequence of slope deposits is very much different from the sequence where the margin is abrupt. We have differentiated between these two types of margins and called them depositional and by-pass margins respectively (Figs. 12 to 15).

The nature of the slope sediments in each case depends on whether the shallow-water margin is formed (1) by reefs, of metazoan, calcareous algal or stromatolitic origin, and occurring either at the edge or slightly downslope below the zone of the most wave movement, or (2) by lime-sand shoals, of pelmatozoan, algal or oolitic origin.

It should be noted that none of the models are mutually exclusive and within a buildup or platform margin all four may be present at any one time, or in the case of a buildup, it may even be possible to have all four occur simultaneously in different places along the buildup margin.

Depositional Margins. The slopes are generally gentle and decrease basinward to merge with the flat basin floor.

A. Shallow-water reef. The zone of peri-platform talus is relatively narrow but the full spectrum of allochthonous deposits is present downslope (Fig. 12). Because most of the allochthonous material comes from the reef or talus pile many of the allochthonous deposits generated high on the slope are deposited far down on the slope or in the basin. Consequently that zone seaward of the peri-platform talus is often composed of hemipelagic limestones and is by-passed by the mass movements. This type of depositional slope occurs most frequently around reef complexes and basinward of platform-margin barrier reef systems along paleotopographic highs, structurally positive elements or hingelines in fairly stable cratonic or miogeosynclinal basins. Examples of this style of slope deposit occur in the Cambrian of Western Canada (McIlreath, 1977a); and the Devonian of the Canning Basin, Australia (Conaghan et al., 1976).

B. Shallow-water lime sands. The slope flanking this style of margin is generally a calcarenite wedge or proximal-to-distal turbidite plain (Fig. 13). These slopes probably represent a depositional equilibrium in that sedimentation controls the slope angle and is active all along the profile. Turbidites and grain flows are the predominant transport mechanisms with debris sheets and breccias rare. Some minor debris sheets composed of cemented lime-sand clasts or other slope-derived lithologies may be present. Hardgrounds and incipient brecciation are common.

Examples of this style of slope deposit includes the Pennsylvanian Dimple Limestone, Texas (Thomson and Thomasson, 1969); Silurian of California and Nevada (Ross, 1965); and the Devonian Fairholme reef complexes of Western Canada (Hopkins, 1977).

By-pass Margins. In these situations the margin is atop a cliff or submarine escarpment so that sediments are transported directly from shallow to deep water, bypassing much of the slope along a wide front or through channels and canyons. The cliffs may result from faulting, large fluctuations

140

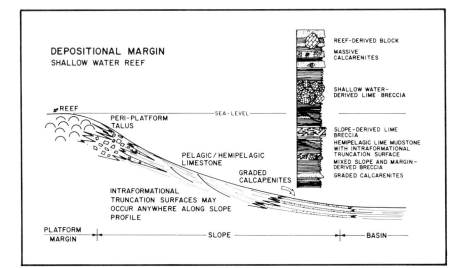

Figure 12
Schematic model for a shallow-water, reef dominated, depositional carbonate margin and illustration of a hypothetical sequence of deposits within the adjacent basin slope.

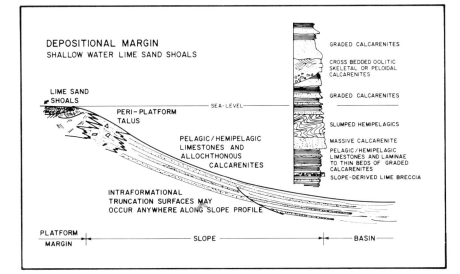

Figure 13
Schematic model for a depositional carbonate margin dominated by shallow-water lime sands and illustration of a hypothetical sequence of adjacent basinal slope deposits.

in sea-level or just rapid upbuilding of the platform as compared to the basinal deposits. This style of margin is particularly common along block-faulted oceanic margins or at the structural hingeline where a basin is subsiding faster than the adjacent platform.

A. Shallow-water reef. Since the reef crowns the escarpment, the most characteristic and spectacular style of accumulation is the wedge of peri-platform talus (Fig. 14). This wedge of material may be enormous, especially if the area is subject to tectonics, with the main transport mechanisms being a combination of rock-fall, sand-streams and gravity-induced

downslope mass movement. If the cliff is dissected by channels or canyons, the peri-platform talus may inter-digitate along strike with carbonate submarine fans similar to those described for siliciclastic deposits (Evans and Kendall, 1977). Slumps, creep and sliding are more active in the deposit than on the adjacent slope due to the variations in lithification. The talus wedge grades downslope into a relatively narrow zone of lime sands and then into pelagic calcilutites to form a debris apron.

This is the style of many modern slope deposits in Belize (Ginsburg and James, 1973); Puerto Rico (Conolly and Ewing, 1967); Jamaica (Goreau and Land, 1974); the Bahamas (Mullins and Newmann, in press); and the Pacific Atolls (Emergy et al., 1954). The most spectacular fossil example is the Cretaceous of Mexico (Enos, 1977).

B. Lime sand shoals. If the shallow-water margin facies is lime sand, then the peri-platform talus will be predominantly lime sand intercalated with calcilutites (Fig. 15) and having fewer limestone blocks than in the previous model, unless there have been substantial movements in sea-level exposing the margin to subaerial diagenesis. Once again away from the escarpment the lime sands grade relatively quickly into slope or basinal pelagic lime muds. There are minor contributions from turbidites.

A fossil example of such a debris apron of calcarenite is the Cambrian Boundary Limestone (McIlreath, 1977b).

The Models as a Norm. In these models we have not consciously placed the slope lithologies in any particular sequence because we feel that the sequence on such a broad scale represents more the complex interactions of sea-level and tectonics at the shallow rim than any secondary sedimentary process on the slope. As a result unusual features are likely to record not so much the style of sedimentation on the slope, as the style of sedimentation and tectonics at the shallow margin.

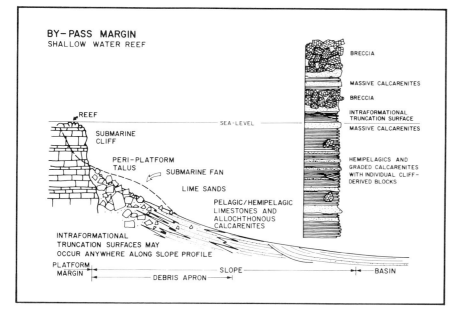

Figure 14
Schematic model for a shallow-water, reef dominated, by-pass type of carbonate margin and illustration of a hypothetical sequence of deposits within the adjacent basin slope.

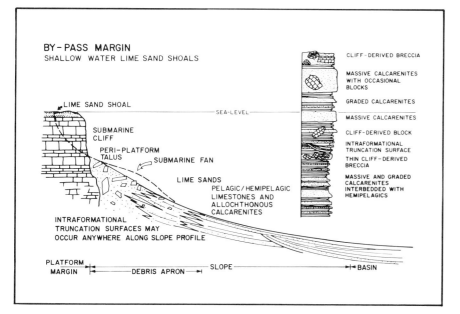

Figure 15
Schematic model for a by-pass type of carbonate margin dominated by shallow-water lime sands and illustration of a hypothetical sequence of adjacent basinal slope deposits.

The Models as a Framework and Guide for Description. The differences in carbonate deposition through time are, in large part, a function of the appearance and disappearance of different types of carbonate-secreting organisms and thus affect the use of these facies models as a framework in two ways: (1) the shallow-water benthic organisms that build massive reefs and so cause relief at the platform margin, as well as contributing major amounts of sediment to the slope, are present only at specific times in geologic history; (2) the pelagic calcareous zooplankton and phytoplankton are insignificant in the early Paleozoic, minor in the middle and late Paleozoic, and prolific in the Mesozoic and Tertiary. As a result, the hemipelagic slope deposition is almost entirely peri-platform ooze in the Precambrian and early Paleozoic and perhaps one-half peri-platform ooze and one-half true pelagic carbonate in the Mesozoic and Tertiary. Interruptions in the fallout of peri-platform ooze in the Paleozoic sometimes resulted in shale interbeds, whereas in the Mesozoic and Tertiary interbeds are thinner but are wholly pelagic carbonate.

The Models as a Predictor for New Situations. Based on a few observations and bearing in mind the age of the deposits as well as their tectonic setting, one can extrapolate and formulate three critical conclusions: (1) examination of the overall sequence indicates the relative position on the slope and possible nearness of the platform; (2) the lithology of the lime-sand beds and relative calcarenite to hemipelagic ratio gives some idea as to the nature of the slope facies, i.e., depositional versus by-pass; and (3) the composition of the clasts indicates the nature of the margin, which is often obliterated or inaccessible.

The Models as a Basis for Hydrodynamic Interpretation. The interpretation of carbonate sediment gravity flows has, to date, been based primarily on an analogy with siliciclastic deposits which have similar sedimentary characteristics. One of the important differences between

142

carbonate and siliciclastic sediment gravity flows, however, is that a dispersal model for the hypothetical evolution of a single flow of carbonate platform-derived debris into deep water has not yet been constructed. In contrast to the relatively unconsolidated sediments on continental shelves, carbonate sediments in similar environments tend to be stabilized by organisms and/or well-lithified. This results in distinctively different slope deposits being produced by a variety of gravity-driven transport processes rather than different types of deposits evolving from the same flow. It should be noted, however, that the concept of a singular flow spawning a series of deposits may apply where slide failure occurs in the lower portion of the slope, remobilizing and transporting these mixed deposits even further basinward.

Summary

Carbonate slope sediments have, in the past, often been either ignored or interpreted as tectonic in origin. Their identification as deposits, separate from tectonically formed mélanges has come largely from a detailed analysis of not only the chaotic deposits but the fine-grained interbeds as well. Our understanding is increasing as more deposits are documented and the first timid steps are being taken beyond the reef into deeper water by submersibles. This latter aspect of carbonate sedimentology is still very much in its infancy.

Refinements of the models presented in this paper must come from two directions, experimentation and observations from modern carbonate slope environments. The hydrodynamic parameters for gravity-induced mass movements involving only carbonate materials must be carefully documented and contrasted with the results from siliciclastic materials. A combination of detailed observations from submersibles, and high resolution seismic and bottom sampling is needed to inventory the spectrum of sediments and structures that make up carbonate slope environments in the modern ocean.

Acknowledgements
Many of the ideas presented here have come about as a result of stimulating discussions with John Harper and Bob Stevens. Hank Mullins and Conrad Neumann kindly gave us a preprint of their paper on Carbonate Slopes in the Bahamas to sharpen our discussion. John Hopkins, Dave Morrow and Federico Krause critically read the manuscript. Gary Allwood drafted all the figures.

References

Basic References on Carbonate Basin Slope Facies

Cook, H.E. and P. Enos, eds., 1977, Deep-water carbonate environments: Soc. Econ. Paleont. Mineral. Spec. Publ. 25, 336p.

Hsu, K.J. and H.C. Jenkyns, eds., 1974, Pelagic sediments: on land and under the sea: Internatl. Assoc. Sedimentologists Spec. Publ. 1, 448p.

Mountjoy, E.W., H.E. Cook, L.C. Pray, and P.N. McDaniel, 1972, Allochthonous carbonate debris flows – worldwide indicators of reef complexes, banks or shelf margins: Proc. 24th Interntl. Geol. Congress, Sect. 6, p. 172-189.

Scholle, P.A., 1977, Deposition, diagenesis and hydrocarbon potential of "deeper-water" limestones: Amer. Assoc. Petrol. Geol. Continuing Education Course Notes, Series No. 7, 25p.

Wilson, J.L., 1969, Microfacies and sedimentary structures in "deeper-water" lime mudstones: in G.M. Friedman, ed., Depositional environments in carbonate rocks: Soc. Econ. Paleont. Mineral. Spec. Publ. 14, p. 4-19.

Examples of Ancient Carbonate Basin Slopes
The following papers contain examples of ancient carbonate basin slopes, with a bias towards examples in Canada. However as with reefs, the location of the country in temperate to boreal latitudes during much of the Mesozoic and, unfortunately, the Cenozoic results in Canadian examples being mainly Paleozoic in age.

Conaghan, P.J., E.W. Mountjoy, D.R. Edgecombe, J.A. Talent, and D.E. Owen, 1976, Nubrigyn algal reefs (Devonian), eastern Australia: allochthonous blocks and megabreccias: Geol. Soc. Amer. Bull., v. 87, p. 515-530.

Cook, H.E., P.N. McDaniel, E.W. Mountjoy, and L.C. Pray, 1972, Allochthonous carbonate debris flows at Devonian bank ("reef") margins Alberta, Canada: Can. Petrol. Geol. Bull., v. 20, p. 439-497.

Davies, G.R., 1977, Turbidites, debris sheets and truncation structures in upper Paleozoic deep water carbonates of the Sverdrup Basin, Arctic Archipelago: in H.E. Cook and P. Enos, eds., Deep-water carbonate environments: Soc. Econ. Paleont. Mineral. Spec. Publ. 25, p. 221-249.

Enos, P., 1977, Tamabra limestone of the Poza Rica trend, Cretaceous, Mexico: in H.E. Cook and P. Enos, eds., Deep-water carbonate environments: Soc. Econ. Paleont.,Mineral. Spec. Publ. 25, p. 273-314.

Hopkins, J.C., 1977, Production of fore-slope breccia by differential submarine cementation and downslope displacement of carbonate sands, Miette and Ancient Wall buildups, Devonian, Canada: in H.C. Cook and P. Enos eds., Deep-water carbonate environments: Soc. Econ. Paleont. Mineral. Spec. Publ. 25, p. 155-170.

Hubert, J.F., R.K. Suchecki, and R.K.M. Callahan, 1977, Cow Head breccia: sedimentology of the Cambro-Ordovician continental margin, Newfoundland: in H.E. Cook and P. Enos, eds., Deep-water carbonate environments: Soc. Econ. Paleont. Mineral. Spec. Publ. 25, p. 125-154.

Kindle, C.H. and H.B. Whittington, 1958, Stratigraphy of the Cow Head Region, Western Newfoundland: Geol. Soc. Amer. Bull., v. 69, p. 315-342.

Mackenzie, W.S., 1970, Allochthonous reef-debris limestone turbidties, Powell Creek, Northwest Territories: Can. Petrol. Geol. Bull., v. 18, p. 474-492.

McIlreath, I.A., 1977a, Stratigraphic and sedimentary relationships at the western edge of the Middle Cambrian Carbonate Facies Belt, Field, British Columbia: Ph.D. dissertation, Univ. of Calgary, Calgary, Alberta, 259p.

McIlreath, I.A., 1977b, Accumulation of a Middle Cambrian, deep-water, basinal limestone adjacent to a vertical, submarine carbonate escarpment, Southern Rocky Mountains, Canada: in H.E. Cook and P. Enos, eds., Deep-water carbonate environments: Soc. Econ. Paleont. Mineral. Spec. Publ. 25, p. 113-124.

Morrow, D.W., 1978, The Prairie Creek Embayment and associated slope, shelf and basin deposits: Geol. Surv. Can. Paper 78-IA, p. 361-370.

Ross, D.C., 1965, Geology of the Independence Quadrangle, Inyo County, California: U.S. Geol. Surv. Bull. 1181-0, 64p.

Srivastava, P.C., C.W. Stearn and E.W. Mountjoy, 1972, A Devonian megabreccia at the margin of the Ancient Wall carbonate complex, Alberta: Can. Petrol. Geol. Bull., v. 20, p. 412-438.

Thomson, A.F. and M.R. Thomasson, 1969, Shallow to deep water facies development in the Dimple Limestone (Lower Pennsylvanian), Marathon Region, Texas: in G.M. Friedman, ed., Depositional environments in carbonate rocks: Soc. Econ. Paleontol. Mineral. Spec. Publ. 14, p. 57-78.

Tyrrell, Jr. W.W., 1969, Criteria useful in interpreting environments of unlike but time-equivalent carbonate units (Tansill-Capitan-Lamar), Capitan Reef Complex, West Texas and New Mexico: in G.M. Friedman, ed., Depositional environments in carbonate rocks: Soc. Econ. Paleontol. Mineral. Spec. Publ. 14, p. 80-97.

Unfortunately a number of excellent papers on Mississippian and Permian basin slope carbonates occurring in New Mexico and Texas exist in difficult-to-obtain guidebooks of local societies. For example, the "Guidebook to the Mississippian Shelf-edge and Basin Facies Carbonates, Sacramento Mountains and Southern New Mexico Region", published by the Dallas Geological Society (1975).

Bein, A. 1977, Shelf basin sedimentation: mixing and diagenesis of pelagic and clastic Turonian carbonates, Israel: Jour. Sed. Petrology, v. 47, p. 382-391.

Bernoulli, D. and H.C. Jenkyns, 1974, Alpine, Mediterranean, and central Atlantic Mesozoic facies in relation to the early evolution of the Tethys: in R.H. Dott, Jr. and R.H. Shaver, eds., Modern and Ancient Geosynclinal Sedimentation: Soc. Econ. Paleont. Mineral. Spec. Publ. 19, p. 129-160.

Cecile, M.P. and B.S. Norford, 1979, Basin to platform transition, Lower Paleozoic strata of Ware and Trutch map areas, northeastern British Columbia: Geol. Survey Can. Paper 79-1A, p. 219-226.

Johns, D.R., 1978, Mesozoic carbonate rudites, megabreccias and associated deposits from central Greece: Sedimentology, v. 25, p. 561-573.

Multer, H.G., S.H. Frost and L.C. Gerhard, 1977, Miocene "Kingshill Seaway" – a dynamic carbonate basin and shelf model, St. Croix, U.S. Virgin Islands: Amer. Assoc. Petrol. Geol., Studies in Geology, no. 4, p. 329-352.

Examples of Modern Carbonate Basin Slopes

Andrews, J.E., 1970, Structure and sedimentary development of the outer channel of the Great Bahama Canyon: Geol. Soc. Amer. Bull., v. 81, p. 217-226.

Andrews, J.E., F.P. Shepard and R.J. Hurley, 1970, Great Bahama Canyon: Geol. Soc. Amer. Bull., v. 81, p. 1061-1078.

Bushby, R.F., 1969, Ocean surveying from manned submersibles: Marine Technol. Soc. Jour., v. 3, p. 11-24.

Conolly, J.R. and M. Ewing, 1967, Sedimentation in the Puerto Rico Trench: Jour. Sed. Petrology, v. 37, p. 44-59.

Emery, K.O., J.I. Tracey, Jr. and H.S. Ladd, 1954, Geology of Bikini and nearby atolls: U.S. Geol. Surv. Paper 260-A, p. 1-262.

Ginsburg, R.N. and N.P. James, 1973, British Honduras by submarine: Geotimes, v. 18, p. 23-24.

Goreau, T.F. and L.S. Land, 1974, Fore-reef morphology and depositional processes, North Jamaica: in L.F. Laporte, ed., Reefs in time and space: Soc. Econ. Paleontol. Mineral. Spec. Publ. 18, p. 77-89.

James, N.P. and R.N. Ginsburg, (in press), The deep seaward margin of Belize barrier and atoll reefs: Internatl. Assoc. Sedimentologists, Spec. Publ. 3.

Land, L.S. and C.H. Moore, 1977, Deep forereef and upper island slope, North Jamaica: in S.H. Frost, M.P. Weiss and J.B. Saunders, eds., Reefs and related carbonates: ecology and sedimentology: Amer. Assoc. Petrol. Geol. Studies in Geology, no. 4, p. 53-67.

Moore, C.H., E.A. Graham, and L.S. Land, 1976, Sediment transport and dispersal across the deep fore-reef and island slope (-55m to -305m), Discovery Bay, Jamaica: Jour. Sed. Petrology, v. 46, p. 174-187.

Mullins, H. and A.C. Neumann, in press, Carbonate slopes along open seas and seaways in the Northern Bahamas: in O. Pilkey and R. Doyle, eds., Slopes: Soc. Econ. Paleont. Mineral. Spec. Publication.

Paulus, F.J., 1972, The geology of site 98 and the Bahama Platform: Initial reports of the deep sea drilling project, v. 11, p. 877, Natl. Sci. Found., Washington, D.C.

Additional References Cited in Text

Ager, D.V., 1973, The nature of the stratigraphical record: New York, J. Wiley and Sons Inc., 114p.

Evans, I. and C.G. St. C. Kendall, 1977, An interpretation of the depositional setting of some deep-water Jurassic carbonates of the central High Atlas Mountains, Morocco: in H.E. Cook and P. Enos, eds., Deep-water carbonate environments: Soc. Econ. Paleont. Mineral. Spec. Publ. 25, p. 249-261.

Hampton, M.A., 1972, The role of subaqueous debris flow in generating turbidity currents: Jour. Sed. Petrology, v. 42, p. 775-793.

Harms, J.C., 1974, Brushy Canyon Formation, Texas: A deep-water density current deposit: Geol. Soc. Amer. Bull., v. 85, p. 1763-1784.

Hay, W.W., J.R. Southam and M.R. Noel, 1976, Carbonate mass balance – cycling and deposition on shelves and in deep sea (Abst.): Amer. Assoc. Petrol. Geol. Bull., v. 60, p. 678.

James, N.P., 1978, Facies models 10. Reefs: Geosci. Canada, v. 5, p. 16-26.

Meischner, K.D., 1964, Allodapische Kalke, Turbidite in riff-nahen Sedimentations-Becken: in A.H. Bouma and A. Brouwer, eds., Turbidites: Elsevier, p. 156-191.

Middleton, G.V. and M.A. Hampton, 1973, Sediment gravity flows; mechanics of flow and deposition: in Turbidites and deep-water sedimentation: Soc. Econ. Paleontol. Mineral., p. 1-38.

Schlager, W. and M. Schlager, 1973, Clastic sediments associated with radiolarites (Tauglboden-Schichten, Upper Jurassic, eastern Alps): Sedimentology, v. 20, p. 65-89.

Schlager, W., and N.P. James, 1978, Low-magnesian calcite limestones forming at the deep-sea floor, Tongue of the Ocean, Bahamas: Sedimentology (in press).

Tucker, M.E., 1974, Sedimentology of Palaeozoic pelagic limestones: the Devonian Griotte (Southern France) and Cephalopodenkalk (Germany): in K.J. Hsu and H.C. Jenkyns, eds., Pelagic sediments: on land and under the sea: Internatl. Assoc. Sedimentologists Spec. Publ. 1, p. 71-92.

Walker, R.G., 1976, Facies models-2, Turbidites and associated coarse clastic deposits: Geosci. Canada, v. 3, p. 25-36.

MS received September 5, 1978.
Reprinted from Geoscience Canada, Vol. 5, No. 4, p. 189-199.

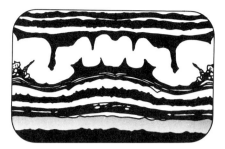

Facies Models 13. Continental and Supratidal (Sabkha) Evaporites

Alan C. Kendall
*Amoco Canada Petroleum
Company Ltd.
444 Seventh Avenue S.W.
Calgary, Alberta T2P 0Y2*

Introduction

Evaporites are a group of rocks that form by precipitation from concentrated brines. Concentration necessary for precipitation is generally attained by evaporation at the air-water interface but can also be achieved by brine freezing or by subsurface processes such as ion-filtration of residual connate fluids. Many evaporites are not strictly primary precipitates, but are diagenetic minerals, emplaced within non-evaporite sediments. Still others are diagenetic replacements of true primary precipitates.

Like iron formations (Dimroth, 1977), evaporites can be viewed in terms of two classes of models: (1) *sedimentary* models that relate structures and textures to hydrodynamic and other depositional parameters, and (2) *post-depositional* models that relate present mineralogical compositions to the physico-chemical environments of diagenetic processes. Because evaporite deposition is controlled mainly by physico-chemical parameters and because many changes occur during early diagenesis (when they are controlled by depositional settings), the distinction between the two classes is blurred. This, and a succeeding paper discuss mainly sedimentary models but mention diagenetic changes when these are early or when a primary or late diagenetic origin is disputed.

Four main factors make evaporites probably the least suitable of sedimentary rocks for facies modelling:

1. Only recently have evaporites been considered as sediments rather than as chemical precipitates. The initial success of the chemical approach caused this to dominate evaporite studies and only in the last decade have sedimentary aspects been stressed. For many evaporite deposits therefore, the basic data upon which facies models are based are lacking. When models have been constructed they all too commonly have been based upon a few occurrences. Thus distillation of essential from local details may be far from complete. The chemical approach has also generated a host of depositional models based upon theoretical concepts of seawater evaporation but which ignore sedimentological evidence. These cannot rightly be considered facies models.

2. Observations upon evaporites may be limited. Only rarely are unaltered evaporites exposed at outcrop. Most evaporite studies are confined to subsurface materials - cores or mine openings. For many poorly sampled evaporite units the gross three-dimensional characteristics are established but internal details (upon which facies modelling depends) are poorly known.

3. Areas of present day evaporite deposition comparable in size with those of the past are absent. It is uncertain whether or not modern small depositional areas (or even artificial salt-pans) are fully representative. Thus the opportunity to utilize modern sediments to construct facies models is either denied to us or is controversial.

4. Lastly, but most importantly, evaporites are most susceptible to extensive post-depositional change. The solubility of evaporite minerals, the tendency for metastable hydrates to be precipitated, and the susceptiblility of many salts to flowage under burial conditions are features unique to evaporites and have the common result of obliterating original sedimentary characteristics during diagenesis. The profound effects of these changes means that some evaporites are better considered metamorphic rocks than sediments. Recognition of primary features and formulation of depositional models for many evaporites may thus be impossible. The situation with respect to many

bittern salt deposits is most extreme for they commonly lack any vestige of original fabrics, structures or mineralogy. There is a corresponding dearth of facies models for these deposits.

In the light of these four factors it is hardly surprising that basic disagreements exist about almost all aspects of evaporite genesis. Most significant amongst them are whether basin-central evaporites were deposited in deep or shallow water, and whether many evaporite structures and textures are of primary or post-depositional origin.

No single facies model can be applied to so heterogeneous a grouping of rocks as the evaporites. The dogma of the decade - supratidal (sabkha) evaporites - has become much too one-sided because there are other evaporite types that clearly are of subaqueous origin. It is probably true that, given the correct environmental conditions, evaporites can mimic most other sediment types. There are evaporite turbidites and oolites; 'reefs' composed of huge gypsum crystals that formed mounds standing proud of the basin floor; and shallow-water clastic evaporites that resemble in texture and sedimentary structure their clastic or carbonate equivalents. Since evaporites may exhibit detrital as well as crystalline precipitate textures, these sediments consititute one of the most variable of sedimentary rock groups.

Evaporite minerals may form only a minor component of some deposits (isolated gypsum crystals in continental redbeds would be an example) and these are best considered part of other facies models.

Of the many possible environments of evaporite precipitation, five major categories (or regimes) were identified by Schreiber *et al.* (1976) with a further subdivision in each category as to whether the evaporites are calcium sulphates or halides (with or without complex sulfates) (Fig. 1). Regimes grade into each other such that identification may depend more upon associated facies than upon internal characteristics. Continental sabkha deposits commonly are internally identical with coastal sabkha deposits, differing only in being inserted within continental deposits. Furthermore, the degree of restriction required to generate halite and/or subaqueous sulphate deposits ensures that all these environments have some of the attributes of the continental regime. Distinction between large hypersaline inland lakes and partially desiccated small seas is a somewhat academic exercise.

Three main environmental groupings are recognized, of which two, continental and coastal sabkha evaporites are considered in this paper. Subaqueous evaporites, whose facies are less clearly defined, are the subject of the next paper.

Continental Evaporites

Evaporites formed exclusively from continental groundwaters are not common in rock record. Many evaporites that formed in continental settings were derived partly from marine input. It is difficult to identify the relative importance of continental and marine influences.

The rarity of continental evaporites also reflects the ephemeral nature of many evaporite minerals in the depositional environment. Many are recycled or move upwards at the same rate as

sediment accretion and are thus non-accumulative. Their former presence may leave evidence in the form of crystal moulds or disrupted lamination.

Continental evaporites occur in saline soils and as sedimentary bodies in central parts of playa (continental sabkha) basins, particularly in association with playa lakes. With the possible exception of gypsum crusts (gypcrete) which form in the same manner as caliche (calcrete) but in more arid areas, the accumulation of pedogenic evaporites is unlikely to be preserved in the rock record. Reference should be made to Cooke and Warren (1973) and Kulke (1974).

Playa (Continental Sabkha) Evaporites

These evaporites, whether precipitated from brine lakes or emplaced within desiccated sediments, are usually precipitated in the lowest areas of enclosed arid drainage basins - environments that are characterized by almost horizontal and largely vegetation-free surfaces of fine-grained sediments. These base-

level plains are a distinctive feature of deserts and are given many different names (sabkha, sebkha, playa, salina, pan, chott, etc.). The name playa is employed here for these features (Fig. 2).

Alluvial fans at basin edges trap most coarse detritus so that only the finest material is carried into the basin. There it is periodically reworked into horizontally laminated sediments by sheetwash associated with storms. Apart from surface flow during storms, water circulation is generally confined to the subsurface.

Some playas have water tables so deep that no groundwater discharge occurs at the surface. These playas possess smooth, hard and dry surfaces and evaporites are commonly lacking. Most evaporites accumulate within playas where groundwater discharge occurs and this may be: 1) indirect, caused by capillary rise, evaporative pumping or evapotranspiration by phreatophytes from a shallow water table, or 2) directly from the water table (perennially or seasonally at the playa surface) or from springs. Many playas are equilibrium deflation-sedimentation surfaces with topography controlled by the water table level and its gradients.

The closeness of the water table to the surface allows great evaporative loss and concomitant concentration of pore fluids. Playas are thus sites of brine formation irrespective of the salinities of peripheral groundwaters that feed into them. The brine type and the mineralogy of evaporites that precipitate are, however, dependent upon the chemical composition of the groundwater supply.

Hydrographic lows on the surface may be occupied by perennial or seasonal bodies of shallow water (playa lakes), fed directly by groundwater seepage, by springs or by accumulation of storm waters. Playa lakes exist only at times when water input (precipitation and inflow) are less than the water lost by evaporation. The latter is dependent upon climate, water salinity and the geometry of the water body.

Continuing evaporation and evapotranspiration generate a pronounced groundwater concentration gradient towards the basin centre or along the flow paths taken by the groundwater. Saturation with respect to calcium and magnesium carbonates is reached at an early stage, causing precipitation of

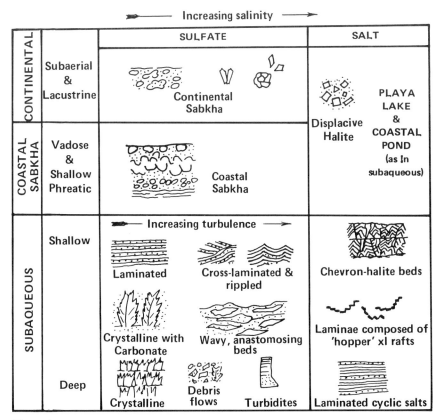

Figure 1
Summary of physical environments of evaporite deposition and the main facies present (modified from Schreiber et al., 1976).

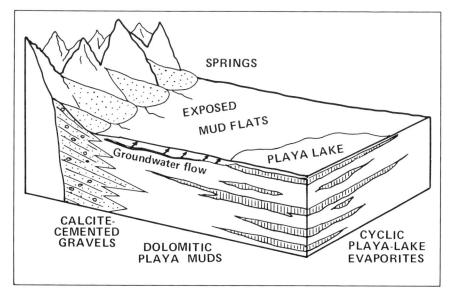

Figure 2
*Schematic block-diagram showing deposi-
tional framework in the Playa Complex model
(after Eugster and Hardie, 1975).*

calcite cement and caliche layers in
alluvial fans, or of soft micron-sized high
Mg-calcite and protodolomite in playa
fringes, or of travertines and pisolitic
caliche when precipitation occurs from
surface waters associated with peripheral springs. Deposition on playa flats
occurs as a mud because sediments
here are kept permanently moist by the
groundwater discharge (Eugster and
Hardie, 1975). The carbonates should
be considered evaporites because they
form in exactly the same manner as
gypsum and more saline minerals
further into the basin. Together with any
detrital sediments, the carbonate muds
are transported toward the basin centre
by storm sheetflood which impart the
laminated or cross-laminated structures. This lamination, however, is also
continuously being disrupted and destroyed by further groundwater discharge
(creating porous 'puffy-ground' surfaces), by the growth and dissolution of
ephemeral evaporite crystals and
crusts, and by episodes of surficial
drying that cause extensive and multiple
mud-cracking. Mine tailings on playas
have been destroyed by these processes in less than 50 years.

Removal of the less soluble mineral
phases (Ca-Mg carbonates and calcium
sulphates) profoundly modifies the
groundwater composition and thus the
sequence and type of saline minerals
that will precipitate in the basin centre. A
mineral zonation is formed with the most

soluble minerals located at the most
distal parts of the groundwater flow and
segregated from the less soluble
phases. In this manner monomineralic
evaporite deposits are formed.

Drying of the playa surface may cause
sediment deflation. Gypsum crystals,
precipitated displacively in the uppermost playa sediments, are concentrated
as lag deposits and may be swept
together to form gypsum dunes.
Surficial gypsum may also dehydrate to
bassanite or anhydrite and, in some
playas, calcium sulphate is emplaced
directly as nodular anhydrite that is
seemingly identical with that in coastal
sabkha environments.

Efflorescent crusts of saline minerals
accumulate on playa surfaces during
groundwater discharge and evaporation, or by the evaporation to dryness of
ponded stormwaters. Because evaporation is rapid and complete, the crusts
include metastable and highly soluble
salts. Rain and storm waters dissolve
these minerals to form concentrated
brines that owe their highly modified
compositons to this fractional dissolution. Ultimately these brines reach the
basin centre.

Evaporite crusts may reach 30 or
more centimetres in thickness. Con-
tinual growth of salt crystals causes
great volume increases and formation of
salt-thrust polygons (and other types of
patterned ground) or highly irregular
surfaces with relief perhaps reaching
several metres.

Even the salts that initially survive
dissolution by storm waters and become
buried are ephemeral if underlying
groundwaters are undersaturated. Upward movement of the less saline water
dissolves the salt crust and reprecipi-
tates it at the new surface. Towards the
basin centre groundwaters become
increasingly saline and calcium sulphate and even halite may become
stable in the sediment. It is important to
note that minerals in surface crusts do
not necessarily reflect the character of
evaporite minerals that are preserved in
underlying sediments. Many fine-
grained dolomitic red-bed sequences,
such as the Keuper of Europe and the
Watrous-Amaranth-Spearfish Formation of the Williston Basin, (Fig. 3),
probably represent deposits of these
evaporitic, but essentially non-
evaporite-preserving, environments.

Halite within playa sediments has not
been specifically described from mod-

Figure 3
*Red dolomitic mudstones with anhydrite
pseudomorphs after gypsum crystals,
Watrous Formation, Saskatchewan. Probable playa flat deposit. Slab 9 cm wide.*

148

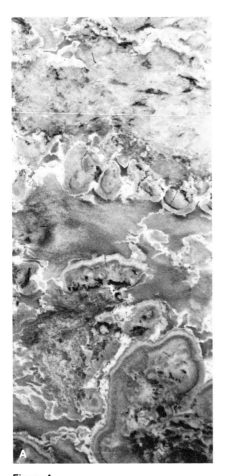
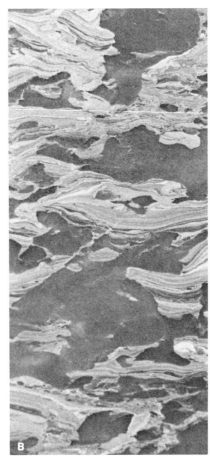

Figure 4
Anhydrite (after gypsum crystals) displacing (and replacing?) pisolitic caliche in (A) and possible laminated dolomite-calcium sulphate

playa flat deposits in (B); Whitkow Anhydrite (Prairie Evaporite Formation; Middle Devonian), Saskatchewan. Slabs about 9 cm wide.

ern environments. Smith (1971) described displacive halite in Permian red mudstones which he interprets as forming in a playa flat. Euhedral to subhedral halite cubes occur in abundant matrix but with increasing halite content the matrix becomes disconinuous, then confined to isolated polyhedral pockets as the halite becomes a near-continuous interlocking mosaic. In more coherent sediments, however, halite occurs interstitially, as veneers around sand grains or as skeletal 'hopper' crystals (Fig. 5) – the last mentioned sometimes assuming extreme forms.

Playa lakes lie at the termination of groundwater flow paths and also accept concentrated brines formed when overland flows dissolve efflorescent crusts on the playa flats. These ponded brines continue to suffer evaporation and saline minerals are precipitated on the brine surface, at the brine-sediment interface and, perhaps also within the

Figure 5
Displacive halite hopper crystal in dolomite. Souris River Formation (Upper Devonian), Saskatchewan. Core is 12 cm across.

bottom sediments to form bedded crystal-brine accumulates. The characteristics of these accumulates are similar to those of marine-derived subaqueous evaporites (see later article).

Evaporation in perennial playa lakes (lasting all year) produces an orderly succession of saline minerals with the more soluble overlying the less soluble. Freshening of lakes after storms or during the 'wet' season dissolves the uppermost, more soluble minerals if water mixing is complete (as in shallow lakes). In deeper playa lakes the brine may become stratified with less saline water overlying a denser, more saline brine which protects the salt-layer from dissolution.

The crystal accumulates become exposed to the air during 'dry' seasons in many shallow playa lakes. The interlocking salt crystals have high porosities and interstices between crystals are occupied by saturated brine. The salt surface is kept moist by evaporative draw and by precipitation of dew on hydrophilic salt surfaces during cold nights. The evaporation rate falls to values as low as 1/170th of the rate from standing bodies of the same brine; thus the brine level rarely drops more than a few metres beneath the surface. The crystalline surface is dissected by salt-thrust polygons and much eolian dust is trapped on the rough and damp surface. During 'wet' seasons, lakes are flooded by storm waters which dissolve surficial salts and introduce clastic material. Since new saline material is introduced during such times, generally less salt is dissolved than was precipitated during preceeding 'dry' seasons. Evaporation during the suceeding 'dry' season creates a new salt layer. Each salt layer is thus largely composed of recycled material. Layers are separated by mud partings composed of detrital material introduced by storm waters and the eolian sediment deposited on the emergent salt surfaces.

The order of salt deposition in seasonal playa lake deposits is commonly not that which would be predicted by the theoretical crystallization sequence from the brine. More soluble salts are found beneath less soluble, forming 'inversely stratified' salt deposits. Such sequences form because: 1) the more soluble salts in surface layers are dissolved during lake-flooding episodes, and 2) concentrated, dense brines created by evaporation during the emergent episodes sink through the crystal accumulate and displace the less concentrated brines, which emerge to the surface, there to cause further dissolution of more saline phases. It is from the descending, dense brines that the permanent, more saline salts precipitate. They must be regarded as early diagenetic additions. Density mixing of brines during emergent phases probably also encourages the replacement of metastable by stable minerals and the recrystallization of earlier formed salts; it thus contributes towards the early diagenetic lithification of the salt deposit. These effects are absent or less efficient within deposits formed from permanent brine lakes.

Variations in the Playa Model

Climate, groundwater source and composition, and the size of the playa complex are the main factors that dictate the type and distribution of evaporites within the playa setting.

Climate. Temperature influences evaporation rate but may also control the type and sequence of salt deposition more direclty. For example, in warm climates brines may precipitate halite before any sodium sulphate is deposited as thenardite (Na_2SO_4). Lakes in colder climates (as in the Prairie Provinces) precipitate mirabilite ($Na_2SO_4 \cdot 10H_2O$) prior to halite.

Water input into the playa basin determines whether evaporites will precipitate and be preserved or not. They accumulate only at times when the water budget is a negative one. The history of playa lake complexes is one of alternating wet (pluvial) and dry (arid) conditions with corresponding transgressive, freshened, non-evaporite-precipitating lakes and regressive (shrinking) saline lake or dry playa stages. Pluvial phases (Fig. 6) are marked by partial to complete dissolution of earlier formed salts, by deposition of basal transgressive conglomerates and beach deposits over former playa flat deposits, and by deposition of non-saline lacustrine sediments (among which oil-shales may be conspicuous). Increasing aridity is recorded by shrinkage of the lake area, a decrease in lake depth, and an increase in salinity ultimately leading to bedded evaporite deposition.

Climatic changes may also be reflected in non-lacustrine playa sediments. Widespread rhythms of increasing evaporite content in red, dolomitic mudstones and siltstones of the Keuper (Upper Triassic) of Europe can be interpreted as indicating gradual reductions in the water influx to the depositional site and a corresponding increase in the persistence of evaporites in the sediments (Wills, 1970).

Groundwater source. It has been assumed that groundwaters move radially from the hinterland, converging toward the basin centre, which also marks the hydrographic low point of the basin. Flow is also assumed to be essentially horizontal and shallow subsurface (except during storms). This produces a concentric pattern of increasing groundwater salinity and a 'bulls-eye' pattern of salt deposition (more saline salts in the centre; Fig. 7). When the deepest part of the basin floor is not centrally located, or when groundwater enters the basin from one side, this ideal pattern is disturbed and becomes asymmetric. The compositions of brines in lakes fed by rivers are not modified by the prior precipitation and retention of less saline salts in peripheral playa flats. These brines retain their carbonate and sulphate contents and low-solubility salts may precipitate within the lake. Consequently there is less mineral segregation in river-fed than in groundwater-fed playa systems.

Chaotic and disturbed sediments with irregularly distributed salt lenses occur beneath playas fed by artesian groundwater. The rise of less saline water dissolves previously deposited evaporites except where they are protected by impermeable clay seals. Removal of deep-lying salts results in localized subsidence, creation of depressions occupied by small playa lakes and pools, and deposition of small, isolated salt deposits.

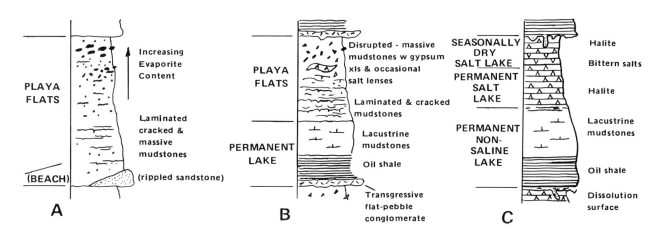

Figure 6
Hypothetical cycles reflecting increasing aridity. A: Distal playa flats, B: Playa flats marginal to playa lake, C: Playa lake.

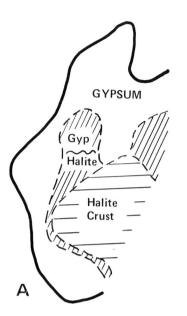

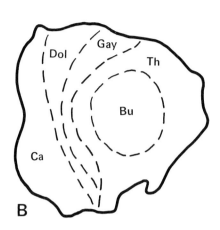

Figure 7
Saline mineral zonation in playas: A: Yotvata Sabkha (Israel) after Amiel and Friedman (1971); B: Deep Spring Lake, California, Ca =

calcite/aragonite; Dol = Dolomite; Gay= gaylussite; Th=Thenardite; Bu= Burkeite (after Jones, 1965).

Playas fed by artesian flow or from rivers may possess water tables that are located higher than those of neighbouring areas. Groundwater moves and becomes compositionally modified towards the basin-edge – in directions opposite to that in the ideal model. Mueller (1960) has shown that saline waters from the Andes evaporate on the floor of the central valley. Residual brines containing nitrates and iodates move upslope through the soil of the coastal mountain slopes by capillary migration and eventually evaporate to complete dryness.

Groundwater composition.
The mineralogy of salts precipitated in closed basins is controlled by the groundwater composition which, in turn, depends mainly upon the rock types in the source area and their mode of weathering. Commonly the evaporite minerals are similar to those precipitated from oceanic waters (hence the difficulty of distinguishing between them) and there is a predominance of alkaline-earth carbonates and various sulphates. This reflects the dominance of the same ions (Ca^{++}, Mg^{++}, Na^+, $CO_3^=$, HCO_3^-, $SO_4^=$, Cl^-), however, these may be in different proportions from those in sea water. Such differences are most evident when the more saline salts are

precipitated. Commonly calcium, sodium, and bicarbonate are present in excess, leading to precipitation of salts such as pirssonite ($CaCO_3 \cdot Na_2CO_3 \cdot 2H_2O$), gaylussite ($CaCO_3 \cdot Na_2CO_3 \cdot 5H_2O$), and trona ($Na_2CO_3 \cdot NaHCO_3 \cdot 2H_2O$). When groundwaters are sulphate-rich then (dependent upon the dominant cations) glauberite ($CaSO_4 \cdot Na_2SO_4$), epsomite ($MgSO_4 \cdot 7H_2O$), bloedite ($MgSO_4 \cdot Na_2SO_4 \cdot 4H_2O$), thenardite ($Na_2SO_4$) and mirabilite ($Na_2SO_4 \cdot 10H_2O$) may be precipitated. Variation in playa and playa-like mineralogy constitutes a vast field of study, one that cannot be discussed here. Reference should be made to Reeves (1968) and Hardie and Eugster (1970).

Size of the playa complex. Much of the surface water introduced during storms will reach the basin centre in small playa basins. Lakes will thus exhibit many cycles of salt dissolution and precipitation and the efficiency of leaching salts from playa flats will be high. In contrast, storm waters may not reach playa lakes that are surrounded by vast playa flats: the water evaporates before it reaches the basin centre. Even during the 'wet' seasons waters may fail to reach lakes, which then become flooded only during

exceptional circumstances. Such 'lakes' spend much of their time with emergent salt surfaces. Salt leaching on adjacent playa flats is inefficient and crystal accumulates will suffer more early-diagenetic changes at depth than those deposited in small playa basins.

Supratidal (Coastal Sabkha) Evaporites
Coastal sabkha evaporites were briefly described in an earlier article in this facies model series under the heading of arid-zone variants of carbonate shallowing-upwards sequences (James, 1977). This style of shallowing-upward sequences (Fig. 8) is composed of (in upwards sequence): (1) carbonates, or less commonly clastics, (2) similar sediments but with angular anhydrite nodules, pseudomorphic after gypsum crystals (Fig. 9) and (3) nodular-mosaic anhydrite, commonly terminated by a sharp erosive contact (Fig. 10). These evaporites are interpreted as diagenetic emplacements within supratidal environments because of their close resemblance to the sequence of lithologies in the progradational wedge along the Abu Dhabi coast of the Persian Gulf (Shearman, 1966; Kinsman, 1969; Butler, 1970; Bush, 1973).

In areas of arid climate and low eolian sand influx the seaward progradation of subtidal and intertidal facies generates broad coastal flats (or sabkhas) that lie just above high tide level and extend between the offshore water body (commonly with coastal lagoons) and regions

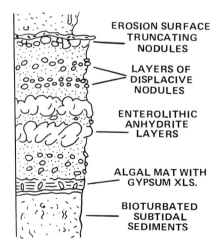

Figure 8
Characteristic features of coastal sabkha evaporites (after Shearman, 1966).

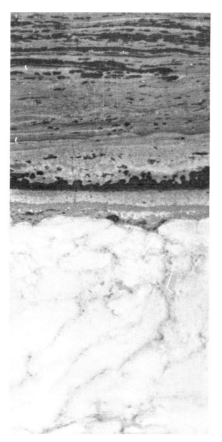

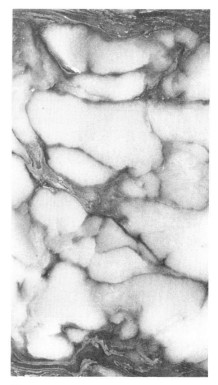

Figure 11

Mosaic anhydrite (displacing subaqeous laminated anhydrite) with individual nodules distorted against each other (a natural consequence of displacive growth). Ordovician Stonewall Evaporite, Saskatchewan. Slab is 8.5 cm wide.

Figure 10

Upper part of sabkha cycle illustrated in Figure 9. Mosaic anhydrite cut across by erosion surface and surmounted by laminated micrites (with late-diagenetic anhydrite) of next lagoonal member. Frobisher Evaporite, Saskatchewan. Slab is 10 cm wide.

Figure 9

Laminated microdolomites (probably hypersaline lagoonal) overlain by algal mat, with gypsum pseudomorphed by anhydrite, and nodular and mosaic coastal sabkha anhydrite. Frobisher Evaporite (Mississippian) Saskatchewan. Core fits together but was rotated during slabbing - slab is 10 cm wide.

of arid continental sedimentation. This environment is a product of both depositional and diagenetic processes, the most important of the latter being the displacive growth of early diagenetic calcium sulphate (or halite). The sabkha is an equilibrium geomorphic surface whose level is dictated by the local level of the groundwater table. Sediment above the capillary fringe dries and is blown away by the wind.

Indigenous sediments of the supratidal flats are a reflection of the offshore sediment mosaic but may contain a substantial proportion of detrital sediment from the hinterland. Offshore sediments are washed over the sabkha during storms that periodically inundate seaward parts with marine floodwaters. Depressions (filled and buried tidal channels) act as conduits for flood and seepage waters.

Groundwaters beneath the sabkha are responsible for transporting materials precipitated as solid phases (eva-

porites, dolomite) and for removing by-products of diagenetic reactions and non-accumulating ions. These waters become progressively concentrated as they advance into the interior of the sabkha and all but the very seaward and landward margins may be saturated with respect to halite. Concentration occurs by evaporation from the capillary fringe and by dissolution of earlier-formed evaporites (particularly halides). Groundwaters lost by evaporation are replenished by: 1) downward seepage of storm-driven floodwaters (flood recharge), 2) gradual intrasediment flow, fluxing from the seaward margin, and 3) intrasediment flow, fluxing from a continental groundwater reservoir that affects landward parts of the sabkha (Fig. 12). Renfro (1974) believes that groundwater flow through continental clastics adjacent to coastal sabkhas (flow induced by evaporative pumping from the sabkha surface) is an important feature in the reddening of these sediments.

The relative importance of the groundwater sources is dependent upon

152

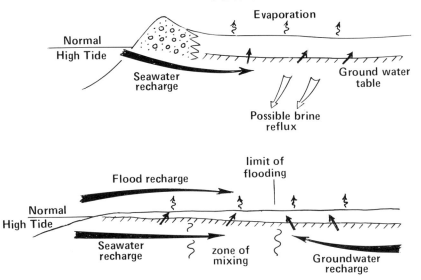

Figure 12
Contrasting water supply in sabkhas. Above, sabkha plain bordered by beach ridge. Seawater recharge is entirely intrasediment flow (based upon a Sinai coastal sabkha;

Gavish, 1974). Below, sabkha groundwaters are replenished by flood recharge and seepage from seawater and continental reservoir (based upon Abu Dhabi sabkha).

local geomorphic conditions. Beach ridges seaward of the sabkha prevent inundation by marine floodwaters, whereas lack of hinterland relief will restrict continental groundwater inflow. Cemented sediment layers and algal mat sediments beneath the sabkha surface inhibit upward movement of deeper-lying groundwaters, thus increasing the importance of marine flooding.

Concentration of groundwater causes precipitation of diagenetic minerals; some as direct precipitates, others as products of reactions between groundwater brines and earlier-deposited sediments. Gypsum is not precipitated on the exposed sediment surface but grows displacively within algal mat or other upper intertidal sediments (forming crystal mushes up to 1 m thick) or grows poikilitically within supratidal sand sediments where it occurs as large, lenticular crystals that include sand grains arranged in herring-bone patterns.

Gypsum precipitation in the intertidal and near-shore supratidal environments causes groundwaters to become depleted in calcium. The increased Mg/Ca ratio of brines induces dolomitization of pre-existing aragonite and the precipitation of magnesite. Dolomitization of

aragonites releases strontium that is precipitated as celestite.

In the Abu Dhabi sabkha, anhydrite first appears one km inland from the normal high water mark, in the capillary zone. It occurs as discrete nodules and as bands of coalesced nodules, some of which may take the form of ptygmatic (enterolithic) layers. Growth of nodules occurs by host sediment displacement. Dilution of the host sediment may occur to such an extent that it is relocated to internodule areas and its fabrics are destroyed. In extreme cases host sediments are confined to mere partings between the anhydrite nodules (mosaic anhydrite: Fig. 11) Some nodules are formed by alteration of earlier formed gypsum crystals (Butler, 1970). Pseudomorphs lose shape because of flowage

Figure 13
Alternations between mosaic anhydrite (top and bottom) and microdolomites much disrupted by growth of halite (now pseudomorphed by anhydrite) and gypsum crystals (now anhydrite). Dolomite intervals probably represent former inter- and subtidal sediments partially obliterated by sulfate growth during reflux dolomitization, Frobisher Evaporite (Mississippian), Saskatchewan. Slab is 10.5 cm wide.

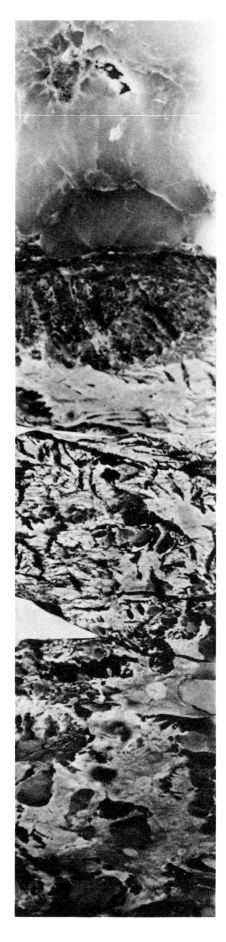

(adjustment during compaction) and the continued growth of primary anhydrite laths in and between pseudomorphs. Composite anhydrite nodules are remnants of gypsum crystal clusters and massive-appearing anhydrite forms from gypsum mush in former upper intertidal sediments. The displacive growth of anhydrite and gypsum in intertidal and supratidal sediments is believed to raise (jack-up) the sediment surface. If the water table does not rise a corresponding amount, then upper parts of the sediment dry out and blow away. Deflation exposes anhydrite and gypsum at the surface, concentrating nodules and crystal fragments as a regolith, or breaking up nodules into laths that become strewn across the sabkha surface. Isolated anhydrite laminae at the top of some ancient sabkha sequences may have formed by such nodule and crystal destruction.

Halite occurs as salt crusts on the surface, as veneers around grains in the upper part of the capillary zone and as solid cubes in sand sediments. Within fine grained sediments the displacive halite cubes assume a skeletal hopper form commonly to extreme degrees (Fig. 5). In most described modern sabkhas halite is not an accumulative phase but is blown away or dissolves in floodwaters. Repeated growth and dissolution of halite can so disrupt the host sediment that all its original fabrics are destroyed.

Variations in the Coastal Sabkha Model
Variations in the nature of the host sediment, the character of the offshore water body, the type of diastrophic control and the effects of differing local topography all may cause profound modifications from the 'norm', as represented by the Abu Dhabi sabkha (Fig. 14).

Nature of the host sediment. This determines the amount of drainage, the subsequent history of sabkha brines and the compactional history of the evaporate deposit.

Impermeable sediments inhibit brine reflux and, by curtailing downward seepage of floodwaters, extend the width of the area affected by flood recharge. This surface flooding, however, causes little dilution of existing groundwaters. Finer grained sediments also allow thicker capillary fringes to

form. Thicker evaporite sequences should be formed in fine-grained sediments because of this but the control has yet to be demonstrated in ancient examples.

Carbonates (particularly aragonite) in host sediments are of major importance. Dolomitization of carbonates releases calcium which reacts with sulphate in groundwaters to form more gypsum and anhydrite. This additional sulphate precipitation and dolomitization reduces the sulphate and magnesium content of brines in carbonate sabkha interiors to

low levels and causes magnesite (precipitated earlier) to redissolve. In non-carbonate sediments, dolomitization does not occur, the sabkha interior brines retain 60 to 70 per cent of their sulphate, much less gypsum and anhydrite is emplaced, and brines remain magnesium rich so that magnesite remains stable. The sulphate and magnesium rich brines formed in non-carbonate sabkha sediments react with earlier-formed gypsum to form polyhalite (Holser, 1966):

$$2CaSO_4 \cdot 2H_2O + 2K + Mg + 2SO_4 = K_2MgCa_2(SO_4)_4 \cdot 2H_2O + 2H_2O$$

gypsum + brine   polyhalite

Reflux of brines capable of dolomitizing deeper-lying carbonates (well beneath the sabkha vadose zone) cause calcium sulphate precipitation (gypsum, perhaps anhydrite) in these deeper-lying sediments. Growth of sulphates in

subtidal carbonate intervals between sabkha evaporites by this reflux dolomitization may obliterate evidence of the cyclic nature of an evaporite deposit and create a single thick, composite unit of nodular anhydrite. Alternations between

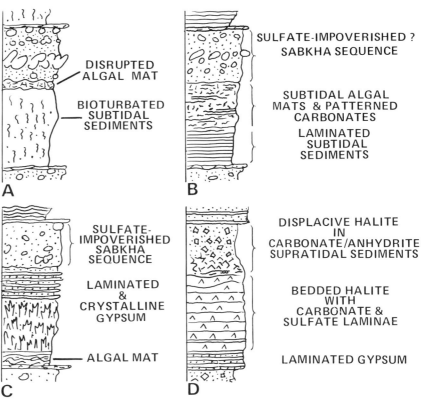

Figure 14
Hypothetical shoaling-upwards cycles: A: marginal to a normal marine to slightly hypersaline water body, B: marginal to a hypersaline water body within which sulphates are bacterially reduced, C: marginal water body precipitates and preserves gypsum, D: marginal water body is salt-precipitating; supratidal sequence largely composed of displacive (and replacive?) halite.

154

nodular and mosaic anhydrites and disrupted dolomite intervals full of gypsum pseudomorphs in parts of the Mississippian Frobisher Evaporite in Saskatchewan may represent such partially obliterated cycles (Fig. 13).

Differences in sediment coherency dictate subsequent compactional history. Lithified or coherent sediments preserve gypsum pseudomorphs or the moulds of dissolved halite crystals. Compressible sediments (particularly organic-rich varieties), on the other hand, allow anhydrite nodules to grow, to coalesce and compact perhaps to form sluggy or even laminar anhydrites (Shearman and Fuller, 1969). Mossop (1978) believes laminar anhydrites in the Ordovician Baumman Fiord Formation (Fig. 15) were originally nodular and have been drastically altered by early diagenetic compaction and flowage.

Nature of the offshore water body. Most commonly the offshore water body is normal marine to slightly hypersaline (well below gypsum saturation). Here subtidal-intertidal sediments are bioturbated and skeletal rich, and algal-mat sediments (if present) are confined to upper intertidal to low supratidal environments (Fig. 14A) where they may become disrupted by subsequent growth of gypsum (James, 1977).

When sabkhas border hypersaline (gypsum precipitating) water bodies, the sediments beneath sabkha evaporites are laminated (burrowing biota absent) and algal mats extend well into subtidal environments where they may be preserved. When precipitated gypsum persists in the bottom sediment, the overlying sabkha sequence forms the uppermost member of a largely subaqueous evaporite sequence (Fig. 14B). However, the abundance of organic matter and dissolved sulphate in hypersaline waters normally induced reducing conditions within which sulphate-reducing bacteria thrive. Their activities cause reduction of gypsum; formation of hydrogen sulphide with precipitation of carbonates and pyrite as by-products (Friedman, 1972) and perhaps formation of patterned carbonates (Dixon, 1976; Kendall, 1977: Fig. 14B). Removal of calcium and sulphate from the offshore water body may severely restrict gypsum and anhydrite formation in adjacent sabkha environments (Fig. 14B and C).

Figure 15
Numerous superimposed sabkha cycles, Baumann Fiord Formation, Ellesmere Island (photo courtesy G. Mossop).

The atmosphere adjacent to large bodies of normal marine water is too humid for halite to persist in the subaerial environment (Kinsman, 1976). If the water body is a concentrated brine, however, its water vapour pressure may be low enough not to increase atmospheric humidity. Halite can thus become an accumulative phase in sabkhas that neighbour hypersaline water bodies (particularly those saturated or near-saturated with respect to halite). Shearman (1966), Friedman (1972) and Smith (1971, 1973) have described halite rocks that appear to have formed by displacing or replacing earlier carbonate-sulphate sabkha sediments. Such sediments form adjacent to halite-precipitating water bodies (Fig. 14D).

Diastrophic control. Shoaling-upwards sequences terminated by coastal sabkha deposits can form as a result of three different events. The most commonly offered interpretation is that each sequence is a separate progradational event. Sabkha plains are generated by sediment accretion with little or no significant sea-level fall. Mossop (in press) and Ginsburg (*in* Bosellini and Hardie, 1973) have independently developed hypotheses which generate successive shoaling-upwards cycles in carbonate-producing areas in a regime of continuous subsidence.

On the other hand, sediment emergence, with formation of supratidal surfaces, can also be achieved by relative falls in sea-level, independently of any sediment up-building. Sea-level changes may be the result of external events (glaciations?) or of restriction of the water body from the world ocean and subsequent removal of water by evaporation (evaporative downdraw). Criteria for distinguishing cycles that form from progradational events from those that reflect episodes of evaporative downdraw do not appear to have been sought.

When greater subsidence occurs towards the basin centre it is possible to recognize distal from proximal locations (Mossop, 1979). Basinwards the cycles are thicker and are dominated by thick subtidal units. Short-lived or less extensive transgressive events may not reach basin margins so that marginal successions contain fewer and thinner cycles that are dominated by supratidal (sabkha) members. Coalescence of several supratidal units may also generate thick evaporite sequences at marginal locations

Topographic control. The Abu Dhabi 'norm' is associated with a relatively simple progradational sediment wedge undissected by active channels, maritime lakes or ridges formed by former beach or offshore spit deposits (Fig. 16). This situation reflects the constant conditions (slightly falling sea-level) that have occurred since the sabkha began to form and the protection afforded by an offshore island chain. When protective barriers are absent, or if sediment supply or rates of sea-level change are variable the accumulation of supratidal sabkha sediments are more discontinuous and parts of the intertidal and subtidal environment are isolated by growth of beach bars and spits. Here we have an arid-zone equivalent of the chenier plain – an environment recently described by Picha (1978) from Kuwait but to date one that has not been recognized in the ancient.

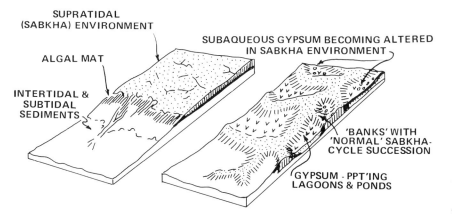

Figure 16
Contrasting patterns of supratidal sedimentation. A: simple sediment wedge, Recent Abu Dhabi sabkha; B: inferred environment for part of Frobisher Evaporite (Mississippian) in southeastern Saskatchewan - numerous shallow maritime lakes isolated by narrow strips of supratidal sabkha.

Drowned valleys or former tidal channels, isolated by spit development or by the formation of beach barrier ridges, may occur within the sabkha environment. If connection is retained with the sea, flow into the former channels occurs in response to a lowered water level caused by evaporation from the standing body of water. Such depressions will also attract groundwaters from beneath the surrounding sabkha and disrupt the more normal pattern of groundwater flow. The Sebkha el Melah (Busson and Perthuisot, 1977) was such a depression but has since been filled with evaporites including a halite sequence 30 metres thick. Beds of subaqueous gypsum, patterned dolomites (representing bacterially-reduced calcium sulphates) or halite beds within 'normal' sabkha deposits may represent the fills of depressions on the sabkha surface.

The evaporite portion of sabkha cycles in the Mississippian Frobisher Evaporite of Saskatchewan (Figs. 9, 10, 13, 17) is dominated by large, subaqueously-precipitated gypsum crystals (now pseudomorphed by anhydrite (also Fig. 10 of Facies Models 14)). They pass laterally into more 'normal' sabkha sequences composed of nodular and mosaic anhydrite. The former gypsum crystals are also deformed by anhydrite nodule growth (Fig. 17) indicating they were transformed to anhydrite or bassanite during early diagenesis. Since more than 90 per cent of the sulphate was precipitated subaqueously a provisional environmental reconstruc-

tion having resemblance to the humid sub-tropical environment of Florida Bay is suggested. Deposition occurred in hypersaline lagoons separated by narrow barriers upon which 'normal' sabkha sequences were formed. The gypsum crystals were precipitated in the lagoons but as progradation of the lagoon complex occurred, older lagoons became more distant from the open sea and dried out to become part of the sabkha plain. In this desiccated environment gypsum dehydrated and new anhydrite grew displacively as nodules.

It is probable that most environments which include supratidal sediments have arid-zone equivalents within which evaporites have formed. We have still to look for them in the rock record.

Acknowledgements
The final manuscript was reviewed by Noel James.

Bibliography
There are numerous papers of merit dealing with evaporite sedimentology but unfortunately few deal with facies models or summarize earlier work. Many facies were first, or are best, described from Canadian deposits but others have yet to be adequately described from Canada. Canadian sources are thus not listed separately but are identified by asterisks.

General

Kirkland, D. W. and R. Evans, 1973, Marine Evaporites: origin, diagenesis and geochemistry: Benchmark Papers in Geology, Stroudsburg, Penn., Dowden, Hutchinson and Ross.
Probably the best starting point. A carefully selected collection of papers (emphasizing calcium sulphate and halite deposits) with informative introductory comments. Now slightly out of date in that subaequeous evaporites are under-represented. This failing is now filled with publication of:

Dean, W. E. and B. C. Schreiber, eds., 1978, Notes for a short course on marine evaporites: SEPM Short Course #4.
The most recent and comprehensive compilation of work upon the evaporites. The paper by Schreiber upon subaqueous sulphates is essential reading and the section upon halite fabrics by Shearman is clearly written and illustrated. Other papers concentrate upon environments, geochemistry and geophysical log evaluation of evaporites.

Busson, G., 1974, Sur les evaporites marines: sites actuels ou Recents de depots d'evaporites et leur transposition dans les series du Passe: Rev. Geog. phys. Geol. dynam., v. 16, p. 189-208.
A critical review of evaporite environments.

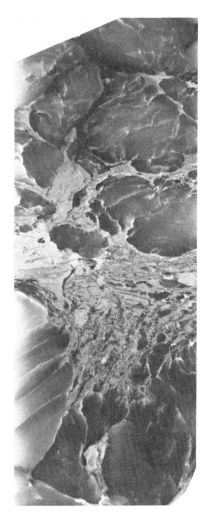

Figure 17
Displacive gypsum crystals (centre) distorted by growth of later but still early-diagenetic anhydrite nodules. Frobisher Evaporite, Saskatchewan. Slab is 10.5 cm wide.

Hsu, K. J., 1972, Origin of saline giants: A critical review after the discovery of the Mediterranean evaporite: Earth Sci. Rev., v. 8, p. 371-396.
Vast bodies of evaporites are reinterpreted as products of desiccated seas.

Shearman, D. J., 1971, Marine Evaporites: the Calcium Sulphate Facies: Alberta Soc. Petrol. Geol. Seminar, Univ. Calgary, 65 p.

Strakhov, N. M., 1970, Principles of Lithogenesis, vol. 3: New York and Oliver and Boyd, Edinburgh, Plenum Publ. Corp., 577 p.
A survey of soviet ideas on arid-zone sedimentation, concentrating upon evaporites. Particularly good in its use of evidence from Recent salt lakes and ancient deposits.

Modern Continental Evaporites

Amiel, A. J. and G. M. Friedman, 1971, Continental sabkha in Arava Valley between Dead Sea and Red Sea: Significance for origin of evaporites: Bull. Amer. Assoc. Petrol. Geol., v. 55, p. 581-592.
Cook, R. U. and A. Warren, 1973, Geomorphology in Deserts: London, B. T. Batsford Ltd.
Parts 2 upon desert surface conditions, 2nd 3.5 upon playa systems provide the essential geomorphic background to continental evaporites.

Glennie, K. W., 1970, Desert Sedimentary Environments: Developments in Sedimentology 14, Amsterdam, Elsevier, 222 p.

Hardie, L. A. and H. P. Eugster, 1970, The evolution of closed-basin brines: Mineral. Soc. Amer. Spec. Publ. 3, p. 273-290.

Jones, B. F., 1965, The hydrology and mineralogy of Deep Springs Lake, Inyo County, California: U.S. Geol Survey Prof. Paper 502-A, 56 p.
Describes concentric zonation of carbonate-sulphate evaporite minerals in a small playa lake.

Kinsman, D. J. J., 1969, Modes of formation, sedimentary associations and diagnostic features of shallow-water and supratidal evaporites: Amer. Assoc. Petrol. Geol. Bull., v. 53, p. 830-840.

Kulke, H., 1974, Zur Geologie und Mineralogie der Kalk- und Gipskrusten Algeriens: Geol. Rundshau, v. 63, p. 970-998.

Reeves, C. C. Jr., 1968, Introduction to Paleolimnology: Developments in Sedimentology 11, Amsterdam, Elsevier, 228 p.

Valyashko, M. G., 1972, Playa lakes - a necessary stage in the development of a salt-bearing basin: in G. Richter-Bernburg, ed., Geology of saline deposits: Proc. Hanover Symposium 1968, UNESCO, Paris, p. 41-51.

Surprisingly, the sedimentology of Recent playa-lake deposits containing sodium sulphate in the Prairie Provinces have not been studied. Most relevant information is to be found within:

*Tomkins, R. V., 1948, Natural sodium sulphate in Saskatchewan: Saskatchewan Dept. Nat. Resources Indust. Tech. Econ. Ser., Rept. 1, 99 p.

Ancient Continental Evaporites

No detailed studies appear to have been made of possible continental evaporites and associated evaporitic sediments in Canada. They occur in the basal Mississippian of the Maritimes, parts of Arctic Canada and in the Juro-Triassic Watrous-Amaranth Formations of Saskatchewan and Manitoba.

Deardorff, D. L. and L. E. Mannion, 1971, Wyoming trona deposits. Wyoming Univ. Contr. Geology, v. 10, p. 25-37.

Dyni, J. R., Hite, R. J. and O. B. Raup, 1970, Lacustrine deposits of bromine-bearing halite, Green River Formation, northwestern Colorado, in J. L. Rau and L. F. Dellwig, eds., Third Symposium on Salt: Northern Ohio Geol. Soc., Cleveland, Ohio, p. 166-180

Euster, H. P. and L. A. Hardie, 1975, Sedimentation in an ancient playa-lake complex: The Wilkins Peak Member of the Green River Formation of Wyoming: Geol. Soc. Amer. Bull., v. 86, p. 319-334.
Although not dealing primarily with evaporites, contains an excellent summary of the playa environment which was used as the basis for the section upon continental evaporites in this paper.

Jacka, A. D. and L. A. Franco, 1974, Deposition and diagenesis of Permian evaporites and associated carbonates and clastics on shelf areas of the Permian Basin: in A. H. Coogan, ed., Forth Symposium on Salt, Northern Ohio Geol. Soc., Cleveland, Ohio, p. 67-89.

Van Houten, F. B., 1965, Crystal casts in Upper Triassic Lockatong and Brunswick Formations: Sedimentology, v. 4, p. 301-313.

Wills, L. J., 1970, The Triassic succession in the central Midlands in its regional setting: Quart. Jour. Geol. Soc. London, v. 126, p. 225-285.

Modern Coastal Sabkhas and Salt-Flats

Bush, P. R., 1973, Some aspects of the diagenetic history of the sabkha in Abu Dhabi, Persian Gulf: in B. H. Purser, ed., The Persian Gulf, Springer-Verlag, Berlin, p. 395-407.

Butler, G. P., 1970, Holocene gypsum and anhydrite of the Abu Dhabi sabkha, Trucial Coast: an alternative explanation of origin: in J. L. Rau and L. F. Dellwig, eds., Third Symposium on Salt, Northern Ohio Geol. Soc., Cleveland, Ohio, p. 120-152.

Gavish, E., 1974, Geochemistry and mineralogy of a recent sabkha along the coast of Sinai, Gulf of Suez: Sedimentology v. 21, p. 397-414.

Holser, W. T., 1966, Diagenetic polyhalite in Recent salt from Baja California: Amer. Mineral., v. 51, p. 99-109.

Kinsman, D. J. J., 1966, Gypsum and anhydrite of Recent age, Trucial Coast, Persian Gulf, in J. L. Rau, ed., Second Symposium on Salt, Northern Ohio Geol. Soc., Cleveland, Ohio, p. 302-326.

Patterson, R. J. and D. J. J. Kinsman, 1976, Marine and continental ground-water sources in a Persian Gulf coastal sabkha: in S. H. Frost, M. P. Weiss, and J. B. Saunders, eds., Reefs and Related Carbonates - Ecology and Sedimentology, p. 381-399.

Picha, F., 1978, Depositional and diagenetic history of Pleistocene and Holocene oolitic sediments and sabkhas in Kuwait, Persian Gulf. Sedimentology, v. 25, p. 427-449.

Phleger, F. B., 1969, A modern evaporite deposit in Mexico: Amer. Assoc. Petrol. Geol. Bull., v. 53, p. 824-829.

Shearman, D. J., 1970, Recent halite rock, Baja California, Mexico: Trans. Instit. Mining Metal., v. 79B, p. 155.

Ancient Coastal Sabkhas and Salt-Flat Deposits

Bosellini, A. and L. A. Hardie, 1973, Depositional theme of a marginal marine evaporite: Sedimentology v. 20, p. 5-27.

*Fuller, J. G. C. M. and J. W. Porter, 1968, Evaporites and carbonates: two Devonian basins of western Canada: Can. Petrol. Geol. Bull., v. 17, p. 182-193.

*Fuzesy, L. M., 1973, The geology of the Mississippian Ratcliffe Beds in south-central Saskatchewan: Saskatchewan Dept. Mineral Resources Rept. 163, 63 p.

*Jansa, L. F., and N. R. Fischbuch, 1974, Evolution of a Middle and Upper Devonian sequence from a clastic coastal plain - deltaic complex into overlying carbonate reef complexes and banks, Sturgeon - Mitsue area, Alberta: Geol. Surv. Canada Bull. 234, 105 p.

Kerr, S. D. and A. Thomson, 1963, Origin of nodular and bedded anhydrite in Permian shelf sediments Texas and New Mexico: Amer. Assoc. Petrol. Geol. Bull., v. 47, p. 1726-1732.

*Mossop, G.D., in press, The Ordovician Baumann Fiord Formation evaporites of Ellesmere Island, Arctic Canada: Geol. Surv. Canada Bull. 298, in press.

*Schenk, P. E., 1969, Carbonate-sulfate-redbed facies and cyclic sedmentation of the Windsorian Stage (Middle Carboniferous), Maritime Provinces: Can. Jour. Earth Sci., v. 6, p. 1037-1066.

Smith, D. B., 1971, Possible displacive halite in the Permian Upper Evaporite Group of northeast Yorkshire: Sedimentology, v. 17, p. 221-232.

Smith, D. B., 1973, The origin of the Permian Middle and Upper Potash deposits of York-shire: an alternative hypothesis: Proc. Yorks. Geol. Soc., v. 39, p. 327-346.

Renfro, A. R., 1974, Genesis of evaporite-associated stratiform metalliferous deposits - a sabkha process: Econ. Geol., v. 69, p. 33-45.

Wood, G. V. and M. J. Wolfe, 1969, Sabkha cycles in the Arab/Darb Formation off the Trucial Coast of Arabia: Sedimentology, v. 12, p. 165-191.

References Cited in Text

Busson, G., and J-P. Perthuisot, 1977, Interêt de la Sebkha el Nelah (sud-tunisien) pour l'interpretation de series evaporitiques anciennes: Sedimentary Geol., v. 19, p. 139-164.

Dimroth, Erich, 1977, Facies Models 5. Models of physical sedimentation of iron formations: Geosci. Canada, v. 4, p. 23-30.

Dixon, James, 1976, Patterned carbonate - a diagenetic feature: Can. Petrol. Geol. Bull., v. 24, p. 450-456.

Friedman, G. M., 1972, Significances of Red Sea in problem of evaporites and basinal limestones: Amer. Assoc. Petrol. Geol. Bull., v. 56, p. 1072-1086.

James, N. P., 1977, Facies Models 8. Shallowing-upward sequences in carbonates: Geosci. Canada, v. 4, p. 126-136.

Kendall, A. C., 1977, Patterned carbonate - a diagenetic feature (by James Dixon): Discussion: Can. Petrol. Geol. Bull., v. 25, p. 695-697.

Kinsman, D. J. J., 1976, Evaporites: relative humidity control of primary mineral facies: Jour. Sed. Petrology, v. 46, p. 273-279.

Mueller, G., 1960, The theory of formation of north Chilean nitrate deposits through 'capillary concentration': Rept. Internatl. Geol. Congress 19th, Norden 1960, Part I, p. 76-86.

Schreiber, B. C., G. M. Friedman, A. Decima and E. Schreiber, 1976, Depositional environments of Upper Miocene (Messinian) evaporite deposits of the Sicilian basin: Sedimentology, v. 23, p. 729-760.

Shearman, D. J., 1966, Origin of marine evaporites by diagenesis: Instit. Mining Met. Trans., v. B75, p. 207-215.

*Shearman, D. J. and J. G. Fuller, 1969, Anhydrite diagenesis, calcitization and organic laminites, Winnipegosis Formation, Middle Devonian, Saskatchewan: Can. Petrol. Geol. Bull., v. 17, p. 496-525.

von der Borch, C. C., 1977, Stratigraphy and formation of Holocene dolomitic carbonate deposits of the Coorong area, South Australia: Jour. Sed. Petrology, v. 46, p. 952-966.

MS received March 8, 1978.
Revised March, 1979.
Reprinted from Geoscience Canada
V. 5, No. 2, p. 66-78.

Facies Models 14. Subaqueous Evaporites

Alan C. Kendall
Amoco Canada Petroleum Company Ltd.
444 Seventh Avenue S.W.
Calgary, Alberta, T2P 0Y2

Introduction

Most studies of marine evaporite deposits have focused attention upon facies that developed along supratidal margins of normal marine basins (see preceeding paper). However, many ancient evaporites were deposited subaqueously within enclosed and hypersaline basins. The primary composition, textures and form of these subaqueous deposits are now only partially understood because, in part, so few hypersaline water bodies occur at the present day for study and there are none comparable in magnitude with those of the past.

The origins of small, thin evaporite deposits and marginal-marine evaporites composed of numerous superimposed sabkha cycles are readily discernable. In contrast, the formation of vast, thick, basin-central evaporites, some of which cover millions of square kilometres and exceed several kilometres in thickness or which may directly overlie oceanic basement, present very different problems.

Some authors suggest that the enormous evaporite deposits form by lateral and vertical accretion in depositional environments similar to those of the present time (in supratidal flats, lagoons and salinas; Shearmen, 1966, Friedman, 1972) whereas others consider that the great difference in scale between Recent and ancient deposits requires drastic departure from the present day settings of evaporites. They suggest

either that precipitation occurred from vast bodies of hypersaline water (Schmalz, 1969; Hite, 1970; Matthews, 1974), or that evaporites were precipitated on the floors of desiccated seas (Hsü *et al.,* 1973).

Theoretical models, which were developed to answer the major compositional problems posed by large evaporite bodies, must be integrated with evidence from rock textures and structures (facies models). Unfortunately this integration is not yet possible because of basic disagreements concerning the depositional palaeogeography of evaporites, and because many evaporite rock characteristics have yet to be studied in detail or have disputed origins. Many evaporites are not just passive chemical precipitates or displacive growth structures, but are transported and reworked in the same ways as siliciclastic and carbonate deposits. For these sediments, sedimentary structures are a major key to unravelling the facies and will be emphasized in this paper.

Facies Models

Internal characteristics of evaporites alone can provide the necessary information about depositional environments. The most pressing environmental concern has been, and still is, the depth of water in which evaporites form.

Schreiber *et al.* (1976) recognize three main environmental settings for subaqueous evaporites. These are identified on the basis of sediment characteristics, believed to reflect the depth at which deposition occurs. Criteria used include: 1) structures indicative of wave and current activity, identifying an intertidal and shallow subtidal environment; 2) algal structures (in the absence of wave and current-induced structures) are believed to identify a deeper environment but one that still resides within the photic zone; and 3) widespread evenly-laminated sediments (rhythmites) that lack evidence of current and algal activity (perhaps associated with gravity-displaced sediments) characterize the deep, subphotic environment.

Considerable difficulty exists in using the presence or absence of algal and current structures as relative depth indicators. Because stromatolites commonly grow in protected, quiet-water, shallow environments, the absence of current structures from algal-bearing

sediments is no criterion of greater depth. In addition, the photic limit in hypersaline waters probably always occurs at shallow depths, because suspended organic residues (preserved because of the poorly-oxygenated nature of brines), surface nucleated and floating evaporite crystals, and numerous anaerobic bacteria (commonly red in colour) all reduce light penetration – sometimes to depths of only a few decimetres. Such turbid brines also trap radiant heat and may reach temperatures of up to 90°C: another adverse environmental factor that will inhibit or curtail algal growth.

For these reasons, only two subaqueous environments are here distinguished: the deep-water environment characterized by laminites and gravity-displaced sediments and the shallow-water environment that represents a plethora of subenvironments which, as yet, are poorly characterized.

Deep Water Evaporite Facies

In this environment the brine is at or near saturation with respect to gypsum and/or halite. Crystal growth probably occurs mainly at the air-water interface and crystals settle through the water column as a pelagic rain. Regular interlamination of minerals of different solubilities (calcite and gypsum, with or without halite) reflect variations in brine influx, annual temperature or evaporation rate. Some calcium sulphate may grow within the upper layers of the bottom sediment and some salt may be precipitated during the mixing of brines in a stratified water body (Raup, 1970). Evaporite turbidites and mass-flow deposits, derived from shallower water carbonate and evaporite accumulations, may also be emplaced within this environment.

The depth of water in which "deep water" evaporites accumulate is difficult to determine. However, where turbidites, (composed of basin-marginal materials) occur at the basin centre, the centre to basin-margin distance combined with a minimal 1° slope suggests a minimum depth. Such a calculation for the Sicilian Basin during the Messinian (Upper Miocene) suggests depths exceeding 175 m (Schreiber *et al.,* 1976).

A minimum water depth can also be obtained by observing the relation of basinal evaporites to topographic elevations. Laminated evaporites at the base

160

of the Muskeg - Prairie Evaporite Formations (Middle Devonian of the Elk Point Basin described by Wardlaw and Reinson, 1971; Davies and Ludlam, 1973) cover flanks of Winnipegosis - Keg River carbonate buildups to heights of at least 20 m (Kendall, *in prep.*). Persistence of laminae up such slopes and the lack of associated lithologic change suggest deposition occurred from a brine body at least 40 m deep.

Geochemical evidence can some-times be employed to calculate water volume and, by implication, water depth. Katz *et al.* (1977) have used the strontium content of aragonite laminae, some interlaminated with gypsum (Begin *et al.*, 1974), from Pleistocene Lisan Formation of the Dead Sea region to establish a water depth of between 400 and 600 m.

Sulphate mm-laminites (Fig. 1). Laminar sulphate (originally gypsum), either alone or in couplets or triplets with carbonate and/or organic matter, is probably the commonest deep-water evaporite facies and occurs in the Permian Castile Formation of Texas and

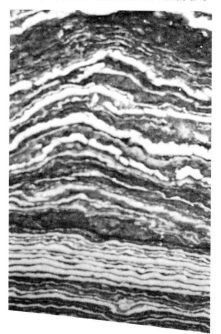

Figure 1
Inter laminated anhydrite (dark) and carbonate, affected by syn-sedimentary slumping. Only carbonate laminae are fractured. Middle Devonian Ratner Member (at base of Prairie Evaporite), Saskatchewan. Laminae such as these are traceable for many kilometres. Slab is 7 cm wide.

New Mexico (Anderson and Kirkland, 1966; Anderson *et al.*, 1972), in the Permian Zechstein group of Germany (Richter-Bernburg, 1957; Anderson and Kirkland, 1966), in the Jurassic Todilto Formation of New Mexico (Anderson and Kirkland, 1966) and in the Middle Devonian Muskeg and Winnipegosis Formations of western Canada (Davies and Ludlam, 1973; Wardlaw and Reinson, 1971).

Laminae are thin (1 to 10 mm thick) and although they are typically bounded by perfectly smooth, flat surfaces they may be uneven, crenulated or plastically disturbed. Over short sections, laminae are nearly of uniform thickness and individual laminae are traceable over long distances (up to several hundred kilometres). The Castile - Lower Salado laminites are 440 m thick, comprise more than 250,000 anhydrite-carbonate couplets and some laminae have been traced laterally for more than 110 km (Anderson *et al.*, 1972).

Some anhydrite laminae exhibit evidence that they were originally composed of small lenticular gypsum crystals, arranged parallel to bedding (Shearman, 1971) and are similar to those in gypsum laminae from the Lisan Formation. Similar lenticular gypsum crystals precipitate from the water column (?) in shallow solar-salt ponds or grow displacively in algal sediments (Schreiber, 1978). Laminae that lack evidence of lenticular gypsum may have accumulated on the basin floor from a rain of fine gypsum needles precipitated at the air-water interface.

Nodular anhydrite intervals occur within the Castile Formation but are the result of a reorganization of pre-existing sulphate laminae, the nodular anhydrite rarely completely losing its laminated appearance. However, some intervals do approach in appearance those formed in supratidal settings. Some authors (notably Friedman, 1972) would use the presence of nodular anhydrite to suggest that the entire Castile succession is shallow water in origin, whereas others (Dean *et al.*, 1975) conclude that this type of nodular anhydrite is not diagnostic of supratidal environments. The occurrence of nodular anhydrite at horizons where anhydrite laminae are thick or immediately beneath halite layers suggests nodule formation was associated with increased salinity.

Laminated sulphates record the precipitation and deposition of sediments in

a water body whose bottom was unaffected by wave action and currents. Such stagnant, permanently stratified water bodies need not be particularly deep and carbonate laminae form in comparatively shallow waters of the Dead Sea (Neev and Emery, 1967) as a result of "whitings" at the brine surface. Lenticular and needle gypsum crystals precipitate in shallow brine ponds and also are non-diagnostic of water depth. The interpretation of some laminated suphate deposits as deep-water thus rests primarily upon: 1) the wide-spread occurrence of individual laminae, 2) the lack of other facies indicative of shallow water, 3) the size of the evaporite unit, and possibly 4) the presence of gravity-displaced sediments.

Laminated halite. Deep-water halite is difficult to recognize because many examples have suffered recrystallization, obliterating original characteristics. Even so, deep-water halite is invariably finely laminated and contains anhydrite-carbonate laminae (Jahresringe) similar to those of deep-water laminated sulphates (which commonly underlie the halite). Lamination within the salt beds bounded by anhydrite-carbonate laminae is common and is defined by variations in inclusion content (liquid inclusions or very fine grained sulphate or pelitic material). Salt layers and laminae have been traced for many kilometres (Schreiber *et al.*, 1976; Richter-Bernburg, 1973; Anderson *et al.*, 1972).

The classic description of a subaqueous basin-central halite deposit is that of the Salina (Silurian of Michigan Basin) by Dellwig (1955) and Dellwig and Evans (1969). Salina salts exhibit a clear salt - cloudy salt banding (in 2 to 9 cm thick couplets) in addition to dolomite-anhydrite laminae. Banding is absent or only poorly developed in some basin-marginal locations where there is additional evidence for shallow-water conditions. Cloudy layers are inclusion-rich and are described by Dellwig as being composed of numerous pyramidal-shaped hopper crystals that grew on the brine surface. When broken or disturbed, these skeletal crystals were swamped, sank and accumulated on the bottom as a sediment. They subsequently developed syntaxial inclusion-free overgrowths, and assumed cubic habits.

Cloudy and clear salt banding was interpreted by Dellwig (1955) as a product of variations in halite saturation on the basin floor. Sinking hopper crystals caused bottom brines to become saturated with respect to halite. A temperature rise in the bottom brines, however, caused undersaturation and some dissolution of previously accumulated hopper crystals. Subsequent cooling of the brine allowed the brine to become supersaturated and promoted growth of clear, inclusion-free halite as overgrowths of surviving bottom hopper crystals and as a new, clear, halite layer. Supply of new hopper crystals from the brine surface formed a new layer of cloudy halite on top of the clear layer. The early diagenetic origin of clear salt layers and of hopper overgrowths is shown where carbonate-anhydrite laminae drape over overgrowth crystal faces or when similar carbonate-anhydrite laminae overlie flat dissolution surfaces that cut across both hoppers and their overgrowths.

If Dellwig's interpretation of the clear halite is correct, then this may account for the recrystallized appearance of other deep-water halites. It is not known, however, whether a deep body of brine would suffer sufficient variation in bottom temperatures to promote this wholesale solution-reprecipitation.

Many of Dellwig's conclusions have been challenged by Nurmi and Friedman (1977). They identify much of the cloudy salt as having grown on the basin floor as crusts of upwardly-directed crystals and infer a shallow-water origin for it (see section on shallow-water halite). However, Dellwig categorically describes some halite crystals as being downwardly-directed so that both bottom-grown and surface-grown (hopper) halite may be present. Nurmi and Friedman also identify some clear halite crystals, interpreted as recrystallized halite by Dellwig, as primary. Such halite occurs as well-developed cubes, is interbedded with stringers of carbonate and anhydrite which drape over underlying halite crystals (Fig. 2). The clear character of these crystals reflects slow precipitation from a brine body that did not suffer rapid compositional changes. This, together with a restriction of this facies to the lower part of the lowest (A-1) Salina salt and to the basin centre, suggests deposition in somewhat deep-water environments. All other salt is reinterpreted to be of shallow-water origin.

Figure 2
Deep-water halite layers, composed of cubic crystals, interbedded with laminated carbonate and anhydrite. Basal A-1 salt (Salina Group) Michigan basin. Vertical scale bar: 1 cm. Photo courtesy R.D. Nurmi.

Gravity-displaced evaporites. Clastic evaporite intervals, interbedded with deep-water laminated evaporites or with non-evaporite sediments, are interpreted as slump, mass-flow and turbidity-current deposits. Their presence is possibly the best indication of a large body of brine during deposition.

Gypsum or anhydrite turbidites are seemingly identical with non-evaporite equivalents. Sometimes the entire Bouma sequence is present (Schreiber *et al.,* 1976) but most beds are only composed of graded units or have poorly developed, parallel laminae in uppermost parts (Schlager and Bolz, 1977). Beds may be entirely evaporite in composition or contain carbonate and other types of clastic material. Gypsum-rich turbidites from the Miocene of the Periadriatic Basin (Parea and Ricci Lucchi, 1972) constitute a thin horizon within a thick siliciclastic flysch sequence, interpreted as a deep-sea fan deposit. Turbidites within some evaporite deposits are entirely carbonate in composition (Davies and Ludlam, 1973) indicating an entirely carbonate upslope source, or that evaporites at such locations contained no coarse-grained

material. Centimetre-thick anhydrite beds (Fig. 3), some exhibiting poorly-developed grading, associated with carbonate turbidites in flanking beds around Winnipegosis banks in Saskatchewan (Kendall, *in prep.*), suggest that the deposits of turbidity currents which only carried fine-grained sulphate may be difficult to distinguish from "normal" basinal evaporites that are deposited as a pelagic rain.

Mass-flow deposits are represented by breccias composed of clasts of reworked sulphate, either alone or with carbonate fragments. They occur in well-defined beds; clasts are tightly packed and large fragments (up to a metre in size) may be concentrated at the base of beds. They are commonly associated with beds affected by slump-folding. Confinement of deformation to certain horizons and possible truncation of deformed beds beneath undisturbed beds are indications that sediment transport down slope was pene-contemporaneous.

The deep-water evaporite model is only just beginning to be understood and to be recognized in ancient sulphate evaporites. When initial basin slopes are gentle, the basin periphery becomes the depositional site of thick shallow-water evaporites that build upwards and outwards into the basin and construct a ramp or platform (Schlager and Bolz, 1977) (Fig. 4). Because evaporite deposition at the deep basin centre is slower than upon the platform, steep depositional slopes develop at the platform edge. Upper parts of slopes are sites of slumping and mass flow, whereas lower parts of slopes contain graded beds that were emplaced by turbidity currents. Laminated sulphates (gypsum) are deposited on basin floors and slopes.

Not all deep-water evaporite basins are flanked by sulphate platforms. Basins flanked by lithified carbonate buildups, for example, possess no source of evaporite detritus and are largely composed of laminites, possibly with minor carbonate turbidites at basin flanks.

Conversely, not all deep-water evaporites need have been emplaced under saturated waters. Some mass-flow deposits described by Parea and Ricci Lucchi (1972) were deposited after evaporites had ceased forming on the platform, and were presumably preserved in the undersaturated waters of their new environment by their fast mode of transport and burial beneath protective non-evaporite sediments.

Deep water halite accumulates in the same general environment as sulphate laminites but the brines became supersaturated with respect to halite. Periodic returns to sulphate precipitation, forming Jahresringe, suggest pulses of seawater entry into the basin – perhaps seasonally. Restriction of clear, cubic halite facies to central parts of the Michigan Basin suggest gradual desiccation of the basin occurred so that deep-water deposition became progressively restricted to the basin centre (Nurmi and Friedman, 1977).

Shallow Water Evaporite Facies

Deposition of shallow-water evaporites occurs in brines that were at or near saturation with respect to gypsum or halite and in environments that may have been subject to strong wave and current action, causing sediment scour, transport and redeposition. Algal activity was significant in more protected (or deeper?) environments and many sedi-

Figure 3

Carbonate and anhydrite turbidites from a facies that flanks M. Devonian (Winnipegosis Fm.) carbonate buildups; Saskatchewan. Graded carbonate turbidites at left are inserted within laminated carbonate and anhydrite. At right, poorly-graded anhydrite layers occur in association with thinner (autochthonous?) anhydrite laminae. Cores are 8.5 cm across.

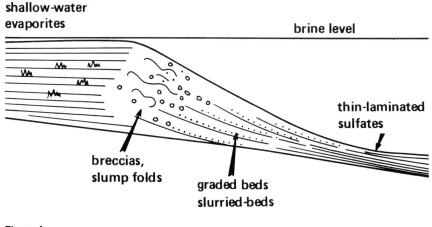

Figure 4

Schematic diagram of deep water and slope evaporite environments (after Schlager and Bolz, 1977).

ments were subject to periodic drying. Whereas water depths may range from a few centimetres to 20 m or more, most facies probably formed in water less than five metres deep. In fact, many evaporites considered subaqueous may have been deposited on evaporitic flats that only became flooded during storm surges or particularly high tides. Evaporite precipitation may occur at the air-water interface, at the sediment-water interface or beneath the sediment surface and varying amounts of continental and marine-derived sediments may be periodically transported into the evaporitic environment.

Laminated sulphates may be similar in character and origin to deep-water sediments but most apparently consisted of current-deposited micrite and clastic gypsum particles in reverse and normally-graded laminae. Laminae were originally composed of silt and sand-sized gypsum crystals or cleavage fragments which grew: 1) as crusts on the depositional surface and so were easily broken and reworked, or 2) as acicular crystals precipitated at the air-water interface which sank and became reworked on the bottom. Other crystals may have grown displacively within the bottom sediment and then were reworked. All crystals and fragments suffer overgrowth on the bottom and laminae become converted into interlocking gypsum mosaics.

In some sediments the gypsum crystals have suffered little if any transport, and in many the crystals displace or poikilitically enclose algal mat carbo-

nate and organic material. Lamination in these sediments is largely a reflection of algal mat lamination (Fig. 5).

Cross-bedding, ripple-drift bedding (Fig. 6), basal scoured surfaces and rip-up breccias testify to environments with periodic high energy events, such as storms. Some small asymmetric ripple-marks with oversteepened sides at the tops of some laminae may represent adhesion ripples and indicate deposition of wind-blown gypsum detritus onto moist surfaces. Shallow-water deposition is also shown by the occurrence of micritic, organic-rich stromatolites between, or within, some laminae; by bird or dinosaur footprints, or by fossil brine shrimp.

Laminae are interpreted as storm deposits. Single laminae form during a storm when evaporitic tidal flats are flooded by sediment-charged water. Blue-green algal mats, which cover the flats, collect and bind evaporite sediment and, as the storm subsides, the coarser load is deposited as a traction layer or as a settle-out to produce a normally graded lamina. Algae grow through the new lamina, re-establish themselves on the surface and protect the underlying sediment from erosion. The analogy may be made with the formation at storm laminae in other tidal-flat sediments.

Reverse-graded laminae are variously interpreted. They may record episodes of brine dilution that induce recrystallization of gypsum in the uppermost parts of laminae, or this feature may be of depositional origin. Upward segregation of coarser particles may have occurred within highly concentrated flowing sand sheets in very shallow waters upon tidal flats during storm surges. The reverse grading may then be emphasized by early diagenetic recrystallization and lithification during quiet periods between storms. Inversely graded layers, adhesion ripples, algal mats and early-diagenetic cementation of gypsum are recorded from evaporitic flats of the Laguna Mormona (Baja California; see Horodyski and Vonder Haar, 1975).

Gypsum laminites that have been altered to anhydrite rarely provide sufficient evidence for environmental reconstruction. Some laminites suffer pervasive recrystallization to coarse gypsum mosaics which transect all earlier fabric elements. If replaced by nodular anhydrite such crystals may yield rocks that superficially resemble sabkha anhydrite.

Shallow-water laminite units may be laterally persistent but typically contain fewer laminae than deep-water units and individual laminae cannot be traced for long distances. Shallow-water laminites may also be distinguished by their association with other facies. Possibly the manner in which laminites deform provides evidence for different environments. Evaporitic flat sediments, which become emergent and suffer extensive early-diagenetic cementation, fracture and become incorporated into rip-up breccias. In contrast, some laminites interpreted as subaqueous, have suffered folding, slumping and plastic stretching (Fig. 7) suggesting that they did not become lithified during early diagenesis.

Coarsely crystalline, selenitic gypsum occurs as single crystals, clusters, crusts and as superimposed beds. This facies is best known from the Miocene of Italy but is also recognized in older sequences, now altered to anhydrite. Similar gypsum has been described from man-made salinas.

Beds of crystalline gypsum are mainly composed of orderly rows of vertically-standing, elongate and commonly swallow-tail twinned crystals that range from a few centimetres to a few metres in height (Figs. 8,9). Crystals are commonly euhedral and in aggregate define a vertical pallisade fabric or may be arranged into radiating-upwards conical clusters (cavoli). Individual crystals are separated from each other by micritic carbonate, fine-grained gypsum or gypsum sands; or secondary overgrowth produces an interlocking crystal mosaic. Other gypsum crystals exhibit more bizarre growth and twinning patterns and suffer crystal splitting to generate palmate to fan-shaped clusters of subparallel crystals (Fig. 10; for details see Schreiber, 1978).

Figure 5
Anhydrite after displacive gypsum crystals which grew within mud-cracked stromatolitic carbonate. Each gypsum crystal is now represented by a small, angular anhydrite 'nodule'. Souris River Formation (U. Devonian, Saskatchewan), Core is 9 cm across.

Figure 6
Laminated anhydrite containing minor amounts of disseminated dolomite that define lamination, ripples, minor cross-stratification and scoured surfaces. Poplar Beds (Mississippian) Saskatchewan. Core is 10.5 cm across.

164

Figure 9
Swallow-tail twinned gypsum crystal (25 cm across) with dissolution surface at arrow. Inset is a cleavage plane surface of the same crystal revealing numerous inclusion-defined growth layers. Miocene of S. E. Spain. Photos courtesy B. C. Schreiber.

Figure 7
Deformed (slumped?) laminated to thin-bedded anhydrite which can easily be confused with displacive nodular anhydrite. Ordovician (Herald Fm.), Saskatchewan. Core is 10.5 cm across.

Figure 8
Coarsely crystalline, selenitic gypsum facies. A: Palisades of gypsum crystals, Miocene of Sicily. Photo courtesy B.C. Schreiber. B: Layered anhydrite with pseudomorphs after gypsum crystals, Otto Fiord Fm. (Pennsylvanian), Ellesmere Island. Photo courtesy N.C. Wardlaw. Scale divisions in cm.

The crystals contain faint lamination (Fig. 9), defined by carbonate and anhydrite inclusions, which passes through the crystalline beds, parallel to bedding. Inclusions lie parallel to crystal facies, recording successive positions of the growing crystal, or defining solution surfaces. Many crystals include algal filaments and appear to have invaded algal mats.

Most authors now conclude that these gypsum crystals are primary and are mostly of very shallow-water origin. Schreiber (1978) notes that in salinas gypsum growth occurs mainly at depths shallower than five metres. The crystals record nucleation and slow incremental growth, presumably in quiet waters. The internal lamination and included algal mats indicate the crystals grew poikilitically, enclosing surficial veneers of sediment. Phases of slight undersaturation create minor dissolution surfaces that truncate the crystals. Renewed precipitation, however, commonly takes place upon the etched surfaces, burying the surface within the crystal (Fig. 9). More severe interruptions may include: 1) a new phase of nucleation, producing a new bed of crystals, 2) lateral dissolution along crystal sides, perhaps with accumulation of residual impurities in the dissolution cavities, or 3) in extreme cases, crystals become disoriented and form residual gypsum breccias.

Beds originally composed of gypsum crystals may be difficult to identify when converted to anhydrite. Inclusions may define crystal faces within massive or mosaic anhydrite, but if the original crystals possessed numerous inclusions that defined laminae, the replacement can be mistaken for laminar sulphate. Gypsum crystals are most easily identified when the pseudomorphs are set within abundant carbonate matrix (Fig. 10). Much of the polyhalite from the Permian of New Mexico appears to have replaced and pseudomorphed selenitic gypsum (see Schaller and Henderson, 1932, pl. 29, 30).

Coarse clastic gypsum. Gypsum sands and pebbly sands, composed of worn gypsum cleavage fragments with variable amounts of carbonate and other materials, may be locally abundant but only rarely constitute major rock units. They do indicate, however, that gypsum may be transported and deposited in the same manner and environments as other clastic sediments, so long as the water body is gypsum-saturated. Such sands exhibit structures indicative of current or wave activity or may be penecontemporaneously disturbed and contain load cast or ball-and-pillow structures (Fig. 11). Clastic gypsum occurs as shoestring sands or in sand sheets; represents channel, beach, offshore shoal or spit deposits or may occur as intercalations between beds of laminar or selenitic gypsum.

Vai and Ricci Lucchi (1977) have interpreted wavy bedded and laminar gypsum (composed of mm-sized gypsum) with accompanying poorly-sorted, broken gypsum-crystal sands, as fluvial deposits that prograded into a basin.

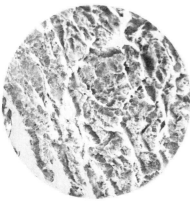

Figure 10
Anhydrite pseudomorphs of extensively twinned and split gypsum crystals within a dolomite matrix. Frobisher Evaporite (Mississippian), Saskatchewan. Cores are 10 cm across.

Figure 11
Parallel laminated, cross-stratified and load-casted gypsum sandstones (gypsarenites). Miocene of Sicily. Photo courtesy B.C. Schreiber. Penknife gives scale.

166

This facies first appears between beds of subaqueous selenite but increases in abundance upwards and includes selenitic nodules interpreted to have been supratidal anhydrite. It grades into a facies of disoriented large gypsum crystals and fragments in a clayey matrix that represents deposits of subaerial debris-flows. Growth of subaqueous gypsum apparently became more and more interrupted by sheet-floods that carried selenite fragments. Progradation caused development of wide supratidal flats composed of this transported material and in this environment sabkha anhydrite was emplaced.

Halite. At least three main facies are present: detrital halite, halite crusts, and halite that grows displacively in pre-existing sediments. It is uncertain what controls which particular facies will be developed.

Halite crusts constitute the best understood facies – one for which there are detailed descriptions from the ancient (Wardlaw and Schwerdtner, 1966), from Recent salt pans (Shearman, 1970) and from experimental studies (Arthurton, 1973). Crusts form: 1) by the foundering of, and continued growth upon, rafts of halite crystals which nucleated on the brine surface,

2) by upward and lateral growth of floor-nucleated crystals, and 3) by accumulation of, and overgrowth upon, detrital halite particles. Various halite growth habits are observed but the most common is layered halite, formed by the superposition of crusts (each crust separated by films or thin beds of detrital carbonate, sulphate or terrigenous sediment) and identified as 'chevron halite' (Fig. 12 A-C). Each halite layer is composed, in part, of vertically elongate crystals that contain abundant brine-filled inclusions. The crystal fabric results from an upward competitive crystal growth on the sea or lake floor such that crystals with coigns uppermost are the most favoured. Inclusions are concentrated in layers parallel to cube faces, so that in the elongate halite crystals with coigns uppermost, the zoning appears as chevrons with upwardly-directed apices. The upper surfaces of halite layers: 1) may exhibit crystal growth faces (interruption in growth caused by only temporary and slight brine undersaturation), 2) are truncation surfaces associated with cavities in the underlying halite crust (recording more extreme episodes of brine undersaturation and halite dissolution), or 3) are flat truncation surfaces

(possible deflation surfaces cut during episodes of emergence). Each halite layer is usually composed of two types of halite; the zoned chevron halite and clear halite which fills former dissolution cavities made in the crust.

Inclusion-rich layers *in zoned halite crystals* form where brines are highly supersaturated and growth is rapid. Reduced brine concentrations (the result of halite precipitation) then allow slower and more perfect (inclusion-free) halite layers to be deposited. Because brine reconcentration (necessary to cause deposition of succeeding inclusion-rich halite layers) can only occur by evaporation of the brine, the numerous alternations between inclusion-rich and inclusion-poor layers in chevron halite, indicate that rapid changes in brine concentration occurred. This can only be achieved in bodies of brine of small volume. The layering in chevron halite is thus indicative of shallow water precipitation and contrasts with the clear halite crystals of deep-water deposits.

Displacive halite has been described previously in connection with playa-flat and sabkha evaporites (Kendall, 1978) but may also be of subaqueous origin. It is recorded from the floor of the Dead Sea where it occurs as large (5 to 10 cm)

Figure 12
Chevron halite. A: Layers of chevron halite interbedded with laminated anhydrite, Souris River Fm. (U. Devonian) Saskatchewan. Core is 10 cm across. B: Thin-section through
chevron halite layer containing cloudy inclusion-rich and clear void-filling crystals, the former truncated by an anhydrite lamina. Prairie Evaporite, Saskatchewan. (Thin sec-
tion loaned by N. C. Wardlaw). C: Isolated crystals from halite crust on bottom of brine pool in a Saskatchewan potash mine. Largest crystal is 3.5 cm long.

cubes with hopper-like pyramidal hollows on each face (Fig. 13). Zoned inclusions of the enclosing sediment, parallel to all cube faces, indicate the crystal grew displacively within the mud. Sediments containing significant quantities of displacive halite cubes are termed Haselgebirge (see Arthurton, 1973) and rock units composed of displacive halite (with host sediment reduced to mere pockets or thin film between crystals) constitute the upper parts of the Prairies Evaporite and other Devonian halites in Saskatchewan.

Detrital halite is probably more important than published studies would suggest, perhaps because this facies seems particularly susceptible to re-crystallization - so that depositional fabrics are lost. Detrital halite is composed of fragmentary surface-grown hopper crystals and small cubes that may represent overgrown hoppers, crystals precipitated during brine-mixing (Raup, 1970) or reworked material from bottom-growing crusts. Detrital halite is commonly ripple-marked and may exhibit cross-bedding and include other detrital material. Crystal growth may continue after deposition, by means of small-scale sediment displacement, and the detrital origin can become obscured.

Weiler et al. (1974) suggest that halite crusts grow preferentially in shallow, quiet-water environments, whereas detrital halite, commonly ripple marked, dominates in higher energy environments because there the sunken surface-grown crystals are subject to bottom movement sufficient to prevent crust development.

Wardlaw (1972) has described crusts of bottom-grown carnallite (KC1-MgC1$_2$-6H$_2$O) interbedded with layers of detrital and surface grown (?) halite. The salts are deformed by synsedimentary folds and by the displacive growth of large carnallite crystals within the sediment. Deformation of the sedimentary layering suggests that the salts were never subaerially exposed or lithified and that subaqueous salts remain unlithified and are capable of being deformed by slumping and differential loading. The occurrence of tachyhydrite (CaC1$_2$.2MgC1$_2$.12H$_2$O) in these evaporites, a mineral that cannot survive exposure to the atmosphere, also indicates evaporite deposition was entirely subaqueous.

Figure 13
Displacive halite crystals (with 'hopper' - faces) that grew in micritic ooze at southern end of the Dead Sea (Photo courtesy B.C. Schreiber).

Shallow water models. Two models are proposed for shallow-water sulphates; both created for the Messinian evaporites of Italy (Fig. 14). Hardie and Eugster (1971) invoke deposition of coarsely crystalline selenite in the quiet waters of a shallow lagoon or gulf, adjacent to a littoral belt of laminated gypsum. Gyp-

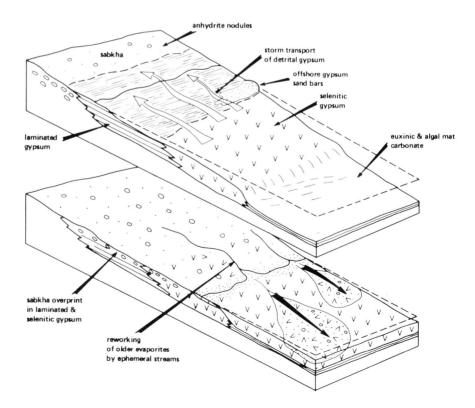

Figure 14
Models for deposition of shallow water sulphate evaporite facies. Above, after Hardie and Eugster (1970); below, Vai and Ricci Lucchi (1977).

sum in the laminites and in associated gypsum sand bodies (beach or offshore-shoal deposits) was derived from the area of selenite deposition and transported shorewards onto the marginal evaporitic flats during storms.

Vai and Ricci Lucchi (1977), on the other hand, working on a sequence that lacked gypsum laminites, suggest transport of gypsum toward the basin centre. Gypsum was reworked from older, emergent beds of selenitic gypsum by ephemeral slope-controlled agents (torrential streams and debris flows) which built up shallow alluvial cones that enchroached the basin. This cannibalistic model can be integrated with Hardie and Eugster's model to obtain a single, more dynamic model: Vai and Ricci Lucchi's interpretation applicable to times of regression, when older evaporites become exposed in marginal areas and subject to reworking, and Hardie and Eugster's interpretation appropriate to times of transgression or when the regression occurs entirely as a consequence of sediment outbuilding (when gradients will be low).

Whereas we possess a reasonable idea about details of the depositional environment of shallow-water halite (since we have a modern-day equivalent: Shearman, 1970), interpretation of the mechanism for depositing the enormous volumes of this material that occur in many evaporite formations remains problematical. As an example, the lower part of the Middle Devonian Prairie Evaporite consists almost entirely of chevron halite with carbonate-anhydrite laminae (Wardlaw and Schwerdtner, 1966) and represents deposition in shallow brine pools and salt flats. This environment apparently stretched across Saskatchewan from central Alberta and the source of the brine was from the northwest and would have had to have travelled more than 1600 km. It is difficult to imagine how this brine could have travelled across brine pools and salt flats without evaporating away before it had travelled for more than a small part of its journey. Interpretations of other large units of shallow-water halite would appear to be afflicted by the same problem.

Evaporite Sequences

Facies models are developed from the characters of individual facies and from the succession and arrangement of these facies. The thickness of many shoaling-upwards subaqueous sedimentary sequences also have commonly been used to estimate a minimum depth of water at the time deposition commenced. It is not possible to use this method for subaqueous evaporite successions (although it has been attempted) because the upper despositional limit, the brine surface, is rarely static. Cycles in subaqueous evaporites can result as much from brine-level lowering (due to evaporation) as from any sedimentary upbuilding. Thus the "minimum" estimate of brine depth (for lower parts of the cycle) can be very much an underestimate. Conversely, because subaqueous evaporite deposition occurs in locations where the brine surface may be lowered by evaporative downdraw, the thickness of cycles need not even record minimal water depths at the start of deposition. Unlike sea-level, which is commonly a more or less static confining surface for marine sedimentation, the level of a brine surface may rise to offset the effect of sediment upbuilding, so maintaining a similar depth of water. In this way, shallow water evaporites may accumulate for many tens of metres without necessarily passing vertically into shallower facies. Shoaling-upwards cycles reflect a gradual decrease in brine depth but this need not occur at the same rate as sedimentary accretion.

It cannot be expected that brine depth will be stable over any great length of time. The very fact of evaporite precipitation means that the brine volume has been depleted by evaporation. In the absence of significant water input, precipitation of evaporites must be accompanied by dramatic lowering of the brine surface (evaporative downdraw) and even when water input offsets this brine-loss it is most unlikely to balance the evaporation rate exactly and brine-level fluctuations will occur. Downdraw and desiccation may cause shallow-water and supratidal sediments to be located only a short distance above those formed in deep water. Conversely, basin refill and dilution may cause episodes of non-evaporite deep-water sedimentation. Thus deep depositional basins may contain both deep and

shallow-water deposits. Application of a single depositional model throughout the history of basin filling is unlikely.

Three types of vertical succession occur within marine evaporite sequences and each corresponds to deposition in a different part of a basin (Fig. 15).

Basin-marginal sequences (which in shallow basins may extend well into the basin) are characterized by sabkha deposits, with evaporites growing *within* the sediments. Calcium sulphate saturation is only achieved in upper intertidal-supratidal environments and shallow water evaporites are either absent or are confined: 1) to existing depressions on the sabkha surface, or 2) to brine-flats that develop if the rate at which the supratidal surface is raised by displacive evaporite growth falls below the rate of subsidence, and the former sabkha surfaces thus becomes flooded with brine. Leeder and Zeidan (1977) interpret laminar sulphates above nodular anhydrite as forming in this last-mentioned situation but, because the brines would have been derived from sabkha groundwaters (therefore calcium and sulphate depleted), salt-pan halite is more likely to be precipitated in such a situation. Halite in the Stettler Formation (Upper Devonian of Alberta; Fuller and Porter, 1969) may have been deposited in such supra-sabkha depressions.

In shallow basin or shelf sequences (which may also be located on the floors of partially desiccated deep basins) gypsum saturation is reached in the shallow subaqueous environment. Vertical variation is caused: 1) by sediment upbuilding, 2) by lowering of brine level by evaporation or drainage (commonly associated with cannibalism of earlier-formed evaporites) or 3) by changes in the rate of brine-recharge that occur as a result of brine-level rise toward the world sea-level, brought about by sediment upbuilding. Decreased recharge will result in desiccation and regressive sequences, whereas increased recharge will produce transgressive sequences. Changes in recharge will also affect the salinity of brines. Salinity increases are recorded by upward transitions from carbonates into sulphates, perhaps via patterned (or pyritic) carbonates that record sulphate precipitation and contemporaneous sulphate-reduction (Fig. 16). By the time halite saturation is reached, the brine-level in

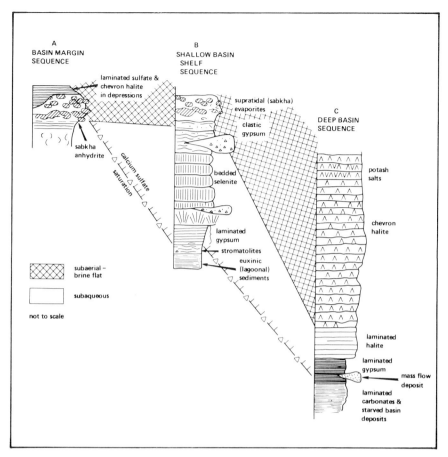

Figure 15
Hypothetical evaporite successions (not to scale) in different parts of evaporite basins.

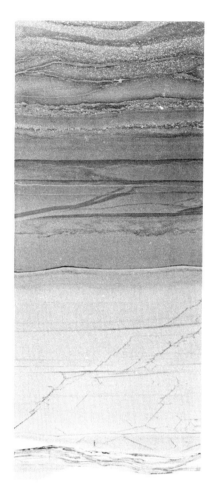

Figure 16
Part of Souris River Fm. (U. Devonian, Saskatchewan) evaporite cycle. Light coloured laminated carbonates at base pass up into pyritic-stained carbonates (marking episode when brines were gypsum saturated but gypsum failed to accumulate because of bacterial reduction in the sediment) and then into laminated anhydrite (deposited when rate of gypsum precipitation exceeded the rate of its removal). A further salinity-increase is revealed by the highly dendritic halite crystals that grew within the laminated carbonates. Core is 8 cm across.

the shallow basin or shelf environment has been lowered to such an extent that only shallow subaqueous or brine-pan salt is deposited.

Calcium sulphate saturation is reached in the central deep-water parts of some basins and laminar sulphates overlie starved basin deposits. Some pyritic basal limestones may represent the by-product of subaqueous sulphate reduction (Friedman, 1972) and the change from carbonate to sulphate marks an increase in the rate of sulphate production so that some of the sulphate survives bacterial reduction. Halite saturation may be reached in shallow or deep-water environments but some basins seem to pass directly from deepwater sulphate laminites into brine-pan halite. The absence of deep-water halite from these basins presents a considerable problem. Upward sequences are produced: 1) by brine-level lowerings, caused by net evaporative loss, which causes precipitation of more saline and shallower water evaporites above deeper, less saline deposits; and 2) by

flooding events that may cause deeper water, less saline deposits to abruptly overlie shallower, more saline evaporites. Most halite appears to have precipitated upon brine-flats, and thick sequences of chevron halite (the lower part of the Prairie Evaporite for example) must record sediment upbuilding coincident with a similar rise in brine-level on the floor of a fairly deep desiccated basin.

Depositional Settings of Evaporites (Depositional Models)
Rather than discuss the classic models that have been applied to subaqueous evaporites (such as the bar model of Ochsenius, the surface reflux model of King and the seepage reflux model developed by Adams and Rhodes) - models that are adequately described and discussed by Stewart (1963), Hsü (1972), and Kirkland and Evans (1973) - attention is directed towards three currently accepted and competing depositional models (Fig. 17).

The deep water, deep basin model is founded upon evidence that giant evaporite basins were deep topographic depressions. Evidence of this comes from: 1) the rate of evaporite deposition compared to possible rates of basin-floor subsidence, and 2) from palaeo-geographic reconstructions.

The Zechstein (Permian) of Germany locally contains almost 2,000 m of evaporites which accumulated at an average rate, based upon varve measurements (assumed to be annual), of 10 mm/year. Even if the Zechstein basin subsided at rates comparable with those of geosynclines (0.1 mm/year) a postulate for which there is no confirming evidence, the initial depth of the Zechstein basin could not have been less than 1,165 m (Schmalz, 1969). The depositional rate may be disputed because the annual nature of the evaporite varves is uncertain (see Shearman, 1970). Other evidence, however, indicates that evaporite depositonal rates can be very high. More than two km of salt in the Messinian of the Mediterranean accumulated in less than two million years and more than 300 m of evaporites in the Muskeg - Prairie Evaporite Formations of the Elk Point Basin (western Canada) were deposited in the time interval corresponding to a fraction of a conodont zone – possibly only 500,000 years. Such rates require that deposition was initiated within a pre-existing deep basin.

Carbonate buildups (reefs or mud-mound complexes) are commonly associated with basin-central evaporites and occur in marginal or basin-central locations. They accumulated either before evaporite deposition was initiated or are (in part) contemporaneous with evaporite deposition. The height of pre-evaporite portions of the buildups can be used to determine depth for the basin and such evidence commonly indicates pre-evaporite basins were at least hundreds of metres deep.

Evaporite deposits also commonly include intercalations of euxinic sediments, such as black shales. These are considered, by some, as evidence for deep-water deposition (Schmalz, 1969).

There is thus considerable evidence for the postulate that many evaporites were deposited on the floors of deep basins. Unfortunately, most of these criteria identify the basin depth but not the depth of water in which the evapor-

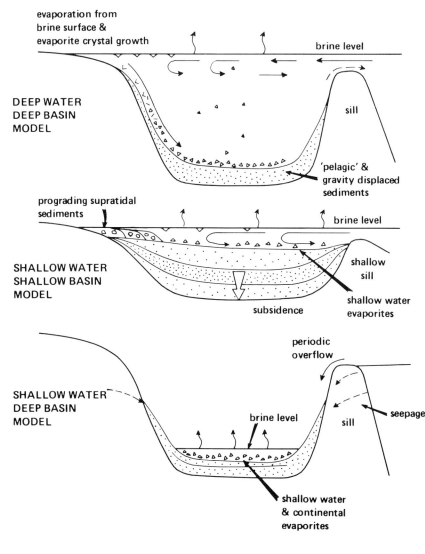

Figure 17
Depositional models for basin-central evaporites.

ites formed. Many evaporites and genetically related sediments exhibit evidence for shallow-water or subaerial deposition. Nevertheless, there are some evaporite deposits *whose internal characteristics* strongly suggest deep water deposition and, as will be seen, the shallow water, deep basin model must progress through a deep water, deep basin stage in its development.

The shallow water, shallow basin model accounts for the sedimentalogic and geochemical evidence for shallow water and/or subaerial depositional environments for basin-central evaporites; evidence that commonly is overwhelming. The main argument against the applicability of the shallow water, shal-

low basin model is the structural and stratigraphic evidence that deposition occurred within pre-existing deep topographic depressions - depressions that, from the evidence of the evaporite facies themselves, must have been only partially filled with brine. There are also difficulties in generating thick evaporite deposits by this model because it must be assumed that the basin floor subsides at approximately the same rate as the evaporites were deposited - unlikely in most tectonic environments. Furthermore, it is also necessary to postulate that the barrier zone, which controls ingress of oceanic water (needed to account for the volume of evaporites precipitated) and egress of refluxing brines (required to prevent isochemical successions from being formed), remains at approximately the same altitude. The barrier can neither subside (as does the basin) nor rise. The first would permit entry of fresh sea-water, which would dissolve previously deposited salts, and reflux out of the basin of the concentrated brines. The second, uplift of the barrier, would permit the basin to dry out and no further deposition would occur. Because the rates of evaporation and corresponding rates of salt deposition are high in arid climates, the rates of basin subsidence would have to be both rapid and constant. Estimated rates of evaporite deposition thus do not support theories of shallow-water deposition with contemporaneous subsidence for the formation of vast bodies of salt. On the other hand, the shallow water, shallow basin model is certainly applicable to evaporite formation in satellite basins and to formation of thin evaporite deposits in basin-central locations. The latter occur at the top of sedimentary cycles which can be numerous and superimposed such that, in aggregate, the evaporites constitute a major proportion of the sedimentary sequence. Examples of such sequences in the Williston Basin occur within the Ordovician (Kendall, 1976), the Upper Devonian (Wilson, 1967; Dunn, 1975) and the Poplar Beds of the Mississippian, although the subaqueous nature of much of these evaporites has not been previously recognized.

Basins bound by active faults are locations where rates of subsidence could be fast enough to keep pace with evaporite deposition. Rates of isostatic readjustment deduced from the Lake Bonneville (Utah) area indicate that such movements can be both rapid and responsive to very small differences in load - in this instance, the weight of water in Pleistocene Lake Bonneville. This leads to an interesting possibility (so far unexplored) that varying climatic conditions control the volume of brine in some basins and induce isostatic subsidence or uplift with the formation of sedimentary cycles that appear to be of purely tectonic origin. Windsor (Mississippian) evaporites of the Maritimes accumulated in graben (Evans, 1970) and could, theoretically, have been deposited in shallow water, shallow basin environments that suffered continuous subsidence.

The shallow water, deep basin model was developed to account for pre-existing deep basins that become filled by evaporites with internal evidence for shallow water and/or subaerial depositional environments.

Calculations made by Lucia (1972) on the degree of basin restriction required to promote gypsum and halite precipitation suggest that the barrier between the open sea and the hypersaline basin must be almost complete. Sea-water-supplying channels into gypsum-precipitating basins can only be of very small dimensions and salt deposition implies complete surface disconnection from the ocean. The source for the halite is from groundwater, sea-water springs or from episodic flooding over the barrier. A corollary to Lucia's argument is that when a deep basin undergoes restriction and loses connection with the world ocean (a requirement needed before evaporites can be precipitated) there may be little to prevent complete desiccation.

The shallow water deep basin model was developed largely to account for two major evaporite deposits - the Middle Devonian Elk Point evaporites of western Canada (Muskeg and Prairie Evaporite Formations) and the Miocene Messinian evaporites of the Mediterranean. Fuller and Porter (1969) and Shearman and Fuller (1969) identified laminated dolomites and anhydrites at the base of the Prairie Evaporite, located between carbonate buildups, as algal mat and sabkha deposits and so postulated desiccation of the Elk Point Basin. The laminated beds are now interpreted as subaqueous deposits (Wardlaw and Reinson, 1971, Davies and Ludlam, 1973) but immediately overlying halites (described by Wardlaw and Schwerdtner, 1977; re-interpreted by Shearman, 1970) testify to deposition of salt flats and imply basin desiccation. Most support for the shallow-water, deep basin model, however, comes from the DSDP program in the Mediterranean (Hsü et al., 1973; Garrison et al., in press). During the Late Miocene the Mediterranean basins were covered by deep marine waters when evaporites were not being formed, but the evaporites were deposited in shallow waters, brine-flats or subaerially on the floor of the basins, thousands of metres below sea level.

Evaporation of the entire Mediterranean Sea would yield only enough salts to reach 60 m in thickness locally. Part of the Messinian evaporites may have been derived from the salt content of waters that drained into the desiccated Mediterranean, but Hsü et al. (1973) estimated that at least 11 flooding-desiccation events would have been required to generate the Messinian evaporite sequence. In this calculation no attention was paid to groundwater contribution. Deep basins must intersect the groundwater pattern of neighbouring areas and, if deep enough, would constitute a major sink for groundwater flow. Much shallow water, deep basin evaporite may thus be derived from groundwaters. The low bromine halite of the Lower Elk Point in Alberta may have been entirely derived from such a groundwater source (Holser et al., 1972).

During basin desiccation and after basin isolation from the world ocean the desiccating sea must pass through a deep water, deep basin stage within which deep water sulphates and halite are precipitated. Hsü (1972) calculates that the Mediterranean would have reached saturation with respect to gypsum while still more than a thousand metres deep. Friedman (1972) however, argues that deep water sulphates will not be preserved because organic matter in the brine promotes bacterial reduction of dissolved sulphate or of already precipitated gypsum. This argument assumes that the rate at which sulphate is bacterially reduced will always exceed the rate at which sulphates are precipitated, itself dependent upon the evaporation rate. The assumption is clearly incorrect in locations where subaqueous sulphates are being precipitated today, like the brine ponds described by Schreiber et al. (1977) and Schreiber and Kinsman (1975).

Upon complete desiccation the floors of basins in which major bodies of evaporite formed (such as the Mediterranean and Zechstein basins) must have lain one or more kilometres below sea level. Such large depressions would provide conditions that are unlike any now present on the earth's surface. Air temperatures would be high (perhaps exceeding 60°C), brine temperatures even higher (80°C or more) and humidities would be very low because of the extreme continentality of basin floors and because of reduced vapour pressures. Such conditions should markedly influence the type of evaporite minerals formed and it is possible that primary subaqueous anhydrite might have been able to form during the extreme desiccation stages of the Mediterranean.

Summary

This paper and an earlier one in the facies model series (Kendall, 1978) have emphasized interpretations based upon evaporite fabrics and structures but have largely ignored geochemical evidence. Potash and similar evaporites have also been little mentioned because few sedimentologic studies have been attempted for these rocks. In part, this sparsity reflects the major changes imposed by diagenesis causing few dispositional characters to survive. The treatment given to other evaporites has also been subject to considerable personal bias. Interpretation of evaporites is still very much an art and many stratigraphic units have been interpreted in very different ways. Important new interpretations seem to appear each year and, probably more than any other paper in this series, this one will be outdated as soon as it is published. Evaporite sedimentology is in a considerable state of flux and probably will remain so for some years to come.

Acknowledgements

Noel P. James read an earlier version of this manuscript and suggested many worthwhile changes. The paper was partially written when the author was a member of the Saskatchewan Geological Survey.

Table I

Sedimentary Aspects of Deep-Water and Shallow-Water Evaporite Deposits.

Mineralogy	Deep Water	Shallow Water
Sulphate Laminites	thin, traceable over long distances.	thicker than in deep water, individual laminae are laterally impersistent.
		evidence of deposition by currents: clastic textures, ripple drift and X-bedding, rip-up breccias, reverse and normal grading.
		associated stromatolites, footprints.
	nodular anhydrite developed from laminated sulphates.	nodular anhydrite developed displacively in inter- inter- and supratidal sediments.
Clastic Sulphates	in form of gravity flows, slumps and turbidites.	in form of offshore bar, channel, beach and sand spit deposits and as intercalations between other facies.
Selenitic Gypsum		layers of swallow-tail twinned crystals.
Halite	finely laminated with carbonate-sulphate laminae. Inclusion-defined laminae traceable over long distances.	layers separated by terrigenous or carbonate-sulphate laminae.
		associated with potash salts.
	clear, inclusion-free halite cubes.	inclusion-rich, 'chevron' halite with clear cavity-filling halite.
		detrital halite, ripple- marked and cross-bedded.
		emersion surfaces, salt-thrust polygons.

Note that recrystallization and other diagenetic changes commonly destroy the evidence necessary to place evaporites into environmental settings. No single criterion is diagnostic.

Bibliography

Canadian sources are not listed separately but are identified by asterisks.

General

Kirkland, D. W. and R. Evans, 1973, Marine Evaporites: origin, diagenesis and geochemistry: Stroudsburg, Penn., Dowden, Hutchinson and Ross, Benchmark Papers in Geology.

Stewart, F. H., 1963, Data of Geochemistry 6th edition: Chapter Y. Marine Evaporites: U.S. Geol. Survey Prof. Paper 440-Y, 52 p.

Although both outdated with respect to subaqueous evaporites, these two works conveniently summarize most classical depositional models that have been developed to explain the major features of evaporites deposits.

*Bebout, D. G. and W. R. Maiklem, 1973, Ancient anhydrite facies and environments, Middle Devonian Elk Point Basin, Alberta: Can. Petrol. Geol. Bull., v. 21, p. 287-343.

Schreiber, B. C., G. M. Friedman, A. Decima and E. Schreiber, 1973, Depositional environments of the Upper Miocene (Messinian) evaporite deposits of the Sicilian Basin: Sedimontology, v. 23, p. 729-760.

Two more recent studies that stress a range of subaqueous evaporite environments.

Schreiber, B. C., 1978, Environments of subaqueous gypsum deposition: in W. E. Dean, and B. C. Schreiber, eds., Notes for a short course on marine evaporites: Soc. Econ. Paleontol. Mineral. Short Course No. 4, p. 43-73.

Depositional Models

Hite, R. J., 1970, Shelf carbonate sedimentation controlled by salinity in the Paradox Basin, southeast Utah. 4th Symposium on Salt: Cleveland, Ohio, N. Ohio Geol. Soc., p. 48-66.

Matthews, R. D. and G. C. Egleson, 1974, Origin and implications of a mid-basin potash facies in the Saline Salt of Michigan. 4th Symposium on Salt: Cleveland, Ohio, N. Ohio Geol. Soc., p. 15-34.

Schmalz, R. F., 1969, Deep-water evaporite deposition: a genetic model: Amer. Assoc. Petrol. Geol. Bull., v. 53, p. 798-823.

Three papers dealing with aspects of the deep basin, deep-water model.

Hsü, K. J., 1972, Origin of saline giants: a critical review after the discovery of the Mediterranean evaporite: Earth Sci. Rev., v. 8, p. 371-396.

Hsü, K. J., M. B. Cita and W. B. F. Ryan, 1973, The origin of the Mediterranean evaporites: in W. B. F. Ryan, K. J. Hsü, et al., Initial Reports of the Dead Sea Drilling Project, Vol. XIII, Washington (U.S. Government Printing Office), p. 1203-1231.

The deep basin, shallow-water model applied to miocene evaporites from the Mediterranean. Hsü's paper contains a useful review of other depositional models.

Lucia, F. J., 1972, Recognition of evaporite-carbonate shoreline sedimentation: in J. K. Rigby, and W. K. Hamblin, eds., Recognition of Ancient Sedimentary Environments: Soc. Econ. Paleontol. Mineral. Spec. Publ. 16, p. 160-191.
Includes calculation of the degree of restriction required for subaqueous evaporite deposition.

Shaw, A.B., 1977, A review of some aspects of evaporite deposition: The Mountain Geologist, v. 14, p. 1-16.
A thought-provoking analysis of deepwater evaporite models which should command more attention than it has.

Deep Water Evaporites

Anderson, R. Y. and D. W. Kirkland, 1966, Intrabasin varve correlation. Geol. Soc. Amer. Bull., v. 77, p. 241-256.

Anderson, R. Y., W. E. Dean, D. W. Kirland and H. I. Snider, 1972, Permian Castile varved evaporite sequence, West Texas and New Mexico. Geol. Soc. Amer. Bull., v. 83, p. 59-86.

*Wardlaw, N. C. and G. E. Reinson, 1971, Carbonate and evaporite deposition and diagenesis, Middle Devonian Winnipegosis and Prairie Evaporite Formations of Saskatchewan: Amer. Assoc. Petrol. Geol. Bull., v. 55, p. 1759-1786.

*Davies, G. R. and S. D. Ludlam, 1973, Origin of laminated and graded sediments, Middle Devonian of western Canada: Geol. Soc. Amer. Bull., v. 84, p. 3527-3546.

For a contrary view see:
*Shearman, D. J. and J. G. Fuller, 1969, Anhydrite diagenesis, calcitization, and organic laminites, Winnipegosis Formation, Middle Devonian, Saskatchewan: Can. Petrol. Geol. Bull., v. 17, p. 496-525.

*Dean, W. E., G. R. Davies and R. Y. Anderson, 1975, Sedimentological significance of nodular and laminated anhydrite: Geology, v. 3, p. 367-372.

Begin, Z. B., A. Ehrlich and Y. Nathan, 1974, Lake Lisan, the Pleistocene procursor of the Dead Sea: Geol. Survey Isreal Bull., v. 63, 30 p.

Dellwig, L. F., 1955, Origin of the Salina Salt of Michigan: Jour. Sedim. Petrol., v. 25, p. 83-110.

Nurmi, R. D. and G. M. Friedman, 1977, Sedimentology and depositional environments of basin-center evaporites. Lower Salina Group (Upper Silurian), Michigan Basin: in J. H. Fisher, ed., Studies in Geology 5: Reefs and Evaporites - Concepts and Depositional Models: Tulsa, Amer. Assoc. Petrol. Geol., p. 23-52.

Parea, G. C. and Ricci Lucchi, F., 1972, Resedimented evaporites in the Periadriatic Trough: Israel Jour. Earth Sci., v. 21, p. 125-141.

Schlager, W. and Bolz, H., 1977, Clastic accumulation of sulphate evaporites in deep-water: Jour. Sedim. Petrol., v. 47, p. 600-609.

Recent Shallow Water Evaporites

Krumbein, W. E. and Y. Cohen, 1974, Biogene, klasticsche und evaporitische Sedimentation in einern mesothermen mono-miktischen ufernahen See (Golf von Aquaba): Geol. Rundschau, v. 63, p. 1035-1065.

Schreiber, B. C. and D. J. J. Kinsman, 1975, New observations on the Pleistocene evaporites of Montallegro, Sicily and modern analog: Jour. Sedim. Petrol., v. 45, p. 469-479.

Schreiber, B.C., R. Catalano and E. Schreiber, 1977, An evaporitic lithfacies continuum: the latest Miocene (Messinian) deposits of the Salemi Basin (Sicily) and a modern analog: in J. H. Fisher, ed., Studies in Geology 5: Reefs and Evaporites - Concepts and Depositional Models: Tulsa, Amer. Assoc. Petrol. Geol., p. 196-180.

Horodyski, R. J. and S. P. Vonder Haar, 1975, Recent calcareous stromatolites from Laguna Mormona (Baja California) Mexico: Jour. Sedim. Petrol, v. 45, p. 894-906.

Neev, D. and K. O. Emery, 1967, The Dead Sea - Depositional processes and environment of evaporites: Geol. Survey Israel, Bull. 41, 147 p.

Shearman, D. J., 1970, Recent halite rock, Baja California, Mexico: Instit. Mining Metallurgy Trans., v. 79B, p. 155-162.

Weiler, Y., E. Sass and I. Zak, 1974, Halite oolites and ripples in the Dead Sea, Israel: Sedimentology, v. 21, p. 623-632.

Ancient Shallow Water Evaporites

Hardie, L. A. and H. P. Eugster, 1971, The depositional environment of marine evaporites: a case for shallow, clastic accumulation: Sedimentology, v. 16, p. 187-220.

Richter-Bernburg, G., 1973, Facies and paleogeography of the Messinian evaporites in Sicily, in C. W. Drooger, ed., Messinian Events in the Mediterranean: North Holland, Amsterdam, p. 124-141.

Via, G. B. and F. Ricci Lucchi, 1977, Algal crusts, autochthonous and clastic gypsum in a cannibalistic evaporite basin: a case history from the Messinian of Northern Appenines: Sedimentology, v. 24, p. 211-244.

Garrison, R., B.C. Schreiber, D. Bernoulli, F.H. Fabricius, R.B. Kidd and F. Melieres, 1978, Sedimentary petrology and structures of Messinian evaporitic sediments in the Mediterranean Sea, Leg 42A, Deep Sea Drilling Project: in K.J. Hsu and L. Montadert et al., Initial Reports of the Deep Sea Drilling Project, v. XLII, p. 571-611; Washington D.C.,

U.S. Government Printing Office. See also Schreiber et al., (1976), Schreiber et al. (1977) and Schreiber (1978).

*Davies, G. R. and W. W. Nassichuk, 1975, Subaqueous evaporites of the Carboniferous Otto Fiord Formation, Canada Arctic Archipelago: A summary. Geology, vol. 3, p. 273-278.

*Wardlaw, N. C. and D. L. Christie, 1975, Sulfates of submarine origin in Pennsylvanian Otto Fiord Formation of Canadian Arctic: Bull. Can. Petrol. Geol., vol. 23, p. 149-171.

*Kendall, A. C., 1976, The Ordovician carbonate succession (Bighorn Group) of southeastern Saskatchewan: Sask. Dept. Mineral Resources Rept. 180, 185 p.

Goto, M., 1967, Oriented growth of gypsum in the Marion Lake gypsum deposit, South Australia: Jour. Faculty Sci., Hokkaido Univ. (Ser. IV) Geology and Mineralogy, v. 13, p. 349-382.

Arthurton, R. S., 1971, The Permian evaporites of the Langwathby Borehole, Cumberland. Rep. Inst. geol. Sci. U.K. 71-17, 18 p.

Dellwig, L. F., 1968, Significant features of deposition in the Hutchinson Salt, Kansas, and their interpretation, in R. B. Mattox ed., Saline Deposits, Geol. Soc. Amer. Spec. Paper 88, p. 421-426.

Dellwig, L. F. and R. Evans, 1969, Depositional processes in Salina Salt of Michigan, Ohio and New York: Amer. Assoc. Petrol. Geol. Bull, v. 53, p. 949-956.
See also Dellwig (1955).

*Wardlaw, N. C. and W. M. Schwerdtner, 1966, Halite-anhydrite seasonal layers in Middle Devonian Prairie Evaporite Formation, Saskatchewan, Canada: Geol. Soc. Amer. Bull., vol. 77, p. 331-342.

Arthurton, R. S., 1973, Experimentally produced halite compared with Triassic layered halite-rock from Cheshire, England: Sedimentology, v. 20, p. 145-160.
Possibly the best illustrated paper written upon evaporites and one that reveals the potential of detailed fabric and experimental studies for environmental interpretation.

References Cited in the Text

*Dunn, C. E., 1975, The Upper Devonian Duperow Formation in southeastern Saskatchewan: Sask. Dept. Mineral Resources, Rept. 197, 151 p.

*Evans, R., 1970, Sedimentation of the Mississippian evaporites of the Maritimes: an alternative model: Can. Jour. Earth Sci., v. 7, p. 1349-1351.

Friedman, G. M., 1972, Significance of Red Sea in problem of evaporites and basinal limestones: Amer. Assoc. Petrol. Geol. Bull., v. 56, p. 1072-1086.

*Fuller, J.G.C.M. and J. W. Porter, 1969, Evaporite formations with petroleum reservoirs in Devonian and Mississippian of Alberta, Saskatchewan, and North Dakota: Amer. Assoc. Petrol. Geol. Bull., v. 53, p. 909-926.

*Holser, W. T., N. C. Wardlaw and D. W. Watson, 1972, Bromide in salt rocks: extraordinarily low content in the Lower Elk Point salt, Canada: in G. Richter-Bernburg, ed., Geology of saline deposits: Paris, UNESCO, p. 69-75.

Katz, A., Y. Kolodny and A. Nissenbaum, 1977, The geochemical evolution of the Pleistocene Lake Lisan-Dead Sea system: Geochim. Cosmochim. Acta, v. 41, p. 1609-1626.

Kendall, A. C., 1978, Facies Models 11: Continental and supratidal (Sabkha) evaporites: Geosci. Canada, v. 5, p. 66-78.

*Kendall, A. C., in prep., The Ashern, Winnipegosis and Lower Prairie Evaporite Formations of the commercial potash area of southern Saskatchewan: Saskatchewan Dept. Mineral Resources Rept. 181.

Leeder, M. R. and R. Zeidan, 1977, Giant late Jurassic sabkhas of Arabian Tethys: Nature, v. 268, p. 42-44.

Raup, O. B., 1970, Brine mixing: an additional mechanism for formation of basin evaporites: Amer. Assoc. Petrol. Geol. Bull., v. 54, p. 2246-2259.

Richter-Berburg, G., 1957, Isochrone Warven in Anhydrite des Zechstein: Geol. Jahrb., v. 74, p. 601-610.

Schaller, W. T. and E. P. Henderson, 1932, Mineralogy of drill cores from the potash field of New Mexico and Texas: U.S. Geol. Survey Bull. 833, 124 p.

Shearman, D. J., 1966, Origin of marine evaporites by diagenesis: Instit. Mining Metallugy Trans., v. 75B, p. 208-215.

Shearman, D. J., 1971, Marine evaporites: the calcium sulfate facies: Amer. Soc. Petrol. Geol., Seminar, University of Calgary, 65 p.

Wardlaw, N. C., 1972, Syn-sedimentary folds and associated structures in Cretaceous salt deposits of Sergipe, Brazil; Jour. Sedim. Petrol., v. 42, p. 572-577.

*Wilson, J. L., 1967, Carbonate-evaporite cycles in Lower Duperow Formation of Williston Basin: Can. Petrol. Geol. Bull., v. 15, p. 230-312.

MS received May 25, 1978.
Reprinted from Geoscience Canada, Vol. 5, No. 3, p. 124-139.

Facies Models 15. Models of Physical Sedimentation of Iron Formations

Erich Dimroth
Sciences de la Terre
Université du Québec à Chicoutimi
Chicoutimi, Qué.

Introduction

A single model of physical sedimentation is sufficient to describe terrigenous sedimentary rocks. Iron formations, on the other hand, must be viewed in terms of two models: 1) A model of physical sedimentation relating the sedimentary structures and textures of the rock to hydrodynamic processes during deposition, and 2) a chemical model that relates the present mineralogy of the rock to diagenetic (and metamorphic) processes. This paper will discuss models of the physical sedimentation of iron formations. Chemical models of the diagenesis of these rocks will be introduced in a later paper.

Any sedimentary rock containing more than 15 per cent iron, except heavy mineral placers, may be called an iron formation. No single facies model can apply to so heterogeneous a rock class. However, we can work with two groups of facies models, since there are two main groups of iron formation (Table I): With few exceptions iron-rich sediments are: 1) detrital chemical sediments analogous to limestones, or 2) iron-rich shales. Models of the physical facies of the first group of iron-rich sediments are based on a comparison of the sedimentary textures and structures of iron formation and limestone. I will discuss here a facies model of shelf iron formations based on the analogy with limestones and will briefly sketch the rudiments of a facies model of pelagic iron formation. Facies models of terrigenous sediments, some of which have been discussed in No. 2 to 4 of this series of articles, can be applied to the iron-rich shales.

Relatively little is known on the sedimentology of iron formations and most previous work (e.g., Borchert, 1952; James, 1954, 1966) has been centered on the mineralogy of these rocks. Therefore, I will have to distill a local model of shelf iron formations from the example of the Sokoman Formation in the Labrador trough, Quebec, Canada. This model has been obtained mainly by lateral correlation of stratigraphic sections by observation of lateral facies changes, and by comparison with analogous recent limestones of the Persian Gulf. I will use some Archean iron formations of the Superior Province to sketch the rudiments of a model of pelagic iron formations. Much work will have to be done to develop and generalize both models.

Sedimentary facies are influenced, to a degree, by biologic processes. Therefore ecologic factors have to be considered, where facies models are based on a comparison of ancient sedimentary rocks and recent sediments. Consideration of ecologic factors is particularly important if Precambrian rocks are compared with recent sediments, as I will do in this paper.

Precambrian sedimentary rocks differ from recent sediments mainly in the following properties: 1) Skeletal and framework building organisms did not exist in the Precambrian. Therefore, Precambrian sedimentary rocks lack skeletal debris, reefs built by framework-builders, and reef debris. 2) Sea-grasses and higher algae that form baffles at the bottom of the present shallow sea did not exist in the Precambrian. However, blue-green algae appparently fulfilled that function in the Precambrian at least to a degree. Blue-green algae also acted as sediment-binders and constructed reef-like mounds. 3) Mud-eating and bottom stirring organisms did not exist in the Precambrian. Therefore delicate laminations are preserved in Precambrian sedimentary rocks whereas they commonly have been destroyed in the Phanerozoic. Thin beds of differently textured sediment survived in the Precambrian, whereas they commonly have been mixed in the Phanerozoic.

Comparison of Iron-Formation and Limestone

The siliceous and aluminous iron formations, variously termed cherty (or banded) iron formations and Minette-type ironstones, are detrital chemical sediments. They have sedimentary textures and structures analogous to the sedimentary textures and structures of limestones. Analogues to all textural types of detrital limestones exist: micritic, intraclastic, peloidal and pelleted, oolitic, pisolitic, stromatolitic. These varieties are briefly described in Table II, but Figures 2 to 8 and 9 to 14 might be better documentation than any description in words would be.

176

Table II
Textural types of cherty iron formation

1. *Micrite-type:* Deposited as a mud whose particles are too fine grained to survive diagenesis. Only lamination and stratification are visible as depositional structures. Small-scale cross-beds here and there prove deposition as particulate, non-cohesive matter. (Fig. 8, 9).

2. *Pelleted:* Fine pellet textures of silt or very fine sand size. (Fig. 2).

3. *Intraclastic:* Containing gravel-size fragments (intraclasts) whose internal textures prove derivation from pene-contemporaneous sediment. Fragments may be embedded in a micrite-type matrix (Fig. 5, 10), or may be bound by a cement that has been introduced during diagenesis.

4. *Peloidal:* Contains sand-size fragments (peloids) without internal textures. Peloids may be embedded in a micrite-type matrix or may be bound by a clear chert cement that has been introduced during diagenesis (Fig. 3).

5. *Oolitic:* Contains concentrically laminated ooids, either set in a micrite-type matrix or, more commonly bound by a clear chert cement introduced during diagenesis (Fig. 10).

6. *Pisolitic:* Contains pisolites (concentrically laminated bodies similar to but larger than ooids, probably produced by blue-green algae) set either in a micrite-type matrix or cemented by clear chert. (Fig. 6).

7. *Stromatolitic:* Wavy, columnar or digitating stromatolites. (Fig. 7).

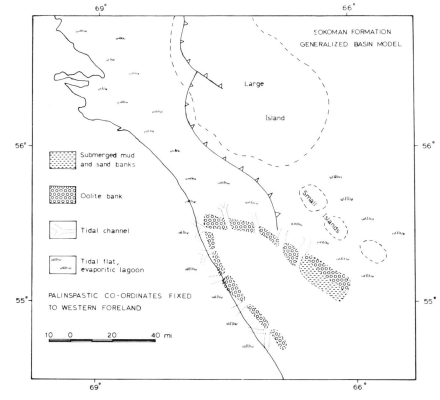

Figure 1
Paleogeography of the Labrador trough during deposition of the Sokoman Formation.

These sedimentary textures and structures permit a number of important conclusions: 1) iron formations, like most limestones, are detrital sediments; they have been mechanically deposited as particulate matter. 2) particles came in various sizes: clay-size (Fig. 8), silt-size (Fig. 2), sand-size (Fig. 3), gravel-size (Fig. 5). 3) particles were relatively hard, and were non-cohesive. The fresh sediment had mechanical properties rather similar to those of lime muds, sands and gravels.

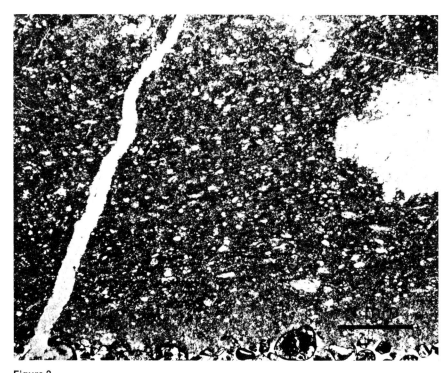

Figure 2
Pelleted iron formation. Relicts of silt-size particles (pellets) are visible. Hematite is the iron mineral. Bar 1 mm long.

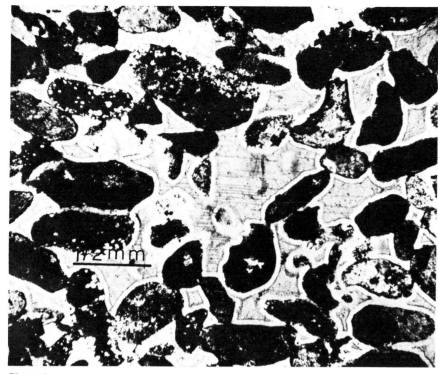

Figure 3
Peloidal iron formation. Structureless particles of sandsize (peloids) are well preserved. Hematite as iron mineral.

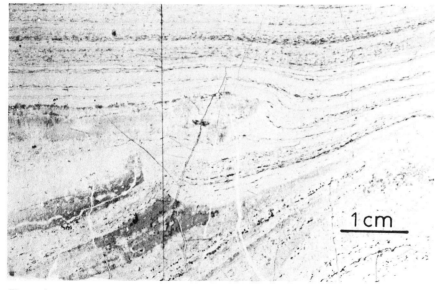

Figure 4
Peloidal iron formation, strongly recrystallized, with crossbedding.

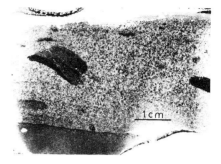

Figure 5
A micrite-type iron formation is overlain by an intraclastic bed. Coarse fragments show internal lamination, thus, are derived from penecontemporaneous sediment. The intraclasts are loosely set in a micrite-type matrix. Minnesotaite is the iron mineral.

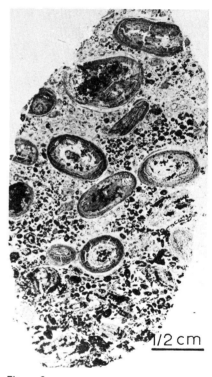

Figure 6
Pisolitic iron formation. Pisolites (P) and ooids (O) cemented by clear chert. Hematite and magnetite predominate.

The Sokoman Formation: Local Model of Shelf Iron Formations

Flat, prograding coastlines away from estuaries, like the south shore of the Persian Gulf (Purser, 1973), generally show a subdivision in four oceanographic domains: 1) a shallow marine shelf sea, 2) a narrow strip of sand bars, tidal deltas, reefs, or island, 3) a lagoonal platform, 4) a tidal flat. The Sokoman iron formation of the Central Labrador trough is very similar to the limestones that are presently being deposited at the south shore of the

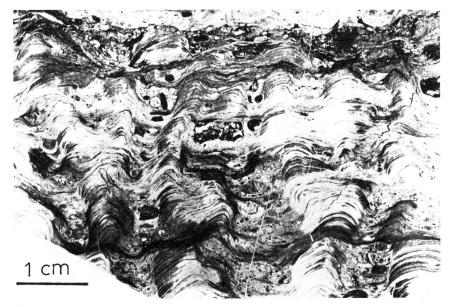

Figure 7
Stromatolitic iron formation. Hematite predominates.

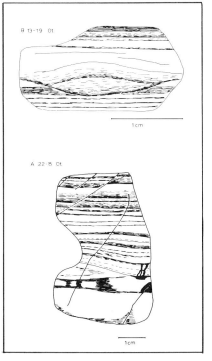

Figure 8
Micrite-type iron formation. Cross-lamination proves that the rock has been deposited as non-cohesive particulate matter, but no remnants of the particles survived diagenesis. Iron minerals are hematite and minnesotaite.

Persian Gulf, and a similar oceanographic subdivision can be recognized (Fig. 1). Each oceanographic domain has its own, characteristic facies. The following facies model is based mainly on a comparison of the Labrador trough iron formation with limestones in the Persian Gulf area, but takes in account certain features of limestones of the Bahama platform and the Florida coast.

Shallow marine shelf: Most parts of the shallow marine shelf, below storm wave base, are covered with micritic muds. Micrite-type shelf iron formation (Fig. 9) of the Labrador trough show delicate parallel laminae; such lamination is absent from shelf muds of the Persian Gulf due to the action of sediment-eating organisms.

Where marine currents are strong, very fine sand (2-3ϕ = 0.1 - 0.2 mm) has been winnowed from coarse foreshore sediment, and has been carried far on to the shelf. Such shelf sands in the Persian Gulf are poorly sorted, due to admixture of terrigenous mud and of shells of foraminifera. In the Labrador trough, shelf sand has not been diluted by other components and is well sorted. Parallel stratification is characteristic.

Foreshore zone. Beds of coarse-grained oolite or peloid sand are intercalated between the shelf muds or sands in the forshore zone. Storm layers formed at the wind-exposed margin of mud shoals

that reached above storm wave base. Such storm layers (Fig. 10) generally are composed of very coarse-grained muddy sand or of muddy gravel. They are poorly sorted and, not uncommonly, show graded bedding. A mud-matrix is absent where the area eroded by the storm lacked muddy sediment.

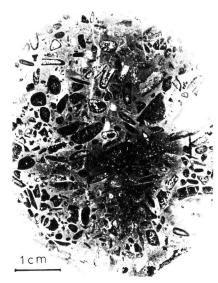

Figure 10
Iron formation deposited as muddy gravel. Intraclasts embedded in a micrite-type matrix. This is a storm layer deposited on top of a mud bank in the foreshore of the shelf. Hematite and magnetite predominate.

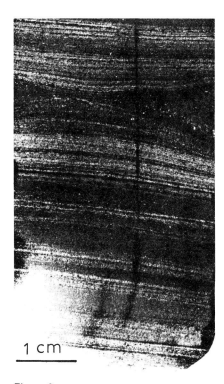

Figure 9
Micrite-type iron formation deposited on the shelf. Minnesotaite predominates.

Shelf-marginal bank: Thick-bedded or massive oolite or peloid sand builds the sand banks, tidal deltas and islands at the shelf margin. The sand (Fig. 11) is coarse grained and very well sorted. Large-scale cross-bedding is common. Oolites formed where sand has been agitated more or less constantly at the sediment surface by alternating tidal currents. Peloids predominate where the sediment was rarely agitated; at present, this is the case where sediment has been stabilized by micro-organic slimes, blue-green algae, algae, or sea-grasses, and where tidal currents are weak.

Lagoonal platform. Sediments of the lagoonal platform are the most variable; they also show relatively great differences between otherwise analogous Precambrian and recent facies. At present, mud, muddy sand, or well sorted sand are deposited on lagoonal platforms, depending on local agitation by tidal currents and waves. Grain size of the sediment increases with increasing intensity of tidal action or wave agitation. Mud deposition is by no means restricted to lagoonal basins; on the contrary, it commonly occurs in water shallower than one metre, on mud banks and tidal flats. Strong storms may carry sand far into areas of mud-deposition, and may produce intraclast gravels. Superficial oolites are common and form where peloids are suspended for brief periods during storms. In the present, mud-eating and bottom-stirring organisms are extremely active and would destroy any small-scale bedding that existed.

In the Labrador trough, sediments deposited on the lagoonal platform are characterized by thin-bedded alternation of micrite-type sediment with oolite or peloid sand (Fig. 12). Surficial ooids are more common than fully developed ooids except in proximity to the basin-marginal oolite banks. Storm layers (Fig. 13) of sandy or muddy gravel are common. Upward-coarsening cycles, 5 to 15 metres thick, have been observed, and probably formed by the gradual infilling of lagoonal basins. Cross-bedding, erosion channels, scours, and flaser bedding are common.

Applications of the Model
Now that we have a local model of shelf iron formations we may ask how this model performs: 1) as a framework and guide to future observations, 2) as a

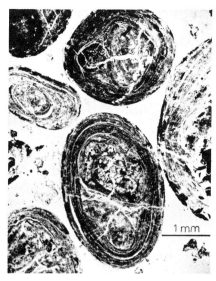
Figure 11
Iron formation deposited as well sorted, coarse-grained ooid sand, of the sand bank at the shelf margin. Hematite and magnetite predominate.

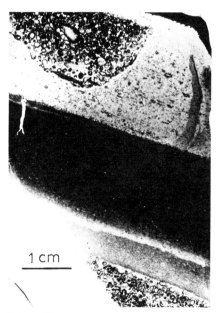
Figure 12
Alternating beds of oolitic and micrite-type iron formations, characteristic of the lagoonal platform. Hematite and magnetite predominate. Note the erosion channel.

norm with which other shelf iron formations may be compared, 3) as a basis for hydrodynamic interpretation, and 4) as a predictor in new geological situations.

(1) *The model as framework and guide to future observations.* The classification of textures (Table II) provides a simple framework for the description of iron formations. Furthermore, all other textures and structures common in limestones are to be expected in iron formations, for example, fenestral textures and solution porosity (Dimroth and Kimberley, 1976, Fig. 14). They have to be recorded carefully as basis for further interpretation.

(2) *The model as norm.* Facies of the Brockman Formation (Hamersley Range, Western Australia) and of the Penge and Kuruman Formation (Transvaal basin, South Africa)

Figure 14
Laminated Algoma-type iron formation.

Figure 13
Intramicrite-type iron formation overlying laminated micrite. Note the graded bedding in the intramicrite, the erosional base of the bed, and loading of the underlying micrite by intraclasts. This is a storm layer deposited on a shallow mud-bank in the lagoonal domain. The iron mineral is siderite.

180

manifestly differ from facies of the
Sokoman Formation (see Trendall,
1968). The two former units are
exclusively or predominantly composed
of laminated micrite-type iron
formations, whereas micrite-type
lithologies do not make up more than
about 30 per cent of the Sokoman
Formation. Based on the model we now
ask why this is so.

(3) *The model as basis for
hydrodynamic interpretations.* As noted
above, the Kuruman formation is mainly
composed of laminated, micrite-type
iron formations (Beukes, 1973; Button,
1976). However, the formation does
contain thin intercalated beds of oolite
sand, of cross-bedded peloid sand, of
intraclast conglomerate and of
edgewise conglomerate. Some of the
laminated rocks documented by Beukes
(1973, Fig. 20B, C, 22C) more closely
resemble peritidal lime muds than deep
basinal lime muds.

These features suggest that facies
differences beween the Sokoman and
Kuruman Formations might be due to
differences in basin hydrodynamics
rather than to differences in basin depth:
The Sokoman formation has been
deposited in a basin exposed to fairly
vigorous tidal action, as the present
Persian Gulf; the Kuruman Formation
possibly has been deposited in a similar
basin from which, however, tidal
currents were more or less excluded. Of
course, this hypothesis has to be verified
by much further work. Button (1976) has
proposed a basin model for the Penge
and Kuruman Formations that very
closely resembles the basin model
proposed for the Sokoman Formation.

The case of the Brockman Formation
is quite different: Micrite-type rocks of
that unit show *only* parallel lamination,
quite similar to the varve-like
laminations of evaporitic limestones
(Trendall and Blockley, 1976). It appears
that the laminated facies grades into
peloidal rocks toward the extreme
margin of the presently preserved part of
the basin (A. D. T. Goode, pers.
commun., 1976). These features
suggest that the Brockman Formation
has been deposited in a fairly deep
basin, permanently below storm wave
base. The Permian Marathon basin of
the US may be a suitable Phanerozoic
analogue; I know of no present-day
equivalent to such a basin.

(4) *The model as predictor.* The
tentative interpretations outlined above
permit us to make some predictions: if
our interpretations are correct, it is to be
expected that micrite-type iron
formations with desiccation cracks and
with fenestral textures, associated with
flat-pebble conglomerates should be
found in the Penge and Kuruman
Formations; such structures are not to
be expected, and have not been found,
in the Brockman Formation.

Pelagic Iron Formations
Our knowledge of Archean pelagic iron
formations is far too limited to permit the
distillation of a facies model that would
fulfill the four functions required by
Walker (1976). Yet, at least a sketch of a
facies model may be presented.
Algoma-type iron formations generally
are thin lenses, rarely more than one or a
few metres thick and rarely continous for
more than a few kilometres. They have
micrite type textures. Parallel lamination
and, rarely, erosion surfaces (Gross,
1972; Schegelski, 1975) are the only
primary sedimentary structures.
Algoma-type iron formations do not
grade into iron formation with different
sedimentary textures and structures
either laterally, or in vertical section.

Algoma-type iron formations here and
there are intercalated between volcanic
flows. Much more common is
intercalation between volcaniclastic
(epiclastic and autoclastic) sediments.
The iron formations are closely
associated with grey or black (graphitic)
cherts, with black shales, and with
argillites all of which show parallel
lamination as their only sedimentary
structure, and with greywakes.
Associated greywackes may be very
fine grained and thin bedded (5 cm or
less), or medium to coarse-grained and
thick-bedded (30 cm and more). Both are
turbidites; the thin bedded greywackes
probably are distal turbidites of Walker's
(1967) type a→e, whereas the thick-
bedded greywackes are proximal
turbidites.

In certain cases, a characteristic
vertical facies change has been
observed: Proximal or distal turbidites
are followed upwards by turbidites that
have thin (one to several mm) laminae of
magnetite-bearing chert at the top of the
Bouma cycle (described in Walker, 1976
b). The proportion of chert gradually
increases upward toward the bed, one to

several metres thick, of magnetite chert
or jasper. Upward, the proportion of iron
formation decreases again. Such a
sequence has been observed in Canada
and in South Africa (compare Beukes,
1972, Fig. 7), but is neither universal nor
is it the only sequence described.

Clearly, these iron formations are
pelagic sediments since they occupy
the uppermost interval of the Bouma
cycle that is normally occupied by
pelagic shale. They are intercalated
indiscriminately between proximal or
distal turbidites. Thus, apparently their
deposition may have taken place on any
part of the turbidite fan. It appears,
therefore, that they were deposited in
any environment from which
volcaniclastic detritus was temporarily
excluded. It is quite unknown whether
the non-deposition of contemporaneous
volcaniclastic material is due to tectonic
causes (subsidence of the source area)
or simply reflects a change in the pattern
of the distributary channels.

This sketch does not fulfill the four
functions of a facies model. It describes
a common case of Algoma-type iron
formation but is not necessarily a norm.
It may serve as guide to future
observations mainly in so far as it is
important to search for, and to describe,
cases where Algoma-type iron
formations occur in a different context.
The model has no predictive value, and
conclusions on hydrodynamic
conditions of deposition are trivial:
obviously these iron formations settled
from very dilute suspensions, in tranquil
water, at a depth that was permanently
below storm wave base.

Origin of Iron Formation
The facies model outlined above is
largely independant of speculations on
the origin of iron formation. Basically,
there are two genetic hypothese:
Cayeux (1922, 1911) suggested that the
cherty iron formations formed by
silicification of aluminous iron
formations and the latter by
penecontemporaneous remplacement
of limestone. On the other hand, Cloud
(1973), Eugster and Chou (1973) and
Lepp and Goldich (1964) believed that
iron formations were precipitated from a
hypothetical "primordial" ocean; that
primordial ocean is thought to have been
devoid of oxygen and saturated with
dissolved iron carbonate. Dimroth and
Kimberley (1976) pointed out that there

is strong evidence against the existence of such a "primordial" ocean even in the Archean.

Iron formations were deposited in the form of non-cohesive particulate matter; this may be interpreted as evidence against direct precipitation of silica and iron (in forms other than iron carbonate) since silica and iron oxide and silicate generally precipitate as gels. Of course, precipitation of iron carbonate is excluded, because of the evidence for deposition of iron formation from an oxygenic environment (Dimroth and Chauvel, 1973; Dimroth and Kimberley, 1976). The limestone-type textures of iron formations point to an origin of iron formation by limestone replacement. Thus, the author believes that iron formations most likely were precipitated as $CaCO_3$, probably as aragonite, organically or inorganically. The original aragonite sediment has been replaced by silica and iron compounds during early diagensis, but the details of the process, and the source of iron and silica (terrigenous clay? tuff?) are still unknown.

Bibliography
This list is very brief on purpose. It is intended as a basic reader for those who wish to become familiar with the models outlined here and with other ideas and concepts on the origin of iron formations.

Sedimentary Textures, Structures, and Facies Models

Cayeux, L., 1911, Comparaison entre les minerais de fer huroniens des Etats-Unis et les minerais de fer oolitiques de France: C.R. Acad. Sci. (Paris), v. 153, p. 1188-1190.

Cayeux, L., 1922, Les minerais de fer oolithiques de France II. Minerais de fer secondaires. Etudes des Gîtes Minéraux de France. Paris, Service de la Carte Géologique. Imprimerie Nationale, 1052 p.

This is the classical account of the Mesozoic iron ores of France by the grand master of sedimentary petrography. It includes a rather detailed comparison of iron ores and limestones; origin of iron ore by pene-contemporaneous replacement of limestone is implied. Cayeux' 1911 paper has great historical interest, as the first comparison of Precambrian and Phanerozoic iron formations.

Chauvel, J.-J. and E. Dimroth, 1974, Facies types and depositional environment of the Sokoman Iron Formation, central Labrador trough, Quebec, Canada: Jour. Sed. Pet. v. 44, p. 299-327.

Dimroth, E., 1975, Depostional environment of the iron-rich sedimentary rocks: Geologische Rundschau, v. 64, p. 751-767.

Dimroth, E., 1976, Aspects of sedimentary petrology of cherty iron formations: in K. H. Wolf, ed., Handbook of Stratabound and stratiform ore deposits: Elsevier, Amsterdam.

Dimroth, E. and J. J. Chauvel, 1973, Petrography of the Sokoman Iron formation in part of the central Labrador trough, Quebec, Canada: Geol. Soc. Amer. Bull., v. 84, p. 111-134.

Dimroth's papers present detailed descriptions of the Labrador trough iron formation based on the limestone-analogy.

Gross, G. A., 1972, Primary features in cherty iron-formations. Sediment. Geol., v. 7, p. 241-261.

First documentation of number of sedimentary structures in Canadian iron formations.

Hofmann, H. J., 1969, Stromatolites in the Proterozoic Animikie and Sibley Groups, Ontario: Geol. Surv. Canada, Paper 68-69.

Detailed documentation of Gunflint stromatolites.

Mukhopadhyay, A. and S. K. Chanda, 1972, Silica diagenesis in the banded hematite jasper and bedded chert associated with the Iron Ore Group of Jamda-Koira Valley, Orissa, India: Sediment. Geol., v. 8, p. 113-135.

Quite detailed comparison of diagenetic textures of quartz in iron formation with calcite textures in limestones.

Trendall, A. F. and J. G. Blockley, 1970, The iron formation of the Precambrian Hamersley Group, Western Australia, with special reference to the associated crocidolite: Geol. Surv. Western Australia, 336. p.

Detailed description of sedimentary textures and structures of an Australian iron formation.

Limestone Literature for Comparison
Bathurst, R. G. C., 1971, Carbonate sediments and their diagenesis: Developments in Sedimentology, Amsterdam, Elsevier, v. 12, 620 p.

Milliman, J. D., 1974, Marine Carbonates: Springer-Verlag, 375 p.

Purser, B. H., ed., 1973, The Persian Gulf - Holocene Carbonate Sedimentation and Diagenesis in a Shallow Epicontinental Sea: Springer-Verlag, New York.

Mineral Facies Concept
Borchert, H., 1952, Die Bildungsbedingungen marin-sedimentärer Eisenerze: Chemie der Erde, v. 16, p. 49-74.

James, H. L., 1954, Sedimentary facies of iron formation: Econ. Geol., v. 49, p. 251-266.

James, H. L., 1966, Chemistry of the iron-rich sedimentary rocks: U.S. Geol. Surv. Prof. Paper 440-W, 61 p.

Genetic Speculations
Cloud, P. E., Jr., 1973, Paleoecological significance of banded iron formations: Econ. Geol., v. 68, p. 1135-1143.

Dimroth, E. and M. M. Kimberley, 1976, Precambrian atmospheric oxygen: Evidence in the sedimentary distributions of carbon, sulfur, uranium and iron. Can. Jour. Earth Sci., v. 13, p. 1161-1185.

Eugster, H. P. and I. Ming, Chou, 1973, The depositional environment of Precambrian banded iron formations. Econ. Geol., v. 68, p. 1144-1168.

Lepp, H. and S. S. Goldich, 1964, Origin of Precambrian iron formations: Econ. Geol., v. 58, p. 1025-1061.

Most hypotheses on the origin of Precambrian iron formations suffer from the defect that they attempt to link deposition of cherty iron formations to a hypothetical, oxygen-free ocean. This is in conflict not only with evidence for the presence of free oxygen in atmosphere and ocean since early Precambrian time, but also with the existence of very large cherty iron formations of late Precambrian age (compare Dimroth and Kimberley, 1976).

Other References Cited

Beukes, N. J., 1973, Precambrian iron-formations of Southern Africa: Econ. Geol., v. 68, p. 960-1004.

Button, A., 1976, Iron formation as end-member in carbonate sedimentary cycles in the Transvaal Supergroup, South Africa: Econ. Geol., v. 71, p. 193-201.

Schegelski, R. J., 1975, Geology and geochemistry of iron formations and their host rocks in the Savant Lake-Sturgeon Lake Greenstone Belts. Univ. Toronto, A progress Report: Superior Geotraverse Workshop 1975, p. 34-1 to 34-18.

Trendall, A. F., 1968, Three great basins of Precambrian banded iron-formation deposition: A systematic comparison: Geol. Soc. Amer. Bull., v. 79, p. 1527-1544.

Walker, R. G., 1967, Turbidite sedimentary structures and their relationship to proximal and distal depositional environments: Jour. Sediment. Petrol., v.37, p. 25-43.

Walker, R. G., 1976a, Facies models-1. General introduction: Geosci. Can., v. 3, p. 21-24.

Walker, R. G., 1976b, Facies models-2. Turbidites and associated coarse clastic deposits: Geosci. Can., v. 3, p. 25-36.

Walker, R. G. and F. J. Pettijohn, 1971, Archean sedimentations: analysis of the Minnitaki Basin, northwestern Ontario, Canada: Geol. Soc. Amer. Bull., v. 82, p. 2099-2130.

MS received November 1, 1976, revised MS received November 15, 1976. Reprinted from Geoscience Canada, Vol. 4, No. 1, p. 23-30.

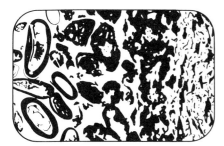

Facies Models 16. Diagenetic Facies of Iron Formation

Erich Dimroth
Sciences de la Terre
Université du Québec à Chicoutimi
Chicoutimi, Qué.

Introduction

Facies is the term used to denote the aspect of part of a sedimentary rock body, as contrasted with other parts. Primary facies are defined by features present at the time of deposition: primary lithology, sedimentary texture and sedimentary structures. Diagenetic facies are defined by diagenetic features. In iron formations, diagenetic facies are primarily defined by the nature of the iron-bearing minerals present.

Information on the distribution of iron minerals in iron formations can be distilled into a facies model in the same way as a model of sedimentary facies is distilled from information on primary lithology, sedimentary texture and structure. In fact, a model of diagenetic facies of iron formation is the oldest facies model available: the models of Borchert (1952) and James (1959). Of course, at that time, the mineralogy of iron formations was thought to be primary, in accordance with the general underestimation of diagenesis in the 1950s, and iron mineral facies of iron formations were therefore thought to be sedimentary facies. This fact cannot detract from the value of the James-Borchert model which I will have to paraphrase.

Just as a model for primary facies, the model of diagenetic facies must fulfill four purposes (Walker 1976).

1) It must act as a *norm*, for purposes of comparison;

2) It must act as a *framework* and *guide* for future observations;

3) It must act as a *predictor* in new geological situations;

4) It must act as a basis for physico-chemical interpretation of the diagenetic environment or system that it represents.

These four purposes are literally taken from Walker's (1976) introduction to the concept of facies models, except for two words. A model of *diagenetic* facies cannot serve as basis for *hydrodynamic* interpretation of the environment, and because *diagenetic* processes are determined by the *physico-chemisty* of the sedimentary system *after* deposition, a model of *diagenetic* facies must permit interpretation of these physico-chemical processes.

Many stages of diagenesis. Iron formations like other chemical sediments, underwent many stages of diagenesis. I define here as *early diagenetic* those processes that took place during the lithification and which are contemporaneous with the infilling of primary porosity, with compaction, and with the consolidation of the sediment. *Late diagenetic* processes took place after lithification. Of course, there is no clearcut limit between late diagenesis (taking place at low temperatures), load metamorphism (taking place at somewhat higher temperatures) and very low grade regional metamorphism (taking place during an orogeny, but not necessarily at higher temperatures than load metamorphism).

Figure 1 diagramatically shows the most important mineral reactions that have been deduced from petrographic observation of iron formation of the Labrador trough. The transitions from limestone into iron formation have been very little studied. Such transitions exist in the Transvaal Supergroup (Button, 1976), in the Hamersley Group (Davy, 1974) and in the Archean of the Slave Province (Kimberley, oral commun., 1976). Therefore, nothing is known about the first step of diagenesis namely the conversion of calcareous sediment to ferriferous chert. However, oolites of iron formations more closely resemble aragonite oolites than calcite oolites; this fact suggests that conversion of calcareous sediment to iron formation occurred during very early diagenesis, and preceded the inversion of aragonite to calcite (Kimberley, 1974).

On the other hand, transformations of iron minerals shown in Figure 1 have

been deduced from petrographic observations like those illustrated by Figures 2 to 5. In Figure 2, concentric texture of oolites is beautifully preserved by sub-microscopic hematite crystallites. Figure 3 shows partial destruction of the concentric texture by coarse-grained magnetite. In Figure 4, nearly all of the concentric texture has disappeared save for a few relicts here and there. Clearly, magnetite formed here at some later stage than hematite. A similar observation is illustrated in Figure 5, where magnetite not only is younger than hematite dust, but where the growth of magnetite is accompanied by differential compaction; consequently, magnetite formed here during a comparatively early stage of diagenesis, before the complete lithification of the sediment. This simple example illustrates the observations that have been used to construct the scheme shown in Figure 1.

The Facies Model

Diagentic facies of iron formation are basically the mineral facies introduced by James (1954). James defined facies by the *predominating* iron mineral and distinguished oxide (hematite, magnetite), silicate, carbonate, and sulphide facies. From this starting point, the mineral facies concept has evolved in two directions. First, Dimroth and Chauvel (1973) defined mineral facies by the relicts of very early diagenetic, iron minerals. These are the very fine grained, submicroscopic, hematite, and fine grained siderite and greenalite. Thus, hematite, siderite, and silicate facies are defined, to which a sulphide facies (mainly pyrite-rich shales) may be added. This scheme fails in formations that do not contain relicts of the very early diagenetic minerals. These have to be defined less precisely as "recrystallized iron formations".

Second, Zajac (1973) and Klein and Fink (1976) defined facies by the *assemblage of iron minerals* now present. This concept can be applied to any iron formation regardless of its diagenetic or metamorphic grade. Generally, it is assumed that all iron minerals present are stably co-existing; however, I will demonstrate below an example where this very likely is not the case.

Obviously, Dimroth and Chauvel's facies represent a relatively early stage of diagenesis and, therefore, show more

184

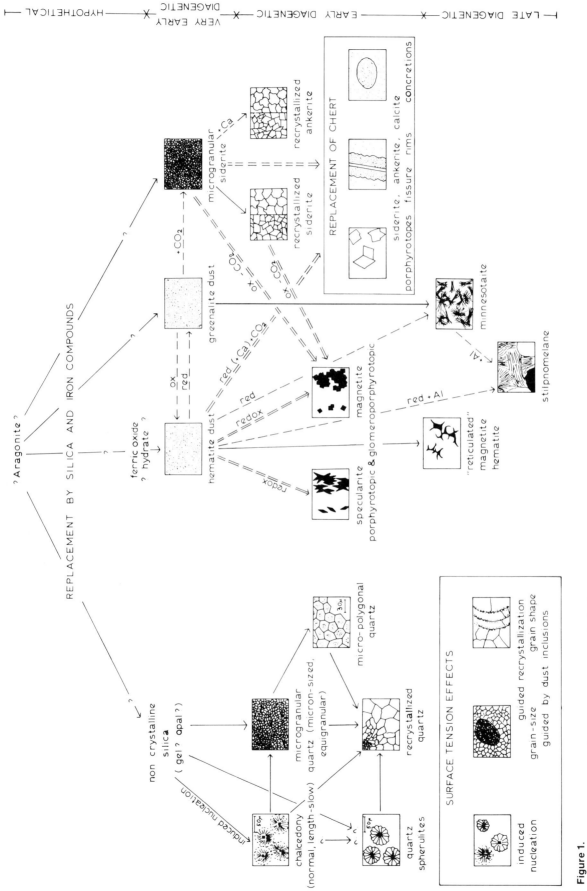

Figure 1.

Mineralogical evolution of the Sokoman Iron Formation, Labrador trough (after Dimroth and Chauvel, 1973); transformations hematite ⇌ greenalite → siderite is from the Gunflint Formation (Goodwin, 1956 and author's observations). Single lines indicate mineral-by-mineral replacement and recrystallization, double lines indicate transformations interpreted essentially as solution-precipitation reactions. Transformations involving redox reactions shown by broken lines (ox = oxidation; red = reduction; redox = reduction of Fe_2O_3 to Fe^{++}, solution and reprecipitation at another place by oxidation).

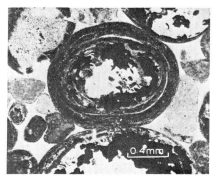

Figure 2.
Hematite oolite. Concentric texture is well preserved by submicroscopic crystallites of hematite. Some recrystallizations of hematite in centre of oolite. Section B13-15, Labrador trough.

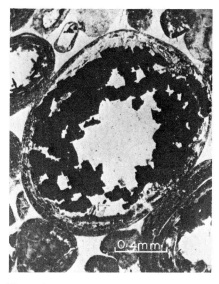

Figure 3.
Same section, about one mm from Figure 2. Most of the concentric texture of the oolite has been destroyed by growth of magnetite.

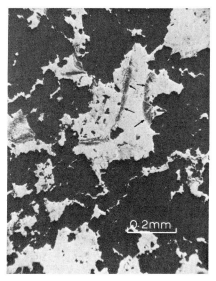

Figure 4.
Same thin section, about 2 mm from Figure 3: All sedimentary textures have been destroyed by growth of magnetite porphyroblasts. A few remnants of oolite skins remain (arrows) and are preserved by submicroscopic crystallites of hematite.

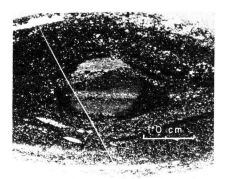

Figure 5.
The spherical grey spot in the center is a nodule of red jasper, stained by submicroscopic hematite crystallites. Lamination and sedimentary textures (ooids, peloids) are beautifully preserved by hematite dust. The dark material surrounding the jasper spot is composed mainly of coarse-grained magnetite and quartz. All sedimentary texture has been destroyed in this material, but some of laminae are continuous from the jasper into the surrounding magnetite-quartz fabric and indicate differential compaction around the jasper nodule. Section C2-S2, Labrador trough.

clearcut relations to sedimentary facies than the assemblage facies of Klein and Zajac which represent a relatively late stage mineralogy. On the other hand, mineral assemblage facies are very important to the economic geologist since the beneficiating properties of taconite iron ore depend largely on the late diagenetic mineral assemblage.

Overprinting mineral facies. Several redox reactions can be shown to have taken place in iron formations (Fig. 1), particularly oxidation of siderite to magnetite and/or hematite and reduction of hematite to magnetite, iron silicate or carbonate. An example of such overprinting is shown in Figures 6-8 hematite oolites and pisolites (Fig. 6), have been partly replaced by iron carbonate, probably siderite (Fig. 7, 8). It should be noted that siderite has never been observed by the writer in contact with the submicroscopic crystallites of hematite staining jasper, suggesting that both minerals do not stably co-exist. On the other hand, grain contacts between siderite and coarsely crystalline hematite are not uncommon; this association of minerals very likely is metastable, metastable preservation of specular hematite being due to its extremely low solubility in natural solutions.

Physico-chemical factors. Which iron mineral is stable during diagenesis depends mainly on Eh, pH, pCO_2 and pS" of the pore solutions. Traditionally, the stability fields of various iron minerals are shown on Eh-pH diagrams at fixed values of pCO_2 and pS" (Garrels and Christ, 1965; James, 1966). However, in *marine* sediments, pH varies little, and is largely determined by pCO_2. Therefore, I follow Berner (1971) in representing the stability fields of the iron minerals in an Eh-pCO_2 diagram at very low pS", and in an Eh-pS" diagram (Fig. 9). The stability fields of ankerite and of iron silicate are unkown. They have been stippled in Figure 9 on the basis of empirical evidence.

Eh depends on the concentration, in the pore solution, of oxygen, or of organic decay products; of course, both are incompatible. pS" depends on the activity of sulphate reducing bacteria, which increases with the proportion of organic matter in the sediment and is limited by the availability of dissolved sulphate. pCO_2 is a measure of the CO_2-

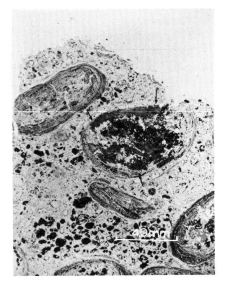

Figure 6.
Oolitic and pisolitic iron formation. Concentric texture of pisolites is beautifully preserved by submicroscopic crystallites of hematite. Carbonate porphyroblasts (probably siderite, dark grey shown by arrows) destroy the concentric texture. Thus, this is a hematite facies rock, overprinted by younger carbonate facies. Section B13-10, Labrador trough.

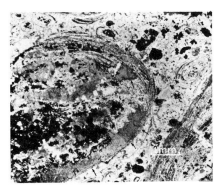

Figure 7.
Detail from Figure 6. Concentric texture of pisolite, preserved by hematite crystallites has been destroyed during subsequent growth of (dark grey) iron carbonate (arrows).

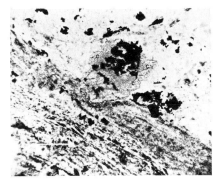

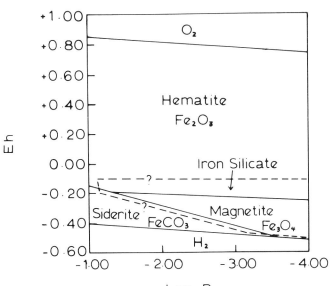

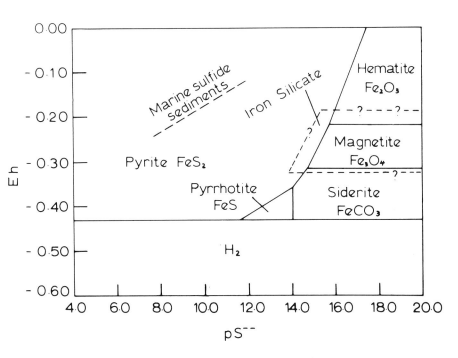

Figure 9.
Stability fields of some iron minerals in Eh-pCO_2 and Eh-pS'' diagrams (after Berner, 1971). The stability field of iron silicate is unknown.

Figure 8.
Detail from Figure 6. A carbonate porphyroblast (grey, c) replaced part of a pisolite and the cementing chert. Thus carbonate facies is late diagenetic.

content of the pore solutions; it depends largely on carbonate precipitation or dissolution during diagenesis and on addition of CO_2 to pore solutions due to decay of organic matter. Therefore, the organic content of the sediment and availability of oxygen during diagenesis are the main factors that determine which iron mineral will form in the sediment. Reduction of Fe^{3+} to Fe^{2+} is caused by the presence of organic matter in the sediment, oxidation of Fe^{2+} to Fe^{3+} by the presence of oxygen.

Sedimentological factors. The organic content of the sediment depends largely on the organic productivity of the environment and on the grain-size of the sediments. Because most organic detritus is plankton, organic matter is enriched in fine grained sediment. During *earliest diagenesis* oxygen diffuses into the sediment, diffusion rates increasing with increasing grain size. During later diagenesis oxygen is excluded from the sediment unless it is transported in decending solutions. On the other hand, ascending diagenetic solutions generally contain decay products of organic matter and, therefore, are reducing.

As an effect of these relations, earliest diagenesis generally takes place under oxidizing or slightly reducing conditions in well-washed sands; it takes place under strongly to moderately reducing conditions in muds, *unless these have been deposited in environments of low productivity and at low rates of sedimentation.* Conditions during later diagenesis depend largely on the direction of pore fluid exchange, pore fluids migrating mainly in well-washed sands.

Iron minerals in recent sediments. Recent iron mineral formation apparently can be understood along the lines outlined above (Berner, 1971). Thus, ferric oxide or oxyhydrates form in the presence of free oxygen, and survive diagenesis in the absence of organic matter. Well washed sands are most likely to contain ferric oxides as are muds deposited in areas of low productivity and at low sedimentation rates (e.g., deep-sea clays). Chamosite forms in overall aerobic environments, in the presence of small amounts of organic matter (Porrenga, 1966). Siderite and sulphides are precipitated in muds relatively rich in organic matter

(Berner, 1970; Ho and Coleman, 1969); which is precipitated seems largely to depend on the availability of SO_4" in the pore solutions.

Iron minerals in ancient iron formations. James (1954, 1966) recognized a similar relationship between the texture, carbon content and mineralogy of iron formations as outlined above. In the Labrador trough, all well washed sands contain either hematite dust or iron silicate as very early diagenetic iron mineral. Most micrite-and intramicrite-type iron formations contain very early diagenetic siderite or iron silicate; however, some do contain hematite dust. As could be expected, relations between later diagenetic mineralogy and sedimentary texture are less clear-cut. A very large class of iron formation is derived from well-washed peloid or ooid sands. Traces of hematite dust prove that these were originally hematite facies rocks; however, this hematite has been replaced nearly completely by coarse grained magnetite, ankerite and siderite during later diagenesis. Thus, they are oxide facies rocks that have been reduced to siderite facies rocks later during their diagenesis.

A correlation between mineralogy and carbon content has also been recognized (James, 1966). Iron formations containing hematite or iron silicate as earliest diagenetic iron mineral are free of carbon, rocks containing fine-grained siderite commonly contain a few tenths of a per cent carbon, and most pyrite-rich rocks are highly bituminous shales.

Applications of the Model
We now will investigate how the model fulfills the four functions defined above.

1) *The model as framework and guide to future observation.* A few simple observations permit ready classification of iron formations into mineral facies. If the reader prefers my definition of mineral facies, he should inspect thin sections for the presence of hematite dust, or microcrystalline iron silicate (greenalite, minnesotaite) or siderite. If he prefers assemblage facies he must determine all iron minerals, paying particular attention to carbonate mineralogy; ankerite, dolomite and calcite are more common than is generally assumed.

2) *The model as norm.* Basically, this model has served as a norm since it was first proposed by James (1954). Of course, deviations from the norm exist: e.g., very rarely, sandstones sandwiched between black shale may contain considerable pyrite. Based on the model, we can try to give an answer to the question why is this so, as I will attempt to do below.
3.) *The model as basis of physico-chemical interpretation.* The Gunflint Formation contains beds, 5 to 10 cm thick, of hematite iron formations intercalated between organic-rich pyritiferous shales (Figs. 10-12). Fine-grained hematite has been replaced at the margins of these beds, first by iron silicate (greenalite), then by carbonate (siderite). Clearly, replacement of earlier hematite by siderite is indicated since the sedimentary textures of the rock are successively destroyed during growth of carbonate.

My tentative interpretation is that solutions containing decay products of organic matter diffused into the bed of hematite iron formation from the adjoining shale layers, and established a reducing diagenetic environment.

Similarly, the sand beds containing pyrite discussed above are sandwiched between organic-rich shales in which sulphate-reducing bacteria were active. Thus, the abnormal precipitation of pyrite in a sandstone probably is due to infiltration of H_2S bearing diagenetic solutions from the adjoining shale into the sandstone bed.

4) *The model as a predictor.* Extremely reducing and extremely oxidizing environments never should be in direct contact. Thus, when we find two outcrops of iron formation, one containing hematite dust as earliest iron mineral, the other with microcrystalline siderite, we can be fairly certain that the two rocks do not belong to stratigraphic units that are directly in contact. However, we are quite unable to predict the thickness of any transitional zone, which may be quite thin.

Earliest Diagenesis of Iron Formation
As noted above, very little is known about the earliest stage of iron formation diagenesis, namely the replacement of calcareous sediment by silica and iron minerals. This is due to two circumstances: 1) Transitions of limestone in iron

188

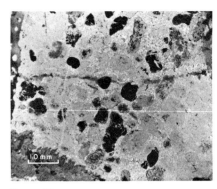

Figure 10.
Part of a jasper layer intercalated between black shale, replaced at the margin by iron silicate (minnesotaite) and carbonate. Here peloids preserved by submicroscopic hematite crystallites are shown. Section G7B, Gunflint Formation.

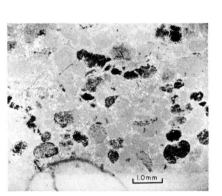

Figure 11.
Same section as Figure 10. Peloids have been replaced by stilpnomelane.

Figure 12.
Same section as Figure 10. Siderite porphyroblasts replaced peloids and part of the cements. Growth of siderite porphyroblasts is related to carbonate-filled fractures, to stylolites, and is most intense at the margin of the jasper bed.

Figure 13
Silicified carbonate porphyroblast in thin-laminated magnetite layer. Hamersley range, Western Australia. Section no. A-2. Photograph I.0 x 1.35 mm. Plane polarized light.

Figure 14
Silicified gypsum nodule in iron formation. Sokoman Formation, Central Labrador Trough. Section H2-10. Photograph. 6.8 x 5.2 mm. Plane polarized light

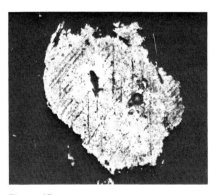

Figure 15
Ooid-moldic carbonate solution-infill, silicified. Sokoman Formation, Central Labrador Trough. Section F2-15. Photograph 5.1 x 4.5 mm. Plane polarized light.

formation have not been studied petrographically. 2) Limestone-derived relict textures have gone unrecognized because most specialists on iron formation are not familiar with limestone petrology. Such relict textures are not uncommon as will be exemplified by Figures 13-15.

Figure 13 shows a small nodule composed of microquartz from the iron-formation of the Hamersley Group, Western Australia. Cleavage planes of a former carbonate mineral are preserved as dust inclusions. There is some differential compaction around the nodule. The texture is characteristic of a laminated micrite partly dolomitized during early diagenesis and then silicified.

Figure 14 documents a silicified gypsum nodule. The fibroradial texture of gypsum crystals is perfectly preserved by inclusions of iron oxide dust. The sample is from the lagoonal platform facies of the Sokoman Formation, Central Labrador trough (Chauvel and Dimroth, 1974). Here deposition of a pellet sand was followed by evaporation and growth of gypsum nodules during earliest diagenesis and, finally by silicification of the gypsum nodule. There is no direct evidence for the original composition of the pellet sand.

Dimroth and Kimberley (1976, Figs. 18-21) documented vuggy solution-porosity and ooid-moldic porosity in cherty iron formations. Solution porosity is common in calcareous sediments that have been exposed to sub-aerial diagenesis; it is unknown from primary silica rocks, like sinters, diatomites, radiolarites. Figure 15 demonstrates a particularly instructive case of solution-infill in iron formations. Here an oolite core has been dissolved. Prismatic crystal grew from the pore-wall into the opening and filled it partly. The prismatic crystals were then replaced by iron oxides (now hematite dust). Finally, coarse-grained quartz grew across all the earlier textures. Crystal shapes suggest that the prismatic crystals were a carbonate, probably iron-rich since they have been replaced by iron oxides.

A complex diagenetic history can be read from Figure 15: after deposition and slight cementation, the rock emerged; it was then percolated by reducing, alkaline, iron-bearing, solutions, the oolite core (aragonite ?) was dissolved and a bladed crust of an iron-rich carbonate grew into the pore; then, solutions became more oxidizing and more acid, s75g215a1oi!!!!!!!!

and carbonate was replaced by iron oxides; finally, the chemistry of pore solution changed again and silicification took place. It is the systematic study of such relict textures which will, in the long run, permit to formulate a well-founded model of the early diagenesis of iron formation.

Bibliography

This list of references has purposely been kept very brief.

Iron Mineral Facies

Borchert, H., 1960, Genesis of marine sedimentary iron ores: Instit. Mining Metallurgy Bull., No. 640, p. 261-279.

James, H. L., 1954, Sedimentary facies of iron formation: Econ. Geol., v. 49, p. 236-293.

James, H. L., 1966, Chemistry of the iron-rich sedimentary rocks: U.S. Geol. Survey Prof. Paper. 440W, 61 p.

Klein, C., Jr. and R. P. Fink, 1976, Petrology of the Sokomon in The Howells River area at the westermost edge of the Labrador trough: Econ. Geol., v. 71, p. 453-487.

Zajac, I. S., 1974, The Stratigraphy and Mineralogy of the Sokomon Formation in the Kuob Lake area, Quebec and Newfoundland: Geol. Survey Canada Bull. 220, 159 p.

In all these papers iron mineral facies is considered a primary, sedimentary facies. Conclusions on the sedimentary environment are drawn that cannot be maintained at the present state of knowledge.

Diagenesis of Iron Minerals in Recent Sediments

Berner, R. A., 1970, Sedimentary pyrite formation: Amer. Jour. Sci., v. 268, p. 1-23.

Berner, R. A., 1971, Principles of Chemical Sedimentology: New York, McGraw Hill, 240 p.

Ho, C. and J. M. Coleman, 1969, Consolidation and cementation of recent sediments in the Atchafalaya Basin: Geol. Soc. Amer. Bull., v. 80, p. 183-192.

Porrenga, D. H., 1966, Glauconite and Chamosite as depth indicators in the marine environment: Marine Geol., v. 5, p. 495-501.

Berner's book should be familiar to any student of iron formations.

Physico-Chemistry of Iron

Garrels, R. M. and C. L. Christ, 1965, Solutions Minerals and Equilibria: Harper and Row, New York, 450 p.

Description of Iron Formations in Canada

Goodwin, A. M., 1956, Facies relations in the Gunflint iron formation: Econ. Geol., v. 51, p. 565-591.

Goodwin, A. M., 1962, Structure stratigraphy and origin of iron formation, Michipicoten area, Algoma District, Ontario: Geol. Soc. Amer. Bull., v. 73, p. 561-586.

Gross, G. A., 1968, Geology of iron deposits in Canada, v. 3, Iron ranges of the Labrador Geosyncline: Geol Survey Canada Econ. Geol. Dept. 22, v. 3, 179 p.

Description of Other Iron Formations

Bayley, R. W. and H. L. James, 1973, Precambrian iron formations of the United States: Econ. Geol., v. 68, p. 934-959.

Beukes, N.J., 1973, Precambrian iron formations of South Africa: Econ. Geol., v. 68, p. 960-1004.

Other References Cited

Button, A., 1976, Iron formation as end-member in carbonate sedimentary cycles in the Transvaal Supergroup South Africa: Econ. Geol., v. 71, p. 193-201.

Chauvel, J-J., and E. Dimroth, 1974, Facies types and depositional environment of the Sokoman Iron Formation, central Labrador Trough, Quebec, Canada: Jour. Sed. Petrol., v. 44, p. 299-327.

Davy, R., 1974, A Geochemical study of a dolomite-BIF transition in the lower part of the Hamersley Group: Geol. Survey Western Australia, Annual Rept. 1974, p. 81-100.

Dimroth, E. and M.M. Kimberley, 1976, Precambrian atmospheric oxygen: Evidence in the sedimentary distributions of carbon, sulfur, uranium and iron: Can. Jour. Earth Sci., v. 13, p. 1161-1185.

Kimberley, M. M., Origin of iron ore by diagenetic replacement of calcareous oolite: Nature, v. 250, p. 319-320.

Walker, R. G., 1976, Facies models 1. General Introduction: Geosci. Canada, v. 3, p. 21-24.

MS received February 10, 1977.
Revised January, 1979.
Reprinted from Geoscience Canada,
Vol. 4, No. 2, p. 83-88.

Facies Models 17. Volcaniclastic Rocks

Jean Lajoie
Département de Géologie
Université de Montréal
Montréal, P.Q. H3C 3J7

Introduction

The realm of volcaniclastic rocks is changing very rapidly due to important work now in progress, but to write on *facies* in volcaniclastites is, to say the least, somewhat ambitious. The subject is in its infancy. There are a few good descriptions of volcaniclastites but in general authors have restricted their work to the petrography, chemistry, and explosive mechanism. Little is known of lateral and vertical variations in these rocks. The study of facies in volcaniclastic sequences is important as it could be a good tool in mineral exploration (Horikoshi, 1969; Sangster, 1972).

In the present paper, I shall try to summarize what we know of lateral and vertical variations and assemblages (facies) in volcaniclastites as seen by a sedimentologist. This will merely whet the appetite of the reader, but I think that to have "summarized" facies in the assemblages that are not well described would have been hazardous, not to say fictitious.

Walker (1976) in his general introduction on facies, suggested that models can be used as frameworks for future observations, as norms and predictors, and as a basis for hydrodynamic interpretation. The volcaniclastic models discussed in the present paper should be seen as general frameworks for future observations. They may be used to initiate hydrodynamic interpretation, but more work needs to be done before they can be considered norms or predictors. Walker (1976) points out that

facies models originate from the distillation of many local examples. In some volcaniclastic rock types there is not much to distill.

The paper is not a review, and therefore the reference list is not exhaustive. I have tried to use general, comprehensive studies of the various subjects treated.

Terminology

When one first begins work on volcaniclastic rocks one is struck by the many definitions there commonly are of the same word. I thus find it necessary to introduce the terminology that will be used in this paper.

Volcaniclastic rocks include all fragmental volcanic rocks that result from any mechanism of fragmentation. The classification most commonly used in North America is that of Fisher (1961, 1966). *Epiclastic* fragments result from the weathering of volcanic *rocks. Autoclastic* fragments are formed by the mechanical breakage or gaseous explosion of lava during movement. *Hyaloclastic* fragments, a variety of autoclastic, are produced by quenching of lava that enters water, water-saturated sediments, or ice. *Pyroclastic* fragments are formed by explosion and are projected from volcanic vents. Showers of pyroclastic fragments produce fall deposits. Clasts that are ejected from vents and that are transported *en masse* on land or in water form pyroclastic-flow deposits. Pyroclastic fragments are primary if ejected from a vent, and secondary if they are recycled from *unconsolidated* primary deposits. In ancient deposits the distinction between primary and secondary pyroclastic debris may be very difficult, if not impossible, to make.

Wentworth and Williams (1932) introduced a grain-size classification for pyroclastic fragments, similar to the

classification used for other clastic sediments. The classification was adopted by Fisher (1961) and is reproduced in Figure 1.

Volcaniclastic sedimentation is such a vast subject, and has so many variables that it will not be possible to treat all of its aspects. Epiclastic rocks which do not differ except in composition from other clastic rocks will not be treated. In this paper, I shall briefly describe the characteristics of autoclastic and pyroclastic rocks. I shall then discuss the observed and possible variations in time and space of the fundamental characteristics (facies). Due to space limitations, fragment petrography will not be discussed. The reader will find in Ross and Smith (1961), and in Heiken (1972, 1974) good descriptions of the morphology and petrography of volcaniclastic fragments.

Autoclastic Rocks

Flow breccias and hyaloclastites are the two autoclastic types that will be discussed in this paper. Both are formed by the fragmentation of lava by friction, or by rapid cooling as it flows in water or under ice, and are therefore commonly associated. The distinction between these two rock types is primarily based on grain size, the breccias being coarse and the hyaloclastites, fine. Flow breccias occur in both subaerial and subaqueous environments. Hyaloclastites are for some authors synonymous with aquagene tuffs, but this usage of the word tuff has been criticized.

Flow breccias and hyaloclastites are common in basaltic sequences, and relatively rare in acid sections as basic lava flows more readily than acid magmas due to its lower viscosity. In these autoclastic rocks, the clasts are generally monolithologic with most fragments being derived from the same parent magma. Lava may scrape vent walls or rip fragments as it flows, but exotic fragments are rarely reported in descriptions of flow breccias. In basaltic sequences, the breccias commonly consist of complete pillows and (or) pillow fragments set in a matrix of devitrified glass shards and lumps (Carlisle, 1963; Dimroth *et al.*, 1978; Figs. 2, 3 and 4). Pillows are exceptional in acid lava. Acid flow breccias are made up of abundant angular blocks, with coarse and fine sand-size fragments, set in a glassy matrix (Pichler, 1965).

PREDOMINANT GRAIN SIZE (mm)	EPICLASTIC FRAGMENTS	PYROCLASTIC FRAGMENTS	PYROCLASTIC ROCKS
	COBBLE	BLOCK AND BOMB	PYROCLASTIC BRECCIA
64	PEBBLE	LAPILLUS	LAPILLISTONE
2	SAND	COARSE ASH	TUFF
1 / 16	SILT	FINE ASH	

Figure 1
Grade size for epiclastic and pyroclastic fragments, and terms for pyroclastic rocks. From Fisher (1961, 1966).

Figure 2
Flow breccia that consists of pillows and pillow fragments set in a matrix of hyaloclas-
tite. Mont Etna, Sicilia. Photo courtesy of John Ludden.

Figure 3
Basaltic flow breccia in the Archean of Rouyn-Noranda, overlain by graded pyroclastic deposits. The lava flowed from left to
right, and the fabric is due to the orientation of the brecciated pillows in shear planes. Photo courtesy of Pierre Trudel.

Figure 4
Brecciated basaltic flow in the Archean of
Rouyn-Noranda. Coin is two cm. Photo courtesy of Pierre Trudel.

Facies in flow breccias and hyaloclastites. Due to their mode of origin one would expect *in situ* flow breccias and hyaloclastites to show little or no systematic lateral and vertical variations in clast content, size, and composition. Autoclastic associations are however much more complex.

The vertical variations in a typical flow breccia of basaltic composition are summarized in Figure 5 and 6. The pillowed lava grades upward into an isolated-pillow breccia that is overlain and transitional with a broken-pillow breccia. In the Archean of the Noranda region, similar sequences are exceptionally overlain by fine-grained hyaloclastites. In this model, clast size decreases from base to top. This grain size variation cannot be called graded-bedding; there is no bedding to begin with as there is no true deposition of clasts since they were formed *in situ.* The distinctive characteristics of this type of breccias are the monolithologic composition of the clasts and the transitional contact with the flow. Since the fragmentation is formed by quenching, the shards are not welded (Carlisle, 1963, p. 68).

Flow breccias are commonly associated with subaerial acid lava flows, where they occur underlying and (or) overlying the flow. There is however no modern example of submarine acid flows. It is a rare and poorly known phenomenon in the rock record. Pichler (1965) described acid subaqueous breccias and hyaloclastites exposed on the island of Ponza in Italy. To my knowledge, this is the only description of acid hyaloclastites in the literature. The lateral and vertical associations of facies that he proposed are summarized in Figure 7. The characteristics of the deposit are somewhat similar to those discussed previously, except for the absence of pillows and of large fragments. The coarse fraction is angular, and found near the parent rhyolitic magma. The clasts are unwelded, and since they are formed *in situ,* the volcaniclastite is massive or shows pseudo stratifications that commonly parallel the breccia-lava contact. Figure 7 has no scale, but Pichler (1965, p. 305) gives an example in which facies 3 overlying the dome reaches 5 m in thickness.

The model presents some difficulties, mostly in the ratio of clastics to the size of the rhyolite dome. On extrusion, acid

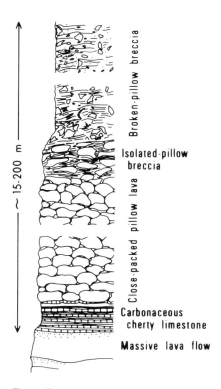

Figure 5
Typical pillow breccia-hyaloclastite sequence in the Triassic of Quadra Island, British Columbia. From Carlisle, 1963.

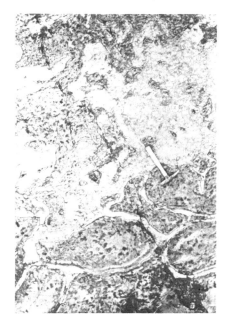

Figure 6
Hyaloclastite transitional with pillow lava. Rouyn-Noranda region. Photo courtesy of Léopold Gélinas.

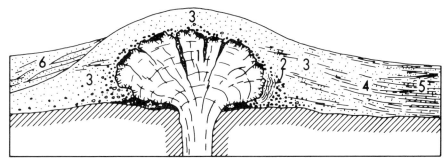

Figure 7
Vertical and lateral associations of rock types in acid hyaloclastites, 1) submarine extruded dome covered by auto-breccias and hyaloclastites. 2) blocky talus with remnants of ancient crust. 3) unstratified, in situ, sand-size hyaloclastites. 4) "stratified", in situ hyaloclastites. 5 and 6) resedimented "hyaloclastites". Modified from Pichler, 1965. For scale, see text.

lavas form a chill margin under which the temperature rapidly decreases. The margin may be brecciated, as well as parts of the underlying dome, but the volume of clasts produced by such a mechanism should be much less than that implied by Pichler. At Lipari, the flow breccias are found under the rhyolitic flow, and the autoclastics overlying it are very thin, where present (Ludden, pers. commun., 1979). The model is not accepted by all workers, and there is some doubt as to the origin of the fragmentation. Pichler (1965) presents little evidence to support the continuous peeling of the chill margin. The large fragments and tongues of felsic magma that one would expect to find in the deposit, are not present. In the example used by Pichler (1965), Santorini, the volcaniclastites overlying the rhyolitic dome are in shallow-marine waters, and unless there are particular conditions which he does not discuss, the volcaniclastic cover should be much thinner than 5 m.

Autoclastic fragments may be reworked by bottom currents or resedimented from density flows. In certain cases the deposit may show some of the characteristics of pyroclastic rocks. These secondary hyaloclastites may be difficult to distinguish from pyroclastites as both types of fragmentation may form similar textures. However, pumiceous fragments should be more common in pyroclastic deposits. Also, fragments in reworked hyaloclastites should normally be monolithologic, and unwelded whereas pyroclastites are commonly polylithologic, and may be welded. Honnorez and Kirst (1975) have shown that shape can be used to distinguish fragments of hyaloclastic origin from those of pyroclastic.

The model proposed for basaltic flow breccias and hyaloclastites is sufficiently well established, and could be used as a norm, but it is safe to say that for acid hyaloclastites we don't yet have a model.

Pyroclastic Rocks
Pyroclastic debris (pumice, shards, crystals, and lithics) ejected from vents, fall or flow in air and (or) water under the influence of gravity. The settling velocities are proportional to fragment size, shape, and density. Therefore, the principles that govern the sedimentation of pyroclastic fragments are identical to those responsible for the deposition of other clastic debris.

The lateral extent and the geometry of pyroclastic deposits are influenced in part by magma composition, and the environment in which the eruption takes place. Basaltic subaerial eruptions generally produce cones of scoria and ash of limited areal extent around or down-wind of the cones. Basaltic eruptions that take place in shallow water or where water has access to the vent are more strongly explosive, and produce ash rings and ash layers that may have considerable areal extent. Due to the higher volatile content of the magma, eruptions of acid and intermediate compositions are generally explosive. These eruptions may project very large volumes of pyroclastic debris to heights well in excess of 25 km, and may produce thick fall and flow deposits.
Facies in pyroclastic fall deposits.
The showering of pyroclastic debris is governed by their settling velocities. It follows that size, composition, and thickness should vary with distance from the vent. Figure 8, modified from Walker (1971), is an example of such lateral size

194

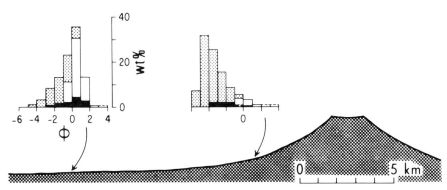

Figure 8
Lateral variations in size and contents of pumice (stippled), crystals (blank), and lithic

(black) in a pyroclastic fall deposit in the Azores. Modified from Walker (1971).

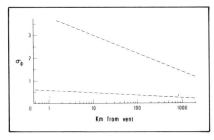

Figure 9
Variation of sorting (σ_ϕ) with distance from the vent for pyroclastic fall deposits. From Walker, 1971, Figure 7.

and composition variations in an air-fall unit. In this example size decreases for all compositions, and the relative abundance of fragment-type varies in the direction of transport. Due to their low density, pumiceous fragments make poor projectiles and are therefore more abundant near the vent, whereas crystals may travel greater distances. This zonation of textures is common in many air-fall deposits of the world (Walker, 1971; Kuno et al., 1964; Lacroix, 1904). Pumices may be lighter than water and thus float if they fall in it. The deposition of pumiceous fragments may only occur when they are waterlogged. It follows that in such depositis, the distribution of pumices may be very erratic.

Any particular fall deposit becomes finer-grained as the distance from the vent increases. However fine-grained deposits may form near the vent as well as far from it. Fine debris accumulates near the vent when the eruption is weak. Even if coarse-grained fall deposits are reliable indicators of vent proximity, fine-grained deposits do not necessarily indicate distance from the vent.

In fall deposits, the dispersion of the graphic standard deviation (σ_ϕ) is considerable near the vent. This results from the differences in settling veloci-

ties; small and dense clasts fall at velocities similar to those of larger and lighter clasts. Figure 9 suggest that a high σ_ϕ could indicate proximity of an eruptive center.

The downwind decrease in thickness of pyroclastic fall deposits is discussed by Eaton (1964). Some windborne ash may be transported over extensive areas and make excellent marker horizons. Fall deposits may be poorly stratified, as stratification is a result of discontinuity in the volcanic activity; beds and laminations may be crudely defined (Fig. 10, A and B). Walker and Croasdale (1971) gave some characteristics of basaltic fall deposits. According to these authors, ashes from subaqueous eruptions are well stratified, and beds are extremely thin (1 mm). Deposits due to rhythmic ejections of incandescent cinder and lapilli are thicker, rarely less than one cm, and commonly more than five cm. Pyroclastic fragments of acid and intermediate compositions may be ejected in very large volumes, and bed thickness may easily reach 4 m near the vent.

There is little systematic variation in the vertical grain sizes within beds. As settling is a function of size and density, grading may be present in fall deposits, but is best observed in the thicker units. The grading may be normal or reverse (Bond and Sparks, 1976). Reverse density-grading has been documented by Koch and McLean (1975). It results from the progressive evacuation of compositionally stratified magma chamber with increasing intensity of eruption. In subaerial falls, all fragments are heavier than the fluid whereas in subaqueous falls as discussed above, pumices may float and may not be present in the deposit, or may only occur at or near the top of the bed.

Because they are controlled by the eruption intensity, vertical variations of facies are unpredictible in pyroclastic fall deposits.

Pyroclastic flow deposits. The characterization and definition of facies in pyroclastic flow deposits is not easy. Most authors have described deposits from flows that are particular to pyroclastites, such as *nuées ardentes*, base surges, ignimbrites, lahars, and others. This terminology of pyroclastic flows is discussed in many standard texts; Macdonald (1972) could be recommended to the neophyte. The major problem in dealing with deposits of these flows is that the criteria used for their definition are not exclusive to pyroclastic rocks. It follows that the characteristics given for some particular facies may also be found in other mass-flow deposits, or the singularities presented for a particular facies of pyroclastic deposit may be found in others. I would like to give two examples to illustrate this particular problem.

Large-scale cross bedding is commonly used to identify base-surge deposits (Figs. 11 and 13; Schmincke et al., 1973; Sparks, 1976; Bond and Sparks, 1976). But base-surge flow as described by Moore (1967) is similar in many respects to the flow of the *nuées ardentes* that Lacroix (1904) described. The eruptive mechanisms may be somewhat different, but the results are similar: turbulent suspensions that move away from the vent, aided by gravity. The descriptions that Lacroix (1904) gave of the *nuées ardentes* are among the best descriptions of turbidity currents in the literature. Large-scale cross bedding does occur in turbidite sequences as well as many other types of sedimentary sequences. It may also be present in *nuées ardentes* deposits although there is no evidence yet since very few deposits have been described. Large-scale cross bedding is closely associated with the *nuées ardentes* of Pelée. Lacroix (1904, p. 469) shows a splendid photograph of more than 1 km of an exposure which he called *dunes de cendres* in which the dunes may reach amplitudes that are in excess of 1 m. Crowe and Fisher (1973) showed the similarity that exists between the primary structures found in base-surge deposits with those found in other turbulent, density-flow deposits. They point out that the structure-forming variables (rate

Figure 11

Base-surge deposit, Gran Canari, near La Calderilla. Two sets of trough cross-bedding are present, one to the right of pencil, and a second one underlying it.

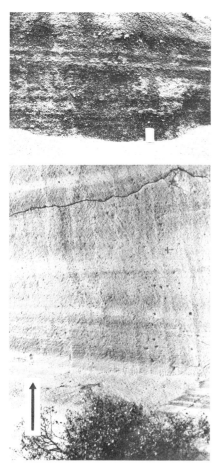

Figure 10

Stratification in air-fall deposits. A: Crude bedding, La Palma, Canary Islands. Notebook is 20 cm long. B: Crude laminations, San Lorenzo de Gran Canaria, Canary Islands. Arrow is 50 cm.

Figure 12

Grading in Archean pyroclastic flow deposits. A: Reversely graded lithics of intermediate composition. Top is to left; notebook is 30 cm long. Reneault, Rouyn-Noranda. B: Reverse overlain by normal grading of rhyolitic lithics. Top is left; notebook is 20 cm long. Don Rhyolite, Noranda.

of deposition, flow power, and grain size) are the same. The point that I wish to make here is that the primary structures of base-surge deposits characterize the flow rather than the eruption mechanism, and as such they may be found in many other types of pyroclastic facies.

Lahar is a second example of ill-defined rock type. It is defined as any unsorted or poorly sorted deposit of volcanic debris that moved and was deposited as a mass that owed its mobility to water. Schmincke (1967) described lahar deposits in the Ellensburg Formation of Washington which he interpreted as being deposited from "watery viscous suspensions", but the primary structures and structure sequences that are described could be interpreted as the result of other types of flow such as turbulent suspensions. There is little difference between these rocks and others found in many turbidite

sequences. The structures characterize the flow, not the environment in which it moves.

Lacroix (1904) showed quite conclusively that gravity controls the movement of pyroclastic flows. It follows that the physics which applies to these flows should be identical to that of other density flows. There is therefore little chance to find *one* model that could characterize subaqueous or subaerial flows. Given similar flow parameters (density, viscosity, velocity, thickness) and grain size, deposits from all types of flow could have similar sedimentary characteristics. It should be clear by now, that in describing pyroclastic flow deposits it was not possible to use the genetic nomenclature that is common in the literature. I may not have much success, but I suggest to workers not to use the genetic ill-characterized pyroclastic-flow terminology.

The deposition of pyroclastic flow ejecta may occur on land or in water. The distinction between subaerial and subaqueous deposits is not always easy because evidence is commonly indirect. According to Fisher and Schmincke (in press) the main problem with subaqueous pyroclastic flows is that the deposits of witnessed *nuées ardentes* that have entered water have not been described. Except for a few occurrences, subaerial flow deposits are not that well described either. Many such deposits are massive (Fig. 13 is an example) probably due, in part, to their relative proximity to the vent. Some subaerial deposits described in the literature are said to be massive, but the photographs that accompany these descriptions commonly show grading and (or) parallel laminations. Workers must then rely on fossils, the presence of pillows in the sequence, and other indirect evidence in order to distinguish

the two types of deposits. The welding of fragments has been used as evidence for subaerial deposition, but there are known occurrences of subaqueous deposits that are welded (Francis and Howells, 1973; Gélinas *et al.*, 1978).

Facies in pyroclastic flow deposits. Most workers accept that pyroclastic flows move under the influence of gravity along topographic depressions. It follows that the mechanics of pyroclastic flows do not basically differ from those of other types of density flows. Grain-size, density, and shape of the fragments affect the rate of sedimentation. However, since pyroclastic flows are the result of complex interactions of many processes some deposits may have anomalous size and thickness characteristics.

Beds deposited from pyroclastic flows are commonly graded (Fig. 12 A and B) or massive (Fig. 13). Normal grading of *lithic* clasts is a frequent feature of many ancient and recent deposits of the world (Sparks *et al.,* 1973). *Pumice* clasts may be either reversely or normally graded. In many subaerial deposits the pumices are reversely graded, and normal grading of pumices seems rare. In such deposits, the pumice size may increase gradually from base to top (true grading), or the larger pumices may be concentrated in an upper horizon within the bed (Sparks, 1976). The reverse grading of pumices is explained by flotation, the matrix being denser than the fragments. In beds of subaqueous deposits, the pumices and the lithic fragments may be concentrated in different horizons but both types of fragment commonly show similar grading trends, reverse or normal (Fiske and Matsuda, 1964; Yamada, 1973; Bond, 1973; Tassé *et al.,* 1978). The grading of pumices in pyroclastic flow deposits is not fully understood. It may be that subaqueous flows have lower densities than their subaerial counterparts, but there is little evidence to support this.

Sparks *et al.* (1973) proposed a "standard" vertical bedding sequence in subaerial pyroclastic flow deposits (Fig. 14, I), derived from observations made in the Azores, Canary Islands, Italy, Greece, Japan, and the West Indies.

Figure 13
Massive pyroclastic flow deposit. Fortaleza Grande, Gran Canaria. Pipe is roughly 1 m in diameter.

This model subdivides the "flow unit" (bed?) in two transitional "layers" (2a and 2b). Layer 2a is relatively fine-grained and reversely graded. In layer 2b, which makes up more than 90 per cent of the bed, the size of the pumiceous fragments increases upward and reaches a maximum at the top, whereas the lithics are concentrated near the base of the bed. This standard sequence contrasts with many sequences observed in subaqueous deposits (Fig. 14, II) described by Fiske and Matsuda (1964), Fiske (1969), Bond (1973) and Tassé *et al.,* (1978), where the entire accumulation is normally graded, by size and (or) density. As already mentioned, reverse grading does occur in subaqueous deposits, but pumices and lithics commonly have similar grain-size variations. The deposits described by Fiske and Matsuda (1964) consist of a lower graded or massive division, overlain and transitional with a stratified division (b, in Fig. 14, II), in which the grading is continuous from the underlying division (a). The authors interpreted the strata in division b as beds. Deposits with similar structures have been described by Tassé *et al.* (1978), in the Archean of Noranda, and by Yamada (1973) in Pleistocene rocks of Japan. They proposed that, since the grading was continuous from the graded division into stratified division, the accumulation was the result of *one event,* that produced *one bed.* The stratifications (Fig. 15) were therefore interpreted as parallel laminations. These proposed standard beds should not be taken as invariable models. The characteristics of the deposit will vary with the flow conditions. Velocity, density, and viscosity of the flow could change such that the characteristics observed in a given bed close to the vent may be very different from those observed in more distal sections. Tassé *et al.,* (1978) have shown that beds of pyroclastic flows in the Noranda region could not be characterized by one standard model.

There is little known on lateral and vertical variations in pyroclastic flow deposits. Grain size and bed thickness may either decrease or increase downflow. There are many examples of increasing bed thickness with distance of transport, particularly close to vent. Schmincke and Swanson (1967) suggested that this was probably due to the relative high velocity and low viscosity of the flow near the vent. Grain size

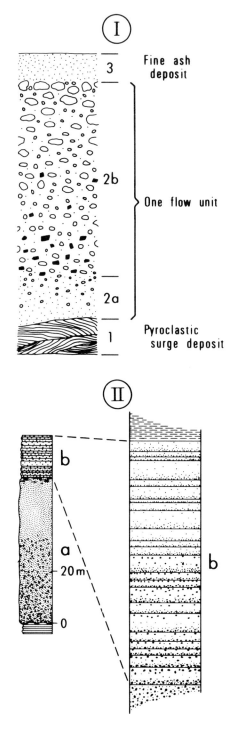

Figure 14
Vertical sequence of primary structures commonly present in subaerial (I) and subaqueous (II) pyroclastic flow deposits. From Sparks et al. *(1973), and Fiske and Matsuda (1964). In Sparks* et al. *(1973), the model has no scale; the thickness of layer 2a varies from a few cms to more than 1 m.*

Figure 15
Parallel stratifications in a Precambrian pyroclastic flow deposit. Noranda, Quebec.

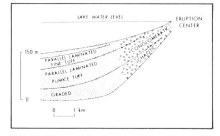

Figure 16
Lateral variations of size and primary structures in a Pleistocene pyroclastic flow deposit in Japan. Modified from Yamada (1973).

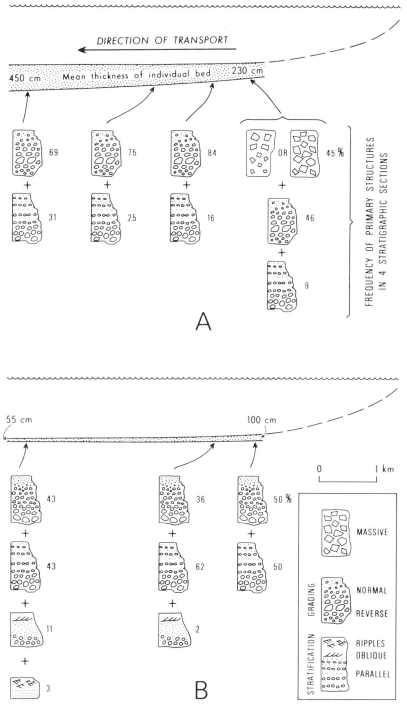

Figure 17
Lateral variations of mean bed thickness and primary structure sequences with distance of transport, in an Archean subaqueous pyroclastic flow deposit, for two different types of accumulation (A and B). The frequency distributions of structure sequences are in percentages, and total 100% for each section. Modified from Tassé et al., (1978).

generally decreases with distance of transport, but in the Precambrian rocks of Noranda grain size both increases and decreases with distance from vent (Tassé *et al.*, 1978). In deposits of turbulent low-density flows the size decreases, but in relatively higher density flows, size may increase before it decreases in the direction of transport. Analogous situations have been described in turbidites.

Primary structures that are controlled both by flow parameters and grain size are perhaps the best tool to establish facies in pyroclastic flow deposits. Yamada (1973) proposed a model (Fig. 16) for lateral variations of primary structures for a subaqueous deposit of Pleistocene age, but the only complete study of lateral variations of facies is that of Tassé *et al.* (1978) for Archean flows of Noranda; Figure 17 is a simplified version of the model that they proposed.

The rocks that were studied are intermediate in composition and not welded. The studied sections show two distinct bed types that have different mean-thickness, grain size, and primary structures. The two types are therefore treated in two different assemblages

(A and B, Fig. 17). Type A beds (Fig. 12, A), are thicker than type B (Fig. 18) and mean bed thickness increases with distance of transport in type A, whereas it decreases in type B. In both bed types, primary structure *sequences* vary systematically away from the vent, but the

198

sequences are somewhat different in the two types. In type A beds, the most abundant structure is grading, and parallel laminations are relatively rare. The proximal section has a high proportion of massive or reversely graded beds whereas the distal section is characterized by normal grading, and a higher number of beds that show parallel laminations. The two intermediate sections suggest a gradation from the proximal to the distal facies. Grading is also very common in type B beds, and is generally normal. Traction structures, such as parallel and oblique laminations (dunes and ripples) are more abundant than in type A, and their proportion increases with distance of transport.

Tassé et al., (1978) interpreted the lateral variation of facies by analogy with sedimentary density flows. Type B beds most probably result from the accumulation of decelerating turbulent suspensions of low density such as turbidity currents. Most of the flows responsible for type A beds appear to have been turbulent in the distal regions, but almost half (45%) of the deposit in the more proximal section probably accumulated from laminar suspensions such as debris flows, as suggested by the reverse grading and massive beds. It follows that different transporting mechanisms acted at different stages of the flows. This may be caused by an increase in flow velocity due to gravity, coupled with a decrease in viscosity.

In the subaqueous deposits of Noranda, the ratio between lithics and pumices decreases with distance from the vent. This relationship has also been observed in some Japanese deposits (Kuno et al., 1964), and is opposite to the pumice-lithic ratio commonly found in pyroclastic fall deposits. This is due to the lower density of the pumice which settle at lower velocities, and are therefore found further down flow.

In ancient pyroclastic flow deposits, grain orientation may be used to help locate vents. The technique could be used with other lateral variations such as size, bed-thickness, and (or) primary-structure sequences. It is described by Elston and Smith (1970), and has been successfully applied in Archean pyroclastic rocks of northwestern Ontario (Teal, 1979).

It is not possible at this stage to use the Noranda model proposed by Tassé et al. (1978) as a norm or a predictor for subaqueous pyroclastic flow deposits.

To my knowledge it is the only study of lateral facies variations, other than grain size and bed thickness, done in this type of deposit, future observations shall have to verify it, and thus make it norm.

Subaerial flow deposits may show facies that are similar to those of subaqueous flows, but because of the higher temperature of the mass and, in certain cases the higher viscosities, they may exhibit different characteristics. Rhyolitic flows have fragments that are commonly deformed, stretched, and welded. The eutaxitic "texture", that results from the parallel arrangement and alternation of layers of different textures or composition may be com-

Figure 18
Primary-structure sequence in a type B bed, Rouyn-Noranda. From Tassé et al. *(1978).*

mon in these rocks. The structures are found in deposits that accumulated from laminar suspensions (Schmincke and Swanson, 1967) but they also occur in turbulent flow deposits (Lock, 1972).

Little is known of vertical variations of facies in pyroclastic flow deposits. Most workers have focused their attention on eruptive cycles. Aramaki and Yamasaki (1963), and Sparks et al. (1973) are examples. The only published systematic work on vertical variations of texture and bed thickness in flow deposits was done in the Archean of Rouyn-Noranda (Tassé et al., 1978; Gélinas et al., 1978). Because pyroclastic flows move under the influence of gravity, they are partly or completely controlled by topographic depressions when the flow is very thick. The channel fills could therefore have characteristics that are similar to those of fills described in other sedimentary sequences, where grain size and bed thickness decrease up-section. We have shown with the Archean examples that these vertical variations are also present in channel fills of pyroclastic flows, and channels have been identified by mapping vertical variations of size and thickness.

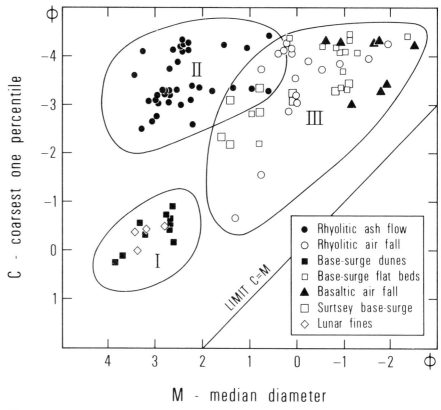

Figure 19
C-M diagram for pyroclastic falls and flows. From Sheridan, 1971.

Distinction between pyroclastic flow and fall. Workers have been interested for quite sometime in finding parameters or characteristics to distinguish pyroclastic flows from falls. Much work has been done on statistical parameters of grain size distribution with some apparent success (Walker, 1971; Sheridan, 1971, Fig. 19). Sheridan (1971) grouped, on Figure 19, rhyolitic base-surge dunes and lunar fines in Field I, and rhyolitic ash flows in Field II. His Field III is said to be composed mostly of air fall deposits. This technique is not infallible and there are many exceptions. For example air falls plot with "base-surge flat beds", and "Surtsey base surges". I would also use primary structures and their sequences to distinguish flow from fall. Our work has shown that flows leave imprints that are different from falls.

Summary

The purpose of this article has been to present what is known of facies relationships in some autoclastic rocks, and in pyroclastic fall and flow deposits. Autoclastic fragments, that are formed *in situ* are monolithologic and not welded. The clastites may be transitional with the parent magma, and have the same composition. The internal structures of these deposits indicate the absence or near absence of transport.

Pyroclastic fall deposits are very well stratified. They commonly show a systematic lateral decrease in grain size and bed thickness. In such deposits, sorting is least close to source due to settling of fragments of different densities. In fall deposits the vertical variations are controlled by eruption intensity, and are therefore unpredictible.

In flow deposits, bed thickness and grain size commonly decrease down flow, but close to vent the variations are not systematic. Beds deposited from flows are commonly graded. In subaqueous deposits the grading of all fragments is generally normal but in many subaerial flow, pumices are commonly reversely graded. The primary structure sequences vary systematically down-flow, and depict the changing flow conditions (density, viscosity, velocity) and the grain size that is transported. Vertical variations of size and thickness are poorly known in these deposits but where they have been studied they are characteristic of channel fills.

Statistical parameters of grain size have been used extensively to distinguish pyroclastic flows from falls. Primary structure sequences are also a powerful tool that could help make this distinction.

The reader must by now be aware that much work remains to be done on volcaniclastic rocks. I hope that this paper has been stimulating.

Acknowledgements

Léopold Gélinas, John Ludden, and Roger G. Walker kindly read an earlier version of this manuscript and offered many helpful suggestions for its improvement.

References

General

Heiken, G.H., 1972, Morphology and petrography of volcanic ashes: Geol. Soc. Amer. Bull., v. 83, p. 1961-1988.

Heiken, G.H., 1974, An atlas of volcanic ash: Smithsonian Contr. Earth Sci., v. 12, 101 p.

Macdonald, G.A., 1972, Volcanoes. Prentice-Hall Inc., Englewood Cliffs, N.J., 510 p.
A good introduction on the subject of volcanoes.

Ross, C.S., and R.L. Smith, 1961, Ash-flow tuffs: their origin, geologic relations and identification: U.S. Geol. Surv. Prof. Paper 366, 81 p.
A good text with abundant photographs. Excellent review of the literature. Excellent descriptions.

Nomenclature and Classification

Fisher, R.V., 1961, Proposed classification of volcaniclastic sediments and rocks: Geol. Soc. Amer. Bull., v. 72, p. 1409-1414.

Fisher, R.V., 1966, Rocks composed of volcanic fragments and their classification: Earth Sci. Rev., v. 1, p. 287-298.

Wentworth, C.K., and H. Williams, 1932, The classification and terminology of the pyroclastic rocks: Rept. Comm. Sed., Bull. Natl. Research Council, v. 89, p. 19-53.

Autoclastic Rocks

Carlisle, D., 1963, Pillow breccias and their aquagene tuffs. Quadra Island, British Columbia: Jour. Geol., v. 71, p. 48-71.
A classic study on basaltic flow breccias.

Dimroth, E., P. Cousineau, M. Leduc, and Y. Sanschagrin, 1978, Structure and organization of Archean subaqueous basalt flows, Rouyn-Noranda, Québec, Canada: Can. Jour. Earth Sci., v. 15, p. 902-918.

Honnorez, J., and P. Kirst, 1975, Submarine basaltic volcanism: morphometric parameters for discriminating hyaloclastites from hyalotuffs: Bull. volcanologique, v. 34, p. 1-25.
A good review of the literature, origin, and characteristics of hyaloclastites.

Pichler, H., 1965, Acid hyaloclastites: Bull. volcanologique, v. 28, p. 293-311.
One of the rare descriptions of acid autoclastic rocks.

Pyroclastic Fall and Flow

Bond, A., and R.S.J. Sparks, 1976, The Minoan eruption of Santorini, Greece: Jour. Geol. Soc., v. 132, p. 1-16.

Crowe, B.M., and R.V. Fisher, 1973, Sedimentary structures in base-surge deposits with special references to cross-bedding, Ubehebe craters, Death Valley, California: Geol. Soc. Amer. Bull., v. 84, p. 663-682.
An excellent description of base-surge deposits.

Elston, W.E., and E.-I. Smith, 1970, Determination of flow direction of rhyolitic ash flow tuffs from fluidal textures: Geol. Soc. Amer. Bull., v. 81, p. 3393-3406.
Grain-orientation measurements are described and used for determination of flow direction.

Fisher, R.V., and H.-U. Schmincke, in press, Subaqueous pyroclastic flow deposits. Unpublished manuscript.
An excellent review of the subject; to be published in 1980 as part of a textbook on pyroclastic rocks.

Fiske, R.S., 1969, Recognition and significance of pumice in marine pyroclastic rocks: Geol. Soc. Amer. Bull., v. 80, p. 1-8.

Fiske, R.S., and T. Matsuda, 1964, Submarine equivalents of ash flows in the Tokiwa Formation, Japan: Amer. Jour. Sci., v. 262, p. 76-106.
One of the rare good descriptions of subaqueous flow deposit.

Gélinas, L., J. Lajoie, M. Bouchard, A. Simard, P. Verpaelst, et R. Sansfaçon, 1978, Les complexes rhyolitiques de la région de Rouyn-Noranda: Min. des Richesses naturelles du Québec, Rapport DPV- 583, 49 p.
One of the rare studies on vertical variation of size and thickness in channelized deposits.

Kuno, H., T. Ishikawa, K. Yagi, M. Yamasaki, and S. Taneta, 1964, Sorting of pumice and lithic fragments as a key to eruptive and emplacement mechanism: Japan Jour. of Geol. and Geography, v. 35, p. 223-238.

Lacroix, A., 1904, La montagne pelée et ses éruptions: Masson et Cie, Paris, France, 662 p.
Anyone seriously involved with pyroclastic flows should have read this reference twice!

Schmincke, H.-U., and D.A. Swanson, 1967, Laminar viscous flowage structures in ash-flow tuffs from Gran Canaria, Canary Islands: Jour. Geol., v. 75, p. 641-664.
Good descriptions of textures and structures.

Sparks, R.S.J., 1976, Grain size variations in ignimbrites and implications for the transport of pyroclastic flows: Sedimentology, v. 23, p. 147-188.

Sparks, R.S.J., S. Self, and G.P.L. Walker, 1973, Products of ignimbrite eruptions: Geology, v. 1, p. 115-122.

Tassé, N., J. Lajoie, and E. Dimroth, 1978, The anatomy and interpretation of an Archean volcaniclastic sequence, Noranda region, Quebec: Can. Jour. Earth Sci., v. 15, p. 874-888.
Perhaps the only published analysis of lateral variations in subaqueous pyroclastic flow deposits.

Teal, P.R., 1979, Stratigraphy, Sedimentology, Volcanology, and Development of the Archean Manitou Group, Northwestern Ontario, Canada: McMaster University, Unpubl. Ph.D. Thesis, 291 p.

Walker, G.P.L., and R. Croasdale, 1971, Characteristics of some basaltic pyroclastics: Bull. volcanologique, v. 35, p. 303-317.
A relatively good description of fall deposits.

Yamada, E., 1973, Subaqueous pumice flow deposits in the Onikobe Caldera, Miyagi Perfecture, Japan: Jour. Geol. Soc. Japan, v. 79, p. 585-597.

Distinction Between Pyroclastic Flow and Fall

Sheridan, M.F., 1971, Particle-size characteristics of pyroclastic tuffs: Jour. Geophys. Res., v. 76, p. 5627-5634.

Walker, G.P.L., 1971, Grain-size characteristics of pyroclastic deposits: Jour. Geol., v. 79, p. 696-714.
An excellent review of size characteristics of more than 1000 samples of flow and fall deposits.

Other References Cited

Aramaki, S., and M. Yamasaki, 1963. Pyroclastic flows in Japan: Bull. volcanologique, v. 26, p 89-99.

Bond, G.A., 1973. A late Paleozoic volcanic arc in the Eastern Alaska Range, Alaska: Jour. Geol., v. 81, p. 557-575.

Eaton, G.P., 1964. Windborne volcanic ash: a possible index to polar wandering: Jour. Geol., v. 72, p. 1-35.

Francis, E.H., and M.F. Howells, 1973. Transgressive welded ash-flow tuffs among Ordovician sediments of N.E. Snowdonia, N. Wales: Jour. Geol. Soc., v. 129, p. 621-641.

Horikoski, E., 1969. Volcanic activity related to the formation of the Kuroko-type deposits in the Kasaka District, Japan: Mineral. Deposita, v. 4, p. 321-345.

Koch, A.J., and H. McLean, 1975. Pleistocene tephra and ash-flow deposits in the volcanic highlands of Guatemala: Geol. Soc. Amer. Bull., v. 86, p. 529-541.

Lock, B.E., 1972. A lower Paleozoic rheo-ignimbrite from White Bay, Newfoundland: Can. Jour. Earth Sci., v. 9, p. 1495-1503.

Moore, J.G., 1967. Base surge in recent volcanic eruptions: Bull. volcanologique, v. 30, p. 337-363.

Sangster, D.F., 1972. Precambrian volcano-genic massive sulphide deposits in Canada: a review: Geol. Surv. Canada, Paper 72-22, 44 p.

Schmincke, H.-U., 1967. Graded lahars in the type sections of the Ellensburg Formation, South-Central Washington: Jour. Sed. Petrol., v. 37, p. 438-448.

Schmincke, H.-U., R.V. Fisher, and A.C. Waters, 1973, Antidune and chute and pool structures in the base-surge deposits of the Laacher Sea Area, Germany: Sedimentology, v. 20, p. 553-574.

Walker, R.G., 1976. Facies models: general introduction: Geosci. Canada, v. 3, p. 21-24.

MS received May 11, 1979.

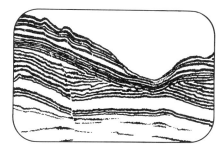

Facies Models 18. Seismic-Stratigraphic Facies Models

T.L. Davis
Department of Geology and Geophysics
University of Calgary
Calgary, Alberta T2N 1N4

Introduction

The reflection seismograph is a geologic tool which produces a geological section or model. Seismic interpretation involves the transformation of seismic data into discrete geologic terms. The interpreter must employ as much geologic information as possible when interpreting seismic data while extracting as much information as possible out of the data to evaluate the "complete" geologic regime. To the geologist the use of reflection seismology and interpretive aspects of "seismic-stratigraphy" can further extend our overall knowledge of depositional systems and facies models within the subsurface, particularly by emphasizing larger scale aspects of the system. An integrated approach is required involving people who are receptive to combining geophysical data with geologic information in order to arrive at an accurate representation of the subsurface.

In order to perform seismic-stratigraphic analysis the interpreter must take on the view that lateral and vertical variations in seismic waveforms, and patterns or configurations of these wiggles on seismic sections have meaning in terms of geology via seismic derived quantities. The problem is one of establishing the meaning and scale of observation that can be attained. The scale of seismic-stratigraphic analyses is commonly orders of magnitude larger than conventional facies modelling - hence seismic stratigraphy can contribute important larger scale relationships that can only be worked out with extreme difficulty by normal sedimentological or stratigraphic methods. Recently, however, the differences in scales of observation have been shrinking.

Although the vertical and horizontal scales of the seismic derived facies models are much larger than the standard level of observation of the geologist working surface sections or subsurface data, the amount of information condensed together within seismic sections is often overwhelming within the total framework of subsurface stratigraphy. The seismic interpreter must focus on selected stratigraphic settings in order to recognize stratigraphic anomalies shown by the data. Anomaly recognition requires complimentary geologic modelling. Often our use or misuse of seismic data to establish stratigraphic relationships is dependent on our understanding or lack of understanding of the geological model itself. Geologic facies modelling is required to accentuate the amount of geologic information derivable from reflection seismic data. Once established, facies may be further documented, refined and expanded upon by proper seismic-stratigraphic analyses.

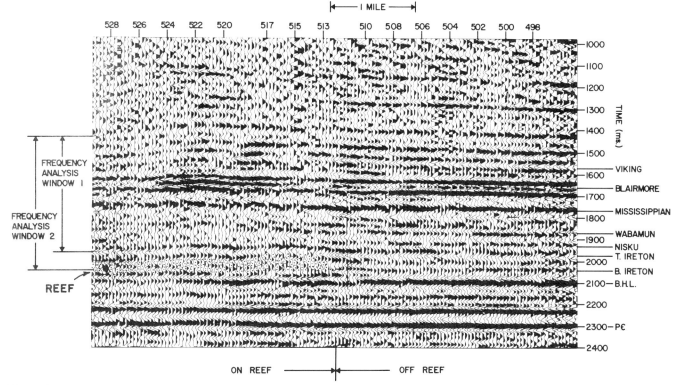

Figure 1
Seismic section over a Devonian Leduc reef (stippled) in western Canada; reef is the lateral equivalent of the Ireton shale. Note the drape of the top of Ireton event over the reef. Note also the change in "character" of the seismic waveform from Top of Ireton to Base of Ireton across section, from reef to off-reef.

Within relatively thin stratigraphic sequence units of interior basins the seismic definition of facies requires a perceptive eye and a substantial knowledge of geologic environments and facies models. A seismic pulse breadth or wavelet displayed in travel time may represent 30 to 50 metres or more of geologic strata. Strata 30 to 50 metres in thickness may encompass substantially different geologic environments. Facies refers to contemporaneous lithologic changes within these environments. As an example of the use of seismic-stratigraphic modelling within interior basin sequences shelf-sedimentation (Spindle) and a fluvial-point bar (Gilby) case studies have been chosen. In each case accurate geologic facies modelling of these systems prior to seismic definition is the primary requisite as to the usefulness of seismic to further delineate facies.

Seismic Reflections—Their Role in Facies Modelling

Elastic waves, generated near the ground surface by some excitation source, travel vertically downward from the vicinity of the source through the adjacent material where some of the wave energy is picked up by vibration detectors distributed in the vicinity. The relative placement of these detectors with respect to the source, and the relative arrival times of seismic events with respect to the initiation time of the energy, by whatever subsurface path they follow, comprise a set of data from which certain structural and lithologic information can be gained. To obtain a seismic section (e.g., Fig. 1), shots are made at equally spaced points along a straight line several miles long. Thus each "wiggle" on Figure 1 represents one shot point, with a spacing of about 220 feet (74 m).

Two variables (*time* and *distance*) are measured at each point in an effort to determine the elastic wave speed (velocity), and the location, configuration, and attitude of structures within the earth's crust. We are guided by two other variables - relative amplitude and "frequency" or frequency content of the seismic pulse. We recognize and attempt to remove the perplexing "noise" which reflects events on the seismogram that are not meaningful in interpreting the geologic significance of the target (Hollister and Davis, 1977).

Elastic waves propagate through crustal material as 1) longitudinal or compressional; 2) transverse or shear; and 3) as surface waves. In general, only the longitudinal waves are useful in subsurface exploration, with the transverse and surface waves (when recorded) being relegated to the category of noise. However, shear wave investigations are now being conducted as a potential stratigraphic tool (Omnés, 1978).

Both longitudinal and transverse waves are body waves, that is, propagating within a material body as opposed to traveling along its surface. The longitudinal wave is characterized by particle motion to-and-fro along the direction of propagation. It travels with a velocity, V_L, which depends on the modulus incompressibility, K, the modulus of rigidity, μ, and the density, δ, related as:

$$V_L = \sqrt{\frac{3K + 4\mu}{3\delta}}$$

In the transverse plane wave, particles oscillate at right angles to the

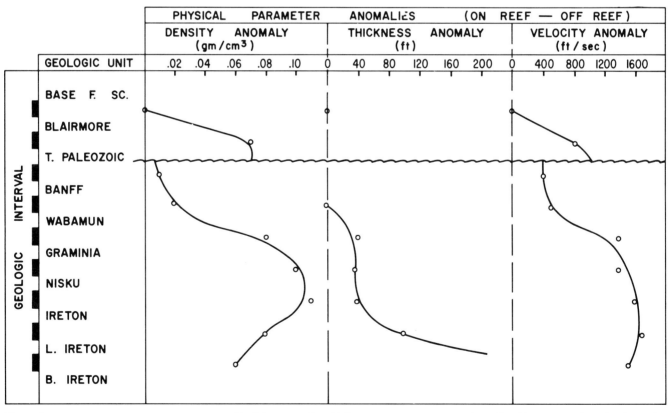

Figure 2
Velocity, density and thickness variations within the geologic section adjacent to and above Leduc reefs in south-central Alberta.

Reefs have a profound influence on petrophysical parameters of rocks not necessarily associated with the reef environment or interval itself. These changes are largely post

depositional in nature and may be due to the influence of reefs as a focus or focal point for fluid movement.

direction of propagation. The propagation velocity, V_T, of the transverse wave is not dependent on the compressibility of the material but only on its rigidity and density and

$$V_L = \sqrt{\frac{\mu}{\delta}}$$

Brief comparison of these two velocity expressions leads to the following conclusions: the longitudinal velocity of a medium is always greater than its transverse; and that fluids, having very low values of rigidity, can be considered as failing to support transverse waves.

In light of the velocity equations, one would think that the increase in density which accompanies greater compaction should decrease rather than increase velocity. Since it is velocity that is observed, it can only be concluded that elastic constants increase at a greater rate than does density.

When, in its travel through the layered crust, a wave encounters an interface between media of different elastic constants and/or densities, it divides into several parts. If the path of an incident wave is normal to the plane interface, only two waves are formed, one reflected and one transmitted, both of the same type as the parent. Under this condition, since longitudinal waves are of concern in seismic prospecting, let us consider the relationships of pulse size which exists among the three waves involved. If a plane longitudinal incident wave of amplitude, M_o, produces a reflected wave of amplitude, M_r, and a transmitted wave of amplitude, M_t, then the amplitude ratios are:

$$\frac{M_r}{M_o} = \frac{\delta_2 V_{L_2} - \delta_1 V_{L_1}}{\delta_2 V_{L_2} + \delta_1 V_{L_1}}$$

and

$$\frac{M_t}{M_o} = \frac{2\delta_1 V_{L_1}}{\delta_2 V_{L_2} + \delta_1 V_{L_1}}$$

where V_{L_1} and V_{L_2} are the velocities of layers 1 and 2 respectively.

The produce of density and velocity (δV_L) is commonly called specific acoustic impedance; and the ratio M_r / M_o is known as the reflection coefficient. Considering the reflected-incident amplitude ratio, it is noted that differences in density, as well as differences in velocity, can produce reflections; also, if $\delta_1 V_{L_1} > \delta_2 V_{L_2}$, a reversal of phase results.

The relatively simple reflection phenomena, occurring at the boundary between two thick media, become extremely complicated when the number of interfaces increases and the

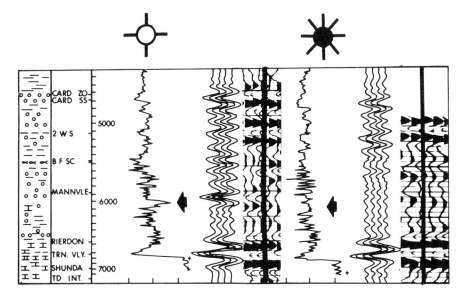

Figure 3
Relation between sonic (velocity) logs plotted as a function of travel time, synthetic seismograms and actual seismic data from two wells a few miles apart in Southern *Alberta. Note the signature (arrow) of approximately 120 feet (30 m) of Mannville present in the 16-24 well which is not present in the other well.*

thickness of the layers becomes less than the breadth of the seismic pulse. Each reflection seen on a seismic record is nearly always a composite of several reflections rather than a distinct event off a single, simple interface. The result is that the interpreter often has to work with waves in the hope that subtle changes in waveform are linked to a change of acoustic impedance rather than noise related phenomena or complicated interference. By tuning the waveform to a symmetrical "zero phase" wavelet during acquisition and processing the amount of information derivable from the wavelet and its potential relation to facies is maximized (Schramm et al., 1977).

Stratigraphic Interpretation Tools
Emphasis within the field of seismic exploration is now being given to seismic stratigraphy; also to the inherent limitations involving resolution of the seismic method (Sheriff, 1977). The only approach to seismic stratigraphic exploration is by experimentation. Often an indirect rather than a direct indication of a stratigraphic trap may exist; for example, seldom is there direct indication of a reef on seismic data because indentifiable reflections from the top and bottom of reefs seldom occur. However, as shown on Figure 1, differential

compaction and velocity anomalies associated with the reef affect the configuration of horizons above and below the feature which, in turn, reflect reef presence and even thickness. Seismic stratigraphy can thus give a broad geometric view of a rock body which might otherwise be hard to determine from the stratigraphic record (see article Number 11 on reefs).

Velocity. Velocity is a stratigraphic tool (Gregory, 1977). Interval velocity variations often reflect changes in stratigraphy (facies). Velocity determination from seismic data is the basis of the inversion technique of Lindseth (1979). Velocity variations on a local and regional scale can help pinpoint stratigraphic anomalies. They can also lead to quantifying physical parameters such as porosity and more abstract parameters such as resolution. Velocity variations around reefs (Davis, 1972, 1973) for example serve to enhance the seismic expression of reefs and offer another tool of searching for reefs and quantifying porosity within the reef from the above-reef section. Figure 2 shows velocity and density variations documented above Leduc reefs. These anomalies extend several hundred feet above the reef. Zones of highest interval velocity above the reefs are often coincident with zones of highest porosity (extensive dolomitization) within the underlying reef.

Synthetic Seismograms. Velocity (and density information, if available) enables construction of synthetic seismograms and models. Synthetic seismograms (Fig. 3) are used to relate seismic data to well information (geology). They also enable an understanding of the compo-

sition of reflections and an insight into the resolution of the seismic method. A step beyond the synthetic seismogram is seismic modelling. Seismic-stratigraphic modelling as described by Meckel and Nath (1977), for example, enables one to examine the influence of a change of geologic, petrophysical or seismic source signature parameters on the seismic expression of stratigraphic features. The interpreter attempts to

match the real seismic data and the synthetic modelled data thereby attempting to converge on a geologic solution. Ambiguity of the seismic method soon becomes apparent when as usual the number of variables exceeds the known parameters.

Frequency. Frequency variations may occur above, within and below stratigraphic anomaly levels. Figure 4 shows

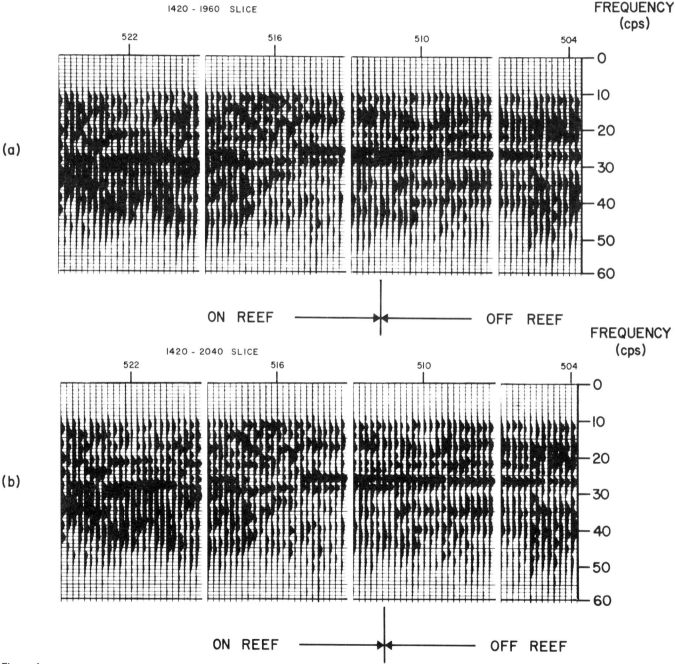

Figure 4
Frequency spectral plots of the data of Figure 1. Seismic data may be evaluated in the time domain or time representation, as illustrated in Figure 1, or in the frequency domain. Any

seismic signal is the sum of a finite number of frequencies each with a unique amplitude and phase shift. Frequency is plotted in cycles per second. The amount of frequency

component present in a particular "window" of seismic-time data is illustrated by the extent of shading on this display of amplitude versus frequency.

frequency analyses of the seismic line of Figure 1. Note the lateral change in dominant frequency bandwidth from the off to on-reef interval. The frequency variation serves to document the reef edge, a stratigraphic objective not easily documented on the time-section of Figure 1.

Waveform. Amplitude variations are controlled by reflection and transmission coefficients which in turn are controlled by velocity and density contrasts. Amplitude variations are usually a very subtle but important stratigraphic tool. Figure 5 shows that the importance of waveform as a stratigraphic tool was recognized long before digital recording and processing of data. Even with our current capability it is often overlooked. Our first reaction often is that amplitude variations or waveform alterations are probably noise related or processing created. Critical

analysis of waves or acoustic impedance signature is a required tool for facies differentiation and determination.

Reflection (Waveform) Configuration. Vail *et al.* (1977) have illustrated the use of seismic sequence analysis and the examination of reflection configurations within these sequences as being indicative of environments of deposition and facies. Sequence and facies analysis utilizing reflection configuration as discussed by Vail and associates requires a large scale analysis or perspective, e.g., a seismic trace consisting of several events comprising a few hundred milliseconds of data while encompassing several hundred feet of geologic strata. Within major seismic-stratigraphic sequences a smaller scale subdivision of units within the framework of depositional systems is required. These units are referred to as genetic units. Genetic units or units of common

genesis have similar lithologies because the environments of deposition are similar (Weimer, 1975). Genetic units will generally be synonomous with facies but it does leave open the possibility of subdividing facies on a smaller scale. They may comprise only a few tens of feet of geologic strata or one pulse breadth or less of seismic reflection time. As a result waveform analysis is our main stratigraphic tool within thin sequences.

Cretaceous Stratigraphic Facies Models
To illustrate the application of genetic unit - facies concepts to seismic stratigraphy two examples have been chosen from the Cretaceous clastic section of two Rocky Mountain foreland basins. The definition of facies is the primary stratigraphic objective of seismic stratigraphy. The concept of facies is taken here in a time-stratigraphic

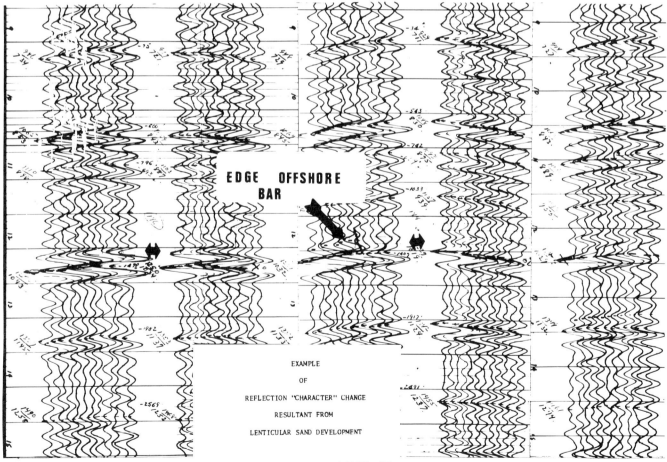

EDGE OFFSHORE BAR

EXAMPLE

OF

REFLECTION "CHARACTER" CHANGE

RESULTANT FROM

LENTICULAR SAND DEVELOPMENT

NATIONAL GEOPHYSICAL COMPANY of CANADA, LTD.

Figure 5
Waveform - a prime stratigraphic facies indicator, displayed on two seismic field records. The central areas of these records shown by the double arrows represent normal incidence (vertical) recording of data. The left hand record has been shot centred on a well that contains no sandstone development at the level of the arrows. The central record has been shot past a well which is in close proximity but contains 80 feet of sandstone reservoir interpreted as Cretaceous offshore bar facies. Note the change in waveshape or "character" at the stratigraphic level illustrated by the arrows.

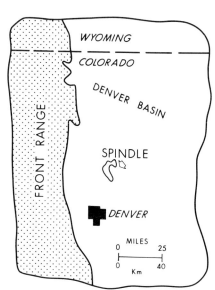

Figure 6
Spindle field location.

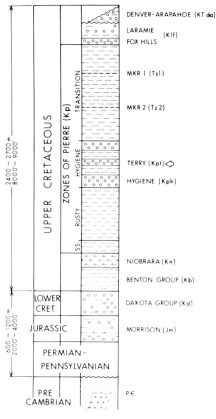

Figure 7
Denver basin stratigraphic column.

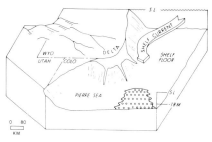

Figure 8
Terry sandstone depositional model (After Porter, 1976).

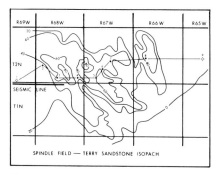

Figure 9
Isopach map of Terry sandstone interval (After Moredock and Williams, 1976). Line of section of Figure 10 and 12 shown.

sense, related to lithologic changes in contemporaneous deposits. It must be determined where seismic reflections are derived from within a stratigraphic sequence in order to maximize the amount of stratigraphic information available from seismic data. In order to understand depositional systems and to perform accurate facies analysis it must be established whether seismic waves or acoustic impedance signatures are "tracking" time-stratigraphic surfaces of deposition, or local prominent changes of lithology which have no time-stratigraphic connotation. Identification of time-stratigraphic units is often a particularly difficult task in interior basins. In some instances log markers can be used. In the Spindle study bentonites were used to establish time-stratigraphic surfaces. Bentonites are derived from volcanic ash which is spread by the wind and settles to the bottom of the marine environment. Upon burial it devitrifies to montmorillonite. Bentonites are often detectable on resistivity, density and gamma ray-neutron logs. In the Gilby study time-surfaces have been associated with high radio-activity markers on the gamma ray-neutron logs. These markers may represent rapid marine incursions and associated shale deposi-

tion or they could represent an unconformity hiatus. These markers are not identifiable on the sonic logs.

Shallow Marine Sedimentation – Spindle. The Spindle field is located in the west-central part of the Denver basin (Fig. 6). The field is producing from the Terry and Hygiene sandstone members of the Pierre Shale (Fig. 7). Producing depths are 4500 to 5000 feet (1300 to 1500 m). The Terry sandstone focused on for this study is thought to represent a shallow marine shelf and offshore bar complex (Fig. 8) (similar to those discussed in Paper 7 of this volume), in accordance with a stratigraphic model developed by Porter (1976) for the Hygiene Zone of the Pierre Shale. A Terry sandstone isopach map is shown on Figure 9. The isopach map adapted by Moredock and Williams (1976) shows lobate trends in a north (northwest) – south (southeast) direction typical of offshore bar sedimentation associated with shelf current activity (Figure 8, and Paper 7 of this volume). Problems arise as to whether the field consists of one bar on a number of bars deposited at different times. A cross-section of the Terry member interval is illustrated on Figure 10. Logs shown are velocity logs derived from resistivity logs (Buchanan, 1978). Time-stratigraphic surfaces T_1 and T_2 determined from bentonite correlations are shown, and rock-stratigraphic surfaces R_1 and R_2 corresponding to the top and bottom of the Terry sandstone are indicted (see isopachs in Fig. 9). Synthetic seismogram sections using a 20 to 80 hertz band-width zero-phase (symmetrical) wave (Fig. 11) illustrate that the acoustic impedance surface (dotted line) parallels the rock-stratigraphic Terry sandstone interval. As a result the Terry reflection, which is a composite waveform encompassing the entire Terry sandstone, "tracks" the Terry sandstone rock-stratigraphic unit. The synthetic section suggests that isochron mapping of the Terry interval will yield no anomalous variation to suggest the presence of the Spindle bar complex. This inability of seismic isochronal mapping to depict the Spindle field area is shown on Figure 12. The synthetics of Figure 11 suggest, however, that careful preservation of amplitude and waveform of the seismic signature may delineate zones (facies boundaries) where the time-stratigraphic surface (T_2) crosses

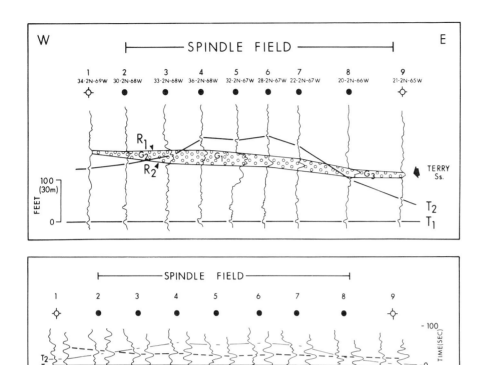

Figure 10
Stratigraphic cross-section of Terry member showing rock-stratigraphic boundaries (R), time-stratigraphic surfaces corresponding to bentonite zones (T) and genetic units G_1, G_2, G_3. Logs are pseudo-velocity logs derived from resistivity logs. Velocity increases to right. Datum is bentonite-time stratigraphic marker T_1.

Figure 11
Spindle synthetic seismogram section utilizing 20-80 hertz zero phase wavelet. Two waveforms are shown at each well location. The waveform on the left corresponds to the pseudo velocity log derived from the resistivity log. The waveform on the right is the synthetic seismogram. The line of reflection correlation corresponds to the rock-stratigraphic Terry sandstone unit.

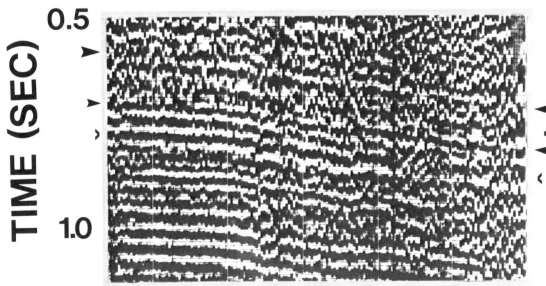

Figure 12
Spindle seismic section. Note lack of stratigraphic definition of the Spindle field. Waveform has not been plotted. Only the peaks have been shaded in a variable density display. No isochronal (time interval between peaks) variance occurs between Transition Zone marker (Tzm) to the Terry, Terry to Hygiene, or Tzm to Hygiene interval.

208

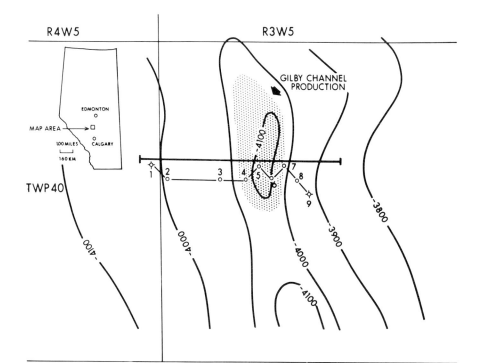

over the rock-stratigraphic Terry sandstone interval. True relative amplitude
wave processing coupled with critical
interpretive waveform analysis and
integration with the well log and geologic
model is required before facies can
accurately be determined in the Hygiene
Zone shallow marine sedimentation
environment. The resolvability in this
case of the Spindle field is dependent on
the identification and use of time-
stratigraphic markers to establish at
least three genetic units along the line of
section of Figure 10 as offshore bar
facies. The oldest, central bar, facies is
the most productive. Bar margins are
relatively non-productive.

*Point-Bar Channel Complex Model –
Gilby.* The Gilby field is located in the
south-central part of the Alberta basin
(Fig. 13). The field is producing
primarily from an Uppermost Jurassic -
Lower Cretaceous (Mannville) channel
complex (Fig. 14). Late Jurassic -
Early Cretaceous channelling cut out
the entire Lower Jurassic section and a
few hundred feet of the underlying
Paleozoic. Uppermost Jurassic and

Figure 13
*Gilby field location, Alberta basin. Structure
contours on top of Mississippian. Basal*
*Cretaceous and Uppermost Jurassic Gilby
channel production superimposed. Line of
section of Figures 14 and 16 shown.*

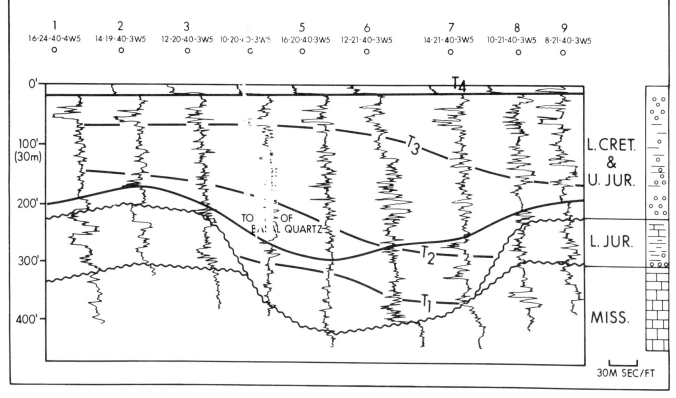

Figure 14
*Gilby sonic log cross-section. Medicine River
coal is taken as datum at 0 feet. Note inferred*
*interpretation of active channel fill on west
(left) side of channel relative to the east side
passive channel margin facies. Note also the*
*progradational infill of Lower Cretaceous.
Time surfaces T_1, T_2, T_3 and T_4 interpreted
from gamma ray logs.*

Figure 15

Gilby synthetic seismogram sections, Medicine River coal datum, utilizing (a) wavelets with a frequency content matched to that of the actual seismic section of Figure 16 and (b) high resolution synthetic seismogram section. Note particularly the lateral waveform change within channel from top of Basal Quartz (BQ) event to the top of the Mississippian (M). T_1, T_2 and T_3 are time-stratigraphic surfaces from Figure 14.

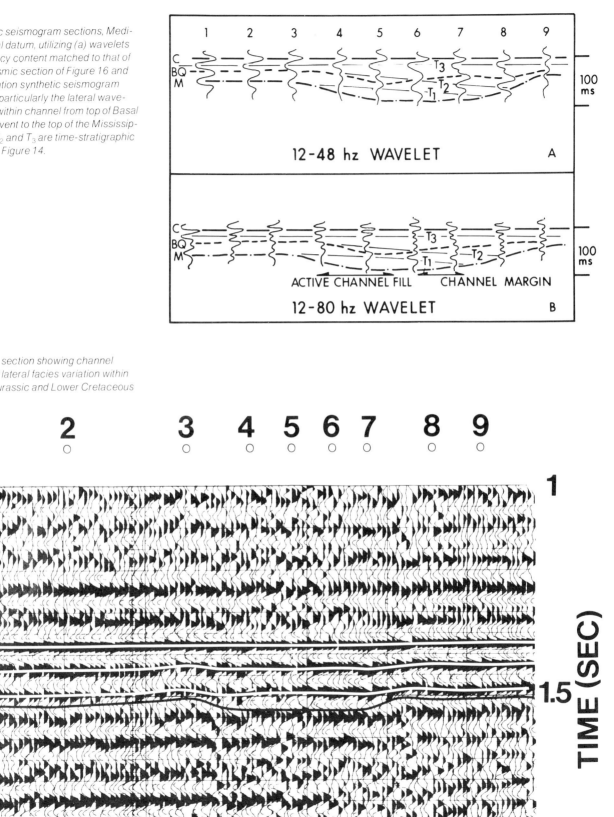

12-48 hz WAVELET A

12-80 hz WAVELET B

ACTIVE CHANNEL FILL CHANNEL MARGIN

Figure 16

Gilby seismic section showing channel definition and lateral facies variation within Uppermost Jurassic and Lower Cretaceous channel fill.

210

Lower Cretaceous channel infill facies vary laterally and vertically. The prime goal of seismic facies modelling in such an environment is no longer to detect the channel but to establish the facies and genetic units within the channel system (e.g., Land and Weimer, 1978; Paper 3 of this volume). Synthetic seismograms (Fig. 15) along the line of section of Fig.14 illustrate that the channel should be identifiable on seismic. An isochronal (time) increase between the coal marker and the top of the Basal Quartz or Mississippian reflector is coincident with the channel. As in the Spindle study, reflections correspond to *rock-stratigraphic boundaries* within the sequence, rather than the time-surfaces identified on the gamma ray logs, making facies definition on the basis of isochronal variations inaccurate. Waveform analysis within the channel sequence is required. Examination of the synthetics (Fig. 15) illustrates that the

waveform from the Lower Cretaceous – Uppermost Jurassic channel fill changes laterally within the channel complex, indicating lateral facies changes within the channel. Waveform analysis on several lines could not only map the spatial position of the channel but also the facies distribution within the channel. Figure 16 illustrates these concepts on a seismic line which traverses the line of section illustrated on Figure 13. Note the channel definition from the sag in the top of the Basal Quartz marker horizon. Note also the change of waveform within the channel between the Basal Quartz and Mississippian marker horizons. The waveform change matches that of the synthetics and marks the facies boundaries (active versus passive channel-fill) within the channel. Proper facies analysis within the Gilby channel is dependent on waveform analysis.

Influence of Tectonics on Facies Models

Our perspective of the influence of structure on stratigraphic depositional systems is often dependent on our scale of observation and perception (Weimer, 1978, 1979). Reflection seismology is one of the most useful tools for monitoring the relation between tectonics and sedimentation as the complete stratigraphic section can be monitored including the basement. Thus a stratigraphic section can be worked from bottom to top in accordance with establishing stratigraphic models rather than from top to bottom as we are often confined and contented to do when working with surface geology and subsurface well control only.

As an example of the role of basement related tectonics on Cretaceous sedimentation and facies of the nature depicted in the foregoing examples consider Figure 17. No wells deeper

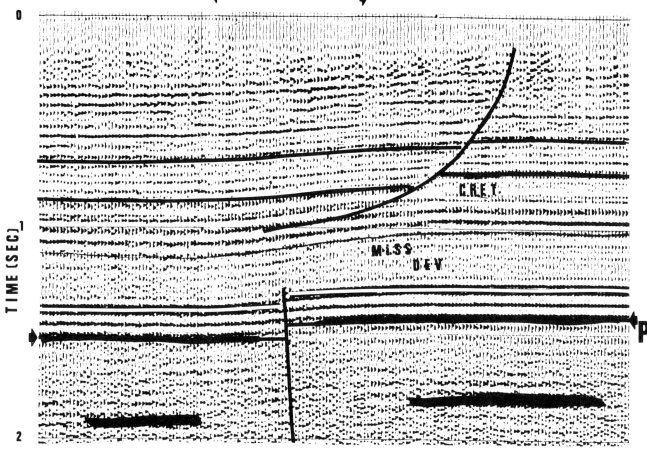

Figure 17
Seismic line from Southern Alberta showing influence of basement tectonics on Cretaceous structure and stratigraphy.

than the Mississippian have been drilled in the area. Basement control of Cretaceous structure and related stratigraphy is obvious on the seismic line provided attention is paid to the basement level. Without seismic control the relationship between basement related fault systems and Cretaceous stratigraphy would go unnoticed. Anomaly recognition and investigations require geophysical as well as geological input.

Summary
Currently logs and seismic data are used to correlate common lithologies. In order to understand depositional systems and perform accurate facies and/or genetic unit analysis lines of lithologic correlation must be separated from time-surfaces of deposition. Often information required to establish time-surfaces and facies is not directly available from seismic data, particularly within "thin" interior basin sequences. Within these "thin" sequences seismic reflections tend to "track" rock-stratigraphic boundaries with subtle variations in waveform being indicative of genetic units within the framework of the depositional sequence. Future success at seismic-stratigraphic facies modelling will be dependent on our ability to integrate seismic with other subsurface information to recognize depositional systems and to establish the influence of regional and local stratigraphic and structural controls on these systems.

Acknowledgements
The author wishes to acknowledge and thank Amoco Production Co., Denver Division, Hudson's Bay Oil and Gas Company and Canexeco for providing and releasing the seismic data used in this paper.

Appreciation is extended to Dr. Roger Walker for reviewing and editing the manuscript. Roger's revisions have helped make the topic a little more palatable for you the reader.

References
Buchanan, P.C., 1978, Seismic stratigraphy of the Terry Hygiene members, Weld County, Colorado: M.Sc. Thesis #T-2110, Colorado School of Mines, 64 p.

Davis, T.L., 1972, Velocity variations around Leduc reefs, Alberta: Geophysics, v. 37, p. 584-604.

Davis, T.L., 1973, Geophysical study of Alberta's Keg River reefs: Oil and Gas Jour., v. 71, no. 43, p. 46-50.

Gregory, A.R., 1977, Aspects of rock physics from laboratory and log data that are important to seismic interpretation: Amer. Assoc. Petrol. Geol., Mem. 26, p. 15-46.

Hollister, J.C. and T.L. Davis, 1977, Seismic prospecting for petroleum, in L.W. Leroy, D.O. Leroy, and J.W. Raese, eds., Subsurface Geology-Petroleum, Mining, Construction: Colorado School of Mines, p. 425-437.

Land, C.B. and R.J. Weimer, 1978, Peoria field, Denver basin, Colorado- J sandstone distributary channel reservoir: Rocky Mountain Assoc. Geol., Symposium Guidebook, p. 81-104.

Lindseth, R.O., 1979, Synthetic sonic logs - a process for stratigraphic interpretation: Geophysics, v. 44, p. 3-26.

Meckel, L.C. and A.K. Nath, 1977, Geologic considerations for stratigraphic modeling and interpretation: Amer. Assoc. Pet. Geol., Mem. 26, p. 417-438.

Moredock, D.E. and S.J. Williams, 1976, Upper Cretaceous Terry and Hygiene sandstone - Singletree, Spindle and Surrey fields - Weld County, Colorado, in R.E. Epis and R.J. Weimer, eds., Studies in Colorado Field Geology: Colorado School of Mines Prof. Contr. no. 8, p. 264-274.

Omnés, G., 1978, Exploring with SH-waves: Can. Soc. Explor. Geophys. Jour., v. 14, p. 40-49.

Porter, K.W., 1976, Marine shelf model, Hygiene member of Pierre Shale, Upper Cretaceous, Denver basin, Colorado, in R.E. Epis and R.J. Weimer, eds., Studies in Colorado Field Geology: Colorado School of Mines Prof. Contr. no. 8, p. 251-263.

Sheriff, R.E., 1977, Limitations on resolution of seismic reflections and geologic detail derivable from them: Amer. Assoc. Petrol. Geol., Mem. 26, p. 3-14.

Schramm, M.W., E.V. Dedman, and J.P. Lindsey, 1977, Practical stratigraphic modeling and interpretation, in C.E. Payton, ed., Seismic Stratigraphy - Applications to Hydrocarbon Exploration: Amer. Assoc. Petrol. Geol., Memoir 26, p. 477-502.

Vail, P.R., R.M. Mitchum, Jr., R.G. Todd, J.M. Widmier, S. Thompson, III, J.B. Sangree, J.N. Bubb, and W.G. Hatlelid, 1977, Seismic stratigraphy and global changes of sea level: Amer. Assoc. Petrol. Geol., Mem. 26, p. 49-212.

Weimer, R.J., 1975, Stratigraphic principles and practices: Energy resources of detrital sequences: Short course lecture notes, Colorado School of Mines, 253 p.

Weimer, R.J., 1978, Influence of Transcontinental Arch on Cretaceous marine sedimentation: a prelimary report: Rocky Mountain Assoc. Geol., Symposium Guidebook, p. 211-222.

Weimer, R.J., 1979, Influence of basement tectonics on depositional systems and seismic stratigraphy: Paper read as part of Society of Exploration Geophysicists Distinguished Lecture series.

MS received May 31, 1979